Dynamics of Natural and Artificial Celestial Bodies

*Proceedings of the US/European Celestial Mechanics Workshop,
held in Poznań, Poland, 3–7 July 2000*

Edited by

**HALINA PRETKA-ZIOMEK and
EDWIN WNUK**

*Astronomical Observatory of the A. Mickiewicz University,
Poznań, Poland*

P. KENNETH SEIDELMANN

*University of Virginia and U.S. Naval Observatory,
U.S.A.*

and

DAVID RICHARDSON

University of Cincinnati, U.S.A.

Partly reprinted from *Celestial Mechanics and Dynamical Astronomy*
Volume 81: 1–2 (2001)

SPRINGER-SCIENCE+BUSINESS MEDIA, B.V.

A C.I.P. Catalogue record for this book is available from the Library of Congress.

ISBN 978-90-481-5865-2   ISBN 978-94-017-1327-6 (eBook)
DOI 10.1007/978-94-017-1327-6

Printed on acid-free paper

All Rights Reserved
© 2001 Springer Science+Business Media Dordrecht
Originally published by Kluwer Academic Publishers in 2001
No part of the material protected by this copyright notice may be reproduced or
utilized in any form or by any means, electronic or mechanical
including photocopying, recording or by any information storage and
retrieval system, without written permission from the copyright owner.

DYNAMICS OF NATURAL AND ARTIFICIAL CELESTIAL BODIES

Preface	1–2
A. ELIPE and M. VALLEJO / On the Attitude Dynamics of Perturbed Triaxial Rigid Bodies	3–12
J. W. MITCHELL and D. L. RICHARDSON / A Simplified Kinetic Element Formulation for the Rotation of a Perturbed Mass-Asymmetric Rigid Body	13–25
W. S. KOON, M. W. LO, J. E. MARSDEN and S. D. ROSS / Resonance and Capture of Jupiter Comets	27–38
D. J. SCHEERES / Changes in Rotational Angular Momentum due to Gravitational Interactions between Two Finite Bodies	39–44
F. MONDÉJAR, A. VIGUERAS and S. FERRER / Symmetries, Reduction and Relative Equilibria for a Gyrostat in the Three-body Problem	45–50
B. S. BARDIN and A. J. MACIEJEWSKI / Quasi-static Motions of Viscoelastic Satellites	51–56
K. T. ALFRIEND, S. R. VADALI and H. SCHAUB / Formation Flying Satellites: Control by an Astrodynamicist	57–62
W. S. KOON, M. W. LO, J. E. MARSDEN and S. D. ROSS / Low Energy Transfer to the Moon	63–73
S. JANCART and A. LEMAITRE / Dissipative Forces and External Resonances	75–80
S. BREITER / Lunisolar Resonances Revisited	81–91
K. MUINONEN, J. VIRTANEN and E. BOWELL / Collision Probability for Earth-Crossing Asteroids Using Orbital Ranging	93–101
I. P. WILLIAMS / The Dynamics of Meteoroid Streams	103–113
V. LAINEY, A. VIENNE and L. DURIEZ / New Estimation of Usually Neglected Forces Acting on Galilean System	115–122
L. BENET / Occurrence of Planetary Rings with Shepherds	123–128
E. LEGA and C. FROESCHLÉ / On the Relationship between Fast Lyapunov Indicator and Periodic Orbits for Symplectic Mappings	129–147
D. VOKROUHLICKÝ, S. R. CHESLEY and A. MILANI / On the Observability of Radiation Forces acting on Near-Earth Asteroids	149–165
J. C. MUZZIO, M. M. VERGNE, F. C. WACHLIN and D. D. CARPINTERO/ Stellar Motions in Galactic Satellites	167–176

General Problems of Celestial Mechanics

C. D. HALL / Attitude Dynamics of Orbiting Gyrostats — 177–186

A. J. MACIEJEWSKI / Non-integrability of a Certain Problem of Rotational Motion of a Rigid Satellite — 187–192

J.-E. ARLOT, W. THUILLOT and Ch. RUATTI / Mutual Events of the Saturnian Satellites: a Test of the Dynamical Models — 193–196

A.-n. SAAD and H. KINOSHITA / An Analytical Theory of Motion of Nereid — 197–204

P. BIDART / MPP01, a New Solution for Moon's Planetary Perturbations. Comparison to Numerical Integration — 205–210

Posters

P. BARTCZAK and S. BREITER / The Improved Model of Potential for Irregular Bodies — 211–212

A. A. VAKHIDOV / An Efficient Algorithm for Computation of Hansen Coefficients and Their Derivatives — 213–214

J. KLACKA / On the Poynting-Robertson Effect and Analytical Solutions — 215–217

V. V. PASHKEVICH / Development of the Numerical Theory of the Rigid Earth Rotation — 219–221

V. A. AVDYUSHEV and T. V. BORDOVITSYNA / Algorithms of the Numerical Simulation of the Motion of Satellites — 223–224

E. Y. TITARENKO, V. A. AVDYUSHEV and T. V. BORDOVITSYNA / Numerical Simulation of the Motion of Martian Satellites — 225–226

Resonances and Stability

A. I. NEISHTADT, C. SIMO and V. V. SIDORENKO / Stability of Long-period Planar Satellite Motions in a Circular Orbit — 227–233

L. FLORIA / Solving a Gylden-Mescerskij System in Delaunay-Scheifele-Like Variables — 235–240

L. E. BYKOVA and T. Y. GALUSHINA / Near-Earth Asteroids Close to Mean Motion Resonances: The Orbital Evolution — 241–246

V. LANCHARES and T. L. MORATALLA / Lyapunov Stability for Lagrange Equilibria of Orbiting Dust — 247–252

P. ROBUTEL and J. LASKAR / Global Dynamics in the Solar System — 253–258

K. GOZDZIEWSKI / Lyapunov Stability of the Lagrangian Libration Points in the Unrestricted Problem of a Symmetric Body and a Sphere in Resonance Cases — 259–260

A. E. ROSAEV / On the Behaviour of Stationary Point in Quasi-central Configuration Dynamics — 261–262

Z. SANDOR and M. H. MORAIS / A Mapping Model for the Coorbital Problem — 263–264

H. UMEHARA / Self-similar Structure Induced by Linear Three-body System — 265–267

Dynamics of asteroids, comets and meteors

S. A. KLIONER / Multidimensional Fourier Transformation of the Planetary Disturbing Function — 269–272

A. A. VAKHIDOV / Asteroid Orbits in Some Three-body Resonances — 273–280

G. MICHALAK / Determination of Masses of Six Asteroids from Close Asteroid-asteroid Encounters — 281–282

O VASILKOVA / Schwarzschield Nonequatorial Periodic Motion about an Asteroid Modeled as a Triaxial Rotating Ellipsoid — 283–288

S. BERINDE / A Quantitative Approach of the Orbital Uncertainty Propagation through Close Encounters — 289–294

J.A. M. FERNANDEZ , A. F. GARCIA and J. P. ERADE / Predictions, Observation and Analysis of Asteroid's Close Encounters — 295–300

ROSAEV A.E. / The Reconstruction of Genetic Relations between Minor Planets, based on their Orbital Characteristics — 301–303

P. A. DYBCZYNSKI and H. PRETKA-ZIOMEK / Dynamically New Comets in the Solar System — 305–306

M. BROZ and D. VOKROUHLICKY / The Peculiar Orbit of Vysheslavia: Further Hints for its Yarkovsky Driven Origin? — 307–312

J. KLACKA and S. GAJDOS / Kuiper-Belt Objects: Distribution of Orbital Elements and Observational Selection Effects — 313–320

O. P. BYKOV / Short Arc CCD Observations of Celestial Bodies: New Approach — 321–326

S. GAJDOS / NEO Observation Program at Astronomical Observatory in Modra — 327–330

P. KANKIEWICZ / Moon-Earth Separation Problem in the Dynamics of Near Earth Asteroids — 331–332

O. M. KOCHETOVA / Determination of Mass of Jupiter and That of Some Minor Planets from Observations of Minor Planets Moving in 2:1 Commensurability with Jupiter — 333–334

N. O. KOMAROVA / Ephemeris Meaning of Parameters of Asteroid's Apparent Motion — 335–337

A. BRUNINI and F. LOPEZ-GARCIA / Primordial Depletion of the Inner Region of the Asteroid Belt, $a < 2.2$ AU — 339–340

V. P. TITARENKO, L. E. BYKOVA and T. V. BORDOVITSYNA / Numerical Simulation of the Motion of Small Bodies of the Solar System by the Symbolic Computation System "Mathematica" — 341–342

I. WLODARCZYK / The Prediction of the Motion of Atens, Apollos and Amors over Long Intervals of Time — 343–345

J. KLACKA and M. GAJDOSIK / Orbital Motion in Outer Solar System 347–349

J. KLACKA / Solar Wind and Motion of Meteoroids 351–353

J. KLACKA and M. KOCIFAJ / On the Stability of the Zodiacal Cloud 355–357

J. KLACKA and M. KOCIFAJ / Interaction of Stationary Nonspherical Interplanetary Dust Particle with Solar Electromagnetic Radiation 359–361

L. NESLUSAN / The Photographically Observed Meteors of (Pegasids?) Stream Associated with Comet 18P/Perrine-Mrkos 363–364

L. NESLUSAN / A Sketch of an Orbital-Momentum-Based Criterion of Diversity of Two Keplerian Orbits 365–366

I. V. SEMENOV and V. P. KOROBEINIKOV / On Modeling of Small Celestial Body Fracture in Planet Atmospheres 367–368

V. P. STULOV / Approximating Trajectories of Observed Bolides with Account for Fragmentation 369–371

Artificial Satellite Dynamics

S. P. RUDENKO, V. K. TARADY, A. V. SERGEEV and N. V. KARPOV / Orbit Determination of the Geosynchronous Space Debris Kupon Satellite Using its CCD Observations at 2-meter Telescope at the Terskol Peak 373–374

A. I. NAZARENKO, V. YURASOV, P. CEFOLA, R. PROULX and G. GRANHOLM / Monitoring of Variations of the Upper Atmosphere Density 375–384

K. KURZYNSKA and A. GABRYSZEWS KA / Influence of the Local Troposphere on GPS Measurements 385–394

A. BOBOJC and A. DROZYNER / Orbit Determination Using Satellite Gravity Gradiometry Observations 395–400

A. I. NAZARENKO and V. YURASOV / Semi-analytical Models of Satellites Motion for Russian Space Surveillance System 401–406

P. H. C. N. LIMA Jr., S. da SILVA FERNANDES and R. V. DE MORAES / Semi-Analytical Method to Study Geopotential Perturbations Considering High Eccentric Resonant Orbits 407–413

G. B. PALMERINI / Coordinated Orbital Control for Satellite Constellations and Formations 415–424

A. I. NAZARENKO, N. N. SMIRNOV and A. B. KISELEV / The Space Debris Evolution Modeling Taking into Account Satellite's Collisions 425–426

W. SCHULZ, A. F. B. A. PRADO and R. V. DE MORAES / Inclination Change Using Atmospheric Drag 427–428

H. UMEHARA / An Optical Scanning Search for GEO Debris 429–431

H. UMEHARA / Co-location of Geostationary Satellites by Imaginary Interaction Maneuvers 433–434

I. WYTRZYSZCZAK, S. BREITER and B. MARCINIAK / Evolution of Super-geostationary Orbits on a Time Span of 200 years   435–436

Stellar and Galactic Dynamics

J. A. LOPEZ, M. J. MARTINEZ and F. J. MARCO / Dynamical Method for the Analysis of the Systematic Errors in Stellar Catalogues   437–442

E. GRIV, M. GEDALIN, D. EICHLER and C. YUAN / Resonant Excitation of Density Waves in Galaxies   443–452

A. TSVETKOV and M. BABADJANIANTS / Constructing of the Stellar Velocity Field Using Hipparcos Data   453–454

T. BORKOVITS / Tidal and Rotational Effects in the Evolution of Hierarchical Triple Stellar Systems   455–456

F. J. MARCO, J. A. LOPEZ, M. J. MARTINEZ / Analysis of Catalog Corrections with Respect to the Hipparcos Reference Frame   457–458

List of Participants

ALFRIEND Kyle T.
Aerospace Engineering
Texas A&M University
3141 TAMU, College Station
TX 77843-3141, USA
alfriend@aero.tamu.edu

ARLOT Jean-Eudes
Institut de mécanique céleste
CNRS, Observatoire de Paris
77, av. Denfert-Rochereau
75014 Paris, France
arlot@bdl.fr

AVDYUSHEV Victor
Applied Mathematics and
Mechanics Institute
Lenin Avenue 36
634050 Tomsk, Russia
astrodep@niipmm.tsu.ru

BABADZHANYANTS Maria
Astronomical Institute
Saint Petersburg University
Bibliotechaya 2, Petrodverets
198904 Saint Petersburg, Russia
masha@lkb.usr.pu.ru

BARDIN Boris S.
Dept. of Theoretical Mechanics
Moscow Aviation Institute
Volokolamskoe Shosse 4
125871 Moscow, Russia
boris.bardin@mailcity.com

BARLIER Francois
Observatoire de la Cote d'Azur
Av. Nicolas Copernic
06130 Grasse, France
francois.barlier@obs-azur.fr

BARTCZAK Przemysław
Astronomical Observatory
A. Mickiewicz University
ul. Słoneczna 36
60-286 Poznań, Poland
przebar@amu.edu.pl

BENET Luis
Max-Planck-Institut für Kernphysik
Postfach 103980
D-69029 Heidelberg, Germany
luis.benet@mpi-hd.mpg.de

BERINDE Ştefan
Astronomical Observatory
Babeş-Bolyai University
Str. Ciresilor No. 19
Cluj-Napoca 3400, Romania
sberinde@math.ubbcluj.ro

BIDART Patrick
DANOF/UMR8630
Observatoire de Paris
61, av. de l'Observatoire
75014 Paris, France
bidart@danof.obspm.fr

List of Participants

BOBOJĆ Andrzej
Institute of Geodesy
University of Warmia and Mazury
ul. Oczapowskiego 1
10-957 Olsztyn, Poland
altair@uwm.edu.pl

BORKOVITS Tamás
Baja Astronomical Observatory
Bács-Kiskun County
Szegedi út, P.O. Box 766
6500 Baja, Hungary
borko@electra.bajaobs.hu

BREITER Sławomir
Astronomical Observatory
A. Mickiewicz University
ul. Słoneczna 36
60-286 Poznań, Poland
breiter@amu.edu.pl

BROZ Miroslav
Institute of Astronomy
Charles University
V Holešovičkách 2
18000 Prague 8, Czech Republic
mira@sirrah.troja.mff.cuni.cz

BUCIORA Marcin
Astronomical Observatory
A. Mickiewicz University
ul. Słoneczna 36
60-286 Poznań, Poland
marcin.buciora@domdata.
depfa-it.com

BYKOV Oleg
Pulkovo Astronomical Observatory
65 Pulkovskoje schosse
196140 St. Petersburg, Russia
oleg@OB3876.spb.edu

DIKOVA Smiliana
Institute of Astronomy
Bulgarian Academy of Sciences
72 Tzarigradsko schossee
1784 Sofia, Bulgaria
skydyn@bas.bg

DROŻYNER Andrzej
Institute of Geodesy
University of Warmia and Mazury
ul. Oczapowskiego 1
10-957 Olsztyn, Poland
drozyner@uwm.edu.pl

DYBCZYŃSKI Piotr A.
Astronomical Observatory
A. Mickiewicz University
ul. Słoneczna 36
60-286 Poznań, Poland
dybol@amu.edu.pl

ELIPE Antonio
Grupo de Mecánica Espacial
Universidad de Zaragoza
50009 Zaragoza, Spain
elipe@posta.unizar.es

FELDMAN Renata
Observatoire de la Côte d'Azur
Av. Copernic
06130 Grasse, France
renata.feldman@obs-azur.fr

FLORÍA Luis
Dept. de Matemática Aplicada
Universidad de Valladolid
Paseo del Cauce s/n
47011 Valladolid, Spain
luiflo@wmatem.eis.uva.es

List of Participants

FROESCHLÉ Claude
Observatoire de Nice
Bv. de l'Observatoire
B.P. 4229
06304 Nice cedex 4, France
claude@obs-nice.fr

GAJDOŠ Štefan
Astronomical Institute
Comenius University
Mlynská dolina 1
84248 Bratislava, Slovak Republic
gajdos@fmph.uniba.sk

GALUSHINA Tatyana
Applied Mathematics and
Mechanics Institute
Lenin Avenue, 36
634050 Tomsk, Russia
astrodep@niipmm.tsu.ru

GOŹDZIEWSKI Krzysztof
Toruń Centre for Astronomy
N. Copernicus University
Gagarina 11
87-100 Toruń, Poland
chris@astri.uni.torun.pl

GRIV Evgeny
Department of Physics
Ben-Gurion University
P.O. Box 653
Beer-Sheva 84105, Israel
griv@bgumail.bgu.ac.il

GUSEV Alexander
Kazan State University
18 Kremljevskaja str.
420008 Kazan, Russia
alexander.gusev@ksu.ru

HALL Christopher D.
Virginia Polytechnic Institute &
State University
Aerospace & Ocean Engineering
VA 24061 Blacksburg, USA
chall@aoe.vt.edu

JANCART Sylvie
Départment de Mathématique
Namur University
8, Rempart de la Vierge
5000 Namur, Belgium
sylvie.jancart@fundp.ac.be

JOPEK Tadeusz
Astronomical Observatory
A. Mickiewicz University
ul. Słoneczna 36
60-286 Poznań, Poland
jopek@amu.edu.pl

KANKIEWICZ Paweł
Astronomical Observatory
A. Mickiewicz University
ul. Słoneczna 36
60-286 Poznań, Poland
ciupo@amu.edu.pl

KIRYUSHKIN Ilya
The Keldysh Institute of
Applied Mathematics
4 Miusskaya Square
125047 Moscow, Russia
kiryush@keldysh.ru

KLAČKA Jozef
Astronomical Institute
Comenius University
Mlynská dolina 1
84248 Bratislava, Slovak Republic
klacka@fmph.uniba.sk

List of Participants

KLIONER Sergei A.
Lohrmann Observatory
Dresden Technical University
Mommsenstr. 13
01062 Dresden, Germany
klioner@rcs.urz.tu-dresden.de

KOCHETOVA Olga
Institute of Applied Astronomy
Russian Academy of Sciences
Zhdanovskaya st. 8
197110 St. Petersburg, Russia
kom@quasar.ipa.nw.ru

KOCIFAJ Miroslav
Astronomical Institute
Slovak Academy of Sciences
Dúbravska cesta 9
84228 Bratislava, Slovak Republic
kocifaj@astro.savba.sk

KOMAROVA Natalya
Pulkovo Astronomical Observatory
65 Pulkovskoje schosse
196140 St. Petersburg, Russia
natali@OB3876.spb.edu

KOON Wang Sang
Control and Dynamical Systems
California Institute of Technology
1200 East California Blvd.
CA 91125 Pasadena, USA
koon@cds.caltech.edu

KURZYŃSKA Krystyna
Astronomical Observatory
A. Mickiewicz University
ul. Słoneczna 36
60-286 Poznań, Poland
kurzastr@amu.edu.pl

LAINEY Valery
IMCCE/Université de Lille
1, Impasse de l'Observatoire
59000 Lille, France
lainey@bdl.fr

LANCHARES Víctor
Universidad de la Rioja
C/Luis de Ulloa s/n
26004 Logroño, La Rioja, Spain
vlancha@dmc.unirioja.es

LEVISON Hal
Southwest Research Institute
1050 Walnut Street, Suite 426
CO 80302 Boulder, USA
hal@boulder.swri.edu

LO Martin W.
Navigation and Mission Design
Jet Propulsion Laboratory
Pasadena, California, USA
Martin.W.Lo@jpl.nasa.gov

LOPEZ Garcia Alvaro
Astronomical Observatory
Valencia University
c/Vicente A. Estelles s/n
46100 Burjassot, Valencia, Spain
Alvaro.Lopez@uv.es

LÓPEZ-GARCIA Francisco
Facultad de Ciencias Exactas
Fisicas y Naturales
Universidad Nacional de San Juan
Av. Alem 368 Sur
5400 San Juan, Argentina
flgarcia@casleo.gov.ar

VIGUERAS Antonio
Departamento de Matematica
Universidad de Cartagena
Aplicada y Estadística
Paseo Alfonso XIII, 44
30203 Cartagena, Spain
Antonio.Vigueras@upct.es

VOKROUHLICKÝ David
Institute of Astronomy
Charles University
V Holešovičkách 2
18000 Prague 8, Czech Republic
vokrouhl@mbox.cesnet.cz

WILKINS Matthew
Aerospace Engineering
Texas A&M University
3141 TAMU, College Station
TX 77843-3141, USA
mwilkins@tamu.edu

WILLIAMS Iwan P.
Astronomy Unit
Queen Mary College
Mile End Rd
London E1 4NS, UK
I.P.Williams@qmw.ac.uk

WŁODARCZYK Ireneusz
Astronomical Observatory
Chorzów Planetarium
41-501 Chorzów 1, Poland
irek@entropia.com.pl

WNUK Edwin
Astronomical Observatory
A. Mickiewicz University
ul. Słoneczna 36
60-286 Poznań, Poland
wnuk@amu.edu.pl

WYTRZYSZCZAK Iwona
Astronomical Observatory
A. Mickiewicz University
ul. Słoneczna 36
60-286 Poznań, Poland
iwona@amu.edu.pl

YAGUDIN Leonid I.
Pulkovo Astronomical Observatory
65 Pulkovskoje schosse
196140 St. Petersburg, Russia
eiya@ipa.rssi.ru

YAGUDINA Eleonora I.
Institute of Applied Astronomy
Russian Academy of Sciences
Naberezhnaya Kutuzova 10
191187 St.Petersburg, Russia
eiya@quasar.ipa.nw.ru

YOUSSEF Manal H.
Astronomy Department
Faculty of Science
Cairo University
Giza, Egypt
manal@frcu.eun.eg

ZHOU Ji-lin
Department of Astronomy
Nanjing University
Nanjing 210093, P.R.China
zhoujl@nju.edu.cn

SEIDELMANN P. Kenneth
Department of Astronomy
University of Virginia
Box 3818 Univ Station
Charlottesville
VA 22903-0818, USA
pks6n@virginia.edu

SEMENOV Ilia
Institute for Computer
Aided Design RAS
2nd Brestskaya 19/18
123056 Moscow, Russia
inapro@glasnet.ru

SIDORENKO Vladislav V.
Keldysh Institute of
Applied Mathematics
Niusskaya Sq. 4
125047 Moscow, Russia
sidorenk@spp.keldysh.ru

STULOV Vladimir
Institute of Mechanics
Moscow Lomonosov University
Michurinsky Ave. 1
119899 Moscow, Russia
stulov@inmech.msu.su

THUILLOT William
Institut de mécanique céleste
CNRS, Observatoire de Paris
77, av. Denfert-Rochereau
74014 Paris, France
thuillot@bdl.fr

TITARENKO Ekaterina
Applied Mathematics and
Mechanics Institute
Lenin Avenue 36
634050 Tomsk, Russia
astrodep@niipmm.tsu.ru

TITARENKO Vadim
Applied Mathematics and
Mechanics Intitute
Lenin Avenue 36
634050 Tomsk, Russia
astrodep@niipmm.tsu.ru

UMEHARA Hiroaki
Communications Research Lab.
Kashima Space Research Center
893-1, Hirai, Kashima
314-0012 Ibaraki, Japan
ume@crl.go.jp

VAKHIDOV Akmal
Lohrmann Observatory
Dresden Technical University
Mommsenstr. 13
01062 Dresden, Germany
vakhidov@rcs.urz.tu-dresden.de

VALSECCHI Giovanni B.
IAS-CNR
Area di Ricerca
via Fosso del Cavaliere 100
00133 Roma, Italy
giovanni@ias.rm.cnr.it

VASHKOVYAK Mikhail
Keldysh Institute of
Applied Mathematics
Niusskaya Sq. 4
125047 Moscow, Russia
vashkov@spp.keldysh.ru

VASILKOVA Olga O.
Institute of Applied Astronomy
Russian Academy of Sciences
Kutuzov quay 10
191187 St.Petersburg, Russia
1021@ita.spb.su

List of Participants

PALMERINI Giovanni B.
Scuola di Ingegneria Aerospaziale
Universitá di Roma
Via Eudossiana 18
00184 Roma, Italy
g.palmerini@caspur.it

PASHKEVICH Vladimir
Pulkovo Astronomical Observatory
65 Pulkovskoje schosse
196140 St. Petersburg, Russia
apeks@gao.spb.ru

PRĘTKA-ZIOMEK Halina
Astronomical Observatory
A. Mickiewicz University
ul. Słoneczna 36
60-286 Poznań, Poland
pretka@amu.edu.pl

RICHARDSON David L.
Dept. of Aerospace Engineering
University of Cincinnati
OH 45221-007 Cincinnati, USA
Walshrd@aol.com

ROBUTEL Philippe
Institut de mécanique céleste
Observatoire de Paris
77, av. Denfert-Rochereau
75014 Paris, France
robutel@bdl.fr

ROSS Shane D.
Control and Dynamical Systems
California Institute of Technology
1200 East California Blvd.
CA 91125 Pasadena, USA
shane@cds.caltech.edu

ROSAEV Alexey
FGUP NPC NEDRA
Svobody 8/38
150000 Yaroslavl, Russia
rosaev@nedra.yar.ru

RUDENKO Sergei P.
Main Astronomical Observatory
National Academy of Sciences
03680 Golosiiv
Kiev-127, Ukraine
rudenko@mao.kiev.ua

SAAD Abdel–Naby S.
The Graduate University
For Advanced Studies (NAOJ)
2-21-1 Osawa, Mitaka-shi
Tokyo 181-8588, Japan
saad@pluto.mtk.nao.ac.jp

SÁNDOR Zsolt
Konkoly Observatory
Hungarian Academy of Scienecs
P.O. Box 67
1525 Budapest, Hungary
sandor@konkoly.hu

SCHEERES Daniel
Dept. of Areospace Engineering
The University of Michigan
1320 Beal Ave., Ann Arbor
MI 48109-2140, USA
scheeres@umich.edu

SCHULZ Walkiria
National Institute for
Space Research (INPE)
Av. dos Astronautas
1758 Sao Jose, Brasil
walkiria@dem.inpe.br

LÓPEZ Jose A.
Departamento de Matemáticas
Univ Jaume I de Castellón
Campus Riu Sec Castellon, Spain
lopez@mat.uji.es

MACIEJEWSKI Andrzej J.
Toruń Centre for Astronomy
N. Copernicus University
Gagarina 11
87-100 Toruń, Poland
maciejka@astri.uni.torun.pl

MARCO Francisco
Departamento de Matemáticas
Univ Jaume I de Castellón
Campus Riu Sec Castellon, Spain
marco@mat.uji.es

MICHALAK Grzegorz
Wrocław University Observatory
Kopernika 11
51-622 Wrocław, Poland
michalak@astro.uni.wroc.pl

MITCHELL Jason W.
Dept. of Aerospace Engineering
University of Cincinnati
OH 45221-007 Cincinnati, USA
Jason.Wm.Mitchell@uc.edu

MORAES Rodolpho Vilhena
DMA-FEG-UNESP
Av. Ariberto P. da Cunha 331
12516-410 Quaratingueta, Brasil
rvm@feg.unesp.br

MORBIDELLI Alessandro
Observatoire de Nice
Bv. de l'Observatoire
B.P. 4229
06304 Nice cedex 4, France
morby@obs-nice.fr

MUINONEN Karri
Observatory
University of Helsinki
Kopernikuksentie 1
FIN-00014 Helsinki, Finland
Karri.Muinonen@Helsinki.Fi

MUZZIO Juan Carlos
Observatorio Astronomico
Universidad Nacional de La Plata
1900 La Plata, Argentina
jcmuzzio@fcaglp.unlp.edu.ar

NAZARENKO Andrey I.
Center for Program Studies
84/32 Profsoyuznaya ul.
117810 Moscow, Russia
nazarenko@iki.rssi.ru

NESLUŠAN Lubos
Astronomical Institute
Slovak Academy of Sciences
05960 Tatranská Lomnica
Slovak Republic
ne@ta3.sk

NESVORNY David
Observatoire de Nice
Bv. de l'Observatoire
B.P. 4229
06304 Nice cedex 4, France
david@obs-nice.fr

PREFACE

This volume contains papers presented at the US/European Celestial Mechanics Workshop organized by the Astronomical Observatory of Adam Mickiewicz University in Poznań, Poland and held in Poznań, from 3 to 7 July 2000.

The purpose of the workshop was to identify future research in celestial mechanics and encourage collaboration among scientists from eastern and western countries. There was a full program of invited and contributed presentations on selected subjects and each day ended with a discussion period on a general subject in celestial mechanics. The discussion topics and the leaders were: Resonances and Chaos – A. Morbidelli; Artificial Satellite Orbits – K.T. Alfriend; Near Earth Objects – K. Muinonen; Small Solar System Bodies – I. Williams; and Summary – P.K. Seidelmann. The goal of the discussions was to identify what we did not know and how we might further our knowledge. The size of the meeting and the language differences somewhat limited the real discussion, but, due to the excellence of the different discussion leaders, each of these sessions was very interesting and productive.

Celestial Mechanics and Astrometry are both small fields within the general subject of Astronomy. There is also an overlap and relationship between these fields and Astrodynamics. The amount of interaction depends on the interest and efforts of individual scientists. With this situation, the teaching, research, students, funding, strength, and collaborations of both astrometry and celestial mechanics can vary in individual countries dependent upon the activities of a few scientists within the country. At this time both celestial mechanics and astrometry lack centers of excellence with critical masses of professors for teaching and research in the United States. Astrometry is primarily centered at the U.S. Naval Observatory and celestial mechanics is a subpart of astrodynamics in various places, with a greatest strength at the Jet Propulsion Laboratory.

There were a number of people drawn into these fields due to the emergence of the space age, with differences in the times due to the variations of when countries started space programs. Many of these people are now approaching, or reaching, retirement age, so the future is dependent upon attracting young scientists into the fields of astrometry and celestial mechanics and encouraging them to undertake new research efforts needed for advancements in those fields.

Astrometry has developed recently due to the success of Hipparcos and the plans for future astrometry missions such as FAME, SIM, and GAIA. All are tied to the applications of astrometry to astrophysics. However, classical ground based astrometry programs need to be redirected away from less accurate observations, that are no longer useful, to programs that can provide observational data to complement the space observations.

Celestial mechanics has historically developed based on the mathematical and computational capabilities available. Today the computational capabilities have grown to the extent that they are generally no longer the limitation, rather the limitations are the methods to be applied and the abilities and cleverness of the scientists.

With that background it appeared desirable to organize a meeting that wasn't based on the older people giving presentations on the research of the past, but rather to try to emphasize what was not known and where research was needed. Also an emphasis was placed on attracting young members of the fields from around the world, so that these people would get to know each other and, hopefully, develop collaborations to investigate the new areas of research discussed at the meeting.

It was felt, in addition, that Poznań, Poland, with a core of scientists covering a range of ages, would provide an example of how a research and educational group could be developed elsewhere. Also, Poznań is a central location convenient to eastern and western countries. Hopefully, the site would be sufficiently attractive to draw young scientists from the U.S. who would have to travel the farthest.

Thus, the gathering of people and the papers presented are to be the bases for building the future of astrometry and celestial mechanics for the future.

We would like to take the opportunity here to thank the National Science Foundation for the idea of this conference and the National Science Foundation, Polish Committee of Scientific Research and European Office of Aerospace Research and Development for supporting the meeting. In addition to them, the following local sponsors made possible the organization of this conference: the telecommunication company Telekomunikacja Polska S.A., the bank Bank Handlowy S.A., the brewery company Kampania Piwowarska and the HARVEST company. Their support is gratefully acknowledged.

Kenneth Seidelmann
David Richardson
Edwin Wnuk
Halina Prętka–Ziomek

ON THE ATTITUDE DYNAMICS OF PERTURBED TRIAXIAL RIGID BODIES

ANTONIO ELIPE[1] and MIGUEL VALLEJO[2]

[1]*Grupo de Mecánica Espacial, Universidad de Zaragoza, 50009 Zaragoza, Spain*
[2]*Real Observatorio de la Armada, 11110 San Fernando, Spain*

Abstract. Attitude dynamics of perturbed triaxial rigid bodies is a rather involved problem, due to the presence of elliptic functions even in the Euler equations for the free rotation of a triaxial rigid body. With the solution of the Euler–Poinsot problem, that will be taken as the unperturbed part, we expand the perturbation in Fourier series, which coefficients are rational functions of the Jacobian nome. These series converge very fast, and thus, with only few terms a good approximation is obtained. Once the expansion is performed, it is possible to apply to it a Lie-transformation. An application to a tri-axial rigid body moving in a Keplerian orbit is made.

Key words: attitude dynamics, rigid body, Lie transform

1. Introduction

Attitude dynamics of rigid bodies is a common problem in both natural and artificial celestial bodies, and has been object of study since the beginning of the celestial mechanics. Indeed, much of the work done in this regard is associated with names of great scientists like Euler, Hamilton, Poinsot, Lagrange, etc. Their work still is valid today, either in the aspect of attitude representation or in attitude dynamics [16].

The Space Age brought to astronomers, among other things, the fact that, apart from the classical bodies like the planets of satellites that are almost spherical, the shape of many natural celestial bodies is far away from the sphere; they present a very irregular shape, which adds a lot of complexity to the motion, that in some cases has been proved to be chaotic [12, 17]. On the other hand, spacecraft are also built under several models, and antennas will modify the distribution of masses, and even will introduce a big component of elasticity. Albeit most of celestial bodies of interest are not perfect rigid bodies, usually, the rigid body model is a good approximation for describing the rotational motion, hence the interest in having good theories for perfect rigid bodies.

The simplest case we meet with rigid body motion is the so called *Euler–Poinsot* problem, that consists of describing the motion of a free rigid body in the space. The attitude is described by elements of the group of rotations SO(3). The Hamiltonian is essentially the kinetic energy, defined on the cotangent space. Since there are no external forces, the angular moment vector is an integral, and

Celestial Mechanics and Dynamical Astronomy **81:** 3–12, 2001.
© 2001 *Kluwer Academic Publishers.*

also are its components and its norm. Thus, the problem is integrable, but unfortunately, Jacobian elliptic functions appear in the solution, which introduces a big complexity when a General Perturbation theory is applied, mainly because of the fact that elliptic functions are not closed over integration.

Of capital importance when integrating or analyzing perturbed problems is the choice of a convenient set of variables. The canonical variables based on the Euler angles reflect the SO(2) symmetry of rotations about the third space axis, thus, the problem is reduced to a two degrees of freedom problem. But the system is also symmetric with respect to the group SO(3), which makes it possible to reduce the system to only one degree of freedom. This is accomplished with the Serret–Andoyer variables (see [7, 9] for details). By denoting with I_1, I_2 and I_3 the principal moments of inertia (that without loss of generality we assume $I_1 < I_2 < I_3$), the Hamiltonian of the Euler–Poinsot problem in these variables (ℓ, g, h, L, G, H) takes the form

$$\mathcal{H} = \frac{1}{2} \left(\frac{\sin^2 \ell}{I_1} + \frac{\cos^2 \ell}{I_2} \right) (G^2 - L^2) + \frac{L^2}{2I_3}, \tag{1}$$

where the angle ℓ still appears, which is not convenient for normalizing a perturbed Hamiltonian.

A complete reduction may be obtained by means of the transformations given by Sadov [14] and Kinoshita [13] that, independently, gave similar sets of action and angle variables. Another transformation due to Deprit and Elipe [9], and based on the Delaunay equation, converts the Hamiltonian into a function depending only on the momenta. In the above three cases the Hamiltonian is independent of the angular variables; then the normalized Hamiltonian, that must belong to the kernel of the Lie derivative associated to the unperturbed Hamiltonian [8], may be taken as the averaged over the coordinates. By doing so, the reduced Hamiltonian depends only on the moments and consequently, is integrable.

Unfortunately, the variables defined in [9, 13, 14] have not a geometric meaning as clear as the Euler or Serret–Andoyer variables, that, usually, are the ones employed to find the potential function. Hence, it is necessary to find expressions of the Serret–Andoyer variables in terms of the angle and action. This is achieved by taking into account that elliptic functions and integrals can be expanded in power series of their nome $q = \exp(-\pi K'/K)$. Thus, we expand the perturbation as the Fourier series of the amplitude, which coefficients are rational functions of the nome. This type of expansion is proved to be very fast convergent and only few terms give enough accuracy [1, 6], and makes possible to normalize the Hamiltonian by means of a Lie transformation [2]. An application to an Eros-like body is made.

2. Sadov versus Serret–Andoyer Variables

To study the attitude of a rigid body about a point C, we set two right-handed frames at that point. One fixed in the space $\mathcal{S} \equiv C s_1 s_2 s_3$, and another one, the system of principal axes of inertia, $\mathcal{B} \equiv C\mathbf{b}_1\mathbf{b}_2\mathbf{b}_3$, fixed in the body. As it is known from Euler, attitude of $\mathcal{B}$ in $\mathcal{S}$ results from three rotations.

In an independent way, Sadov [14] and Kinoshita [13] obtained sets of action- and angle-variables for the rotational motion of a triaxial rigid body. Essentially, they are obtained by solving the Hamilton–Jacobi equation from the Hamiltonian (1). Sadov's transformation is given in terms of the Legendre elliptic functions of the first and third kind, whereas in the Kinoshita one, the Heuman Lambda function appears. Both transformation are equivalent [5] and here, we shall use Sadov's ones.

The first step consists in obtaining the generating function $\mathcal{S}$ of the transformation $(\ell, g, h, L, G, H) \longrightarrow (\beta_1, \beta_2, \beta_3, \alpha_1, \alpha_2, \alpha_3)$, by solving the Hamilton–Jacobi equation corresponding to the Hamiltonian (1). Since the system is autonomous and the variables g and h are cyclic, the generating function may be put as

$$\mathcal{S} = -\alpha_1 t + \alpha_2 g + \alpha_3 h + \mathcal{U}(\ell; \alpha_1, \alpha_2, \alpha_3),$$

with the function $\mathcal{U}$ obtained by a quadrature from

$$\left(\frac{\partial \mathcal{U}}{\partial \ell}\right)^2 = I_3 \left(\frac{a + b \sin^2 \ell}{c + d \sin^2 \ell}\right),$$

with

$$\begin{array}{ll}
a = I_1(\alpha_2^2 - 2\alpha_1 I_2), & c = I_1(I_3 - I_2) > 0, \\
b = (I_2 - I_1)\alpha_2^2 > 0, & d = I_3(I_2 - I_1) > 0.
\end{array}$$

After this transformation, Sadov proceeds to obtain a second one

$$(\beta_1, \beta_2, \beta_3, \alpha_1, \alpha_2, \alpha_3) \longrightarrow (\varphi_\ell, \varphi_g, \varphi_h, I_\ell, I_g, I_h), \tag{2}$$

such that the new actions are

$$\begin{aligned}
I_\ell &= \frac{1}{2\pi} \oint L \, d\ell, \\
I_g &= \frac{1}{2\pi} \oint G \, dg = G = \alpha_2, \\
I_h &= \frac{1}{2\pi} \oint H \, dh = H = \alpha_3.
\end{aligned}$$

To compute these quadratures, Sadov defines a parameter κ and a state function λ by the relations

$$\kappa^2 = \frac{I_3(I_2 - I_1)}{I_1(I_3 - I_2)} \geqslant 0, \qquad \lambda^2 = \kappa^2 \frac{I_1}{I_3} \frac{2 I_3 \alpha_1 - \alpha_2^2}{\alpha_2^2 - 2 I_1 \alpha_1} \geqslant 0. \tag{3}$$

With these quantities the angular moment L may be put as

$$L^2 = \alpha_2^2 \frac{\kappa^2}{\kappa^2 + \lambda^2} \frac{1 - \lambda^2 + (\kappa^2 + \lambda^2)\sin^2 \ell}{1 + \kappa^2 \sin^2 \ell} = I_g^2 L_1^2(\ell, \lambda), \tag{4}$$

and by means of the change

$$\sin \ell = \frac{\cos z}{\sqrt{1 + \kappa^2 \sin^2 z}}, \qquad \cos \ell = -\frac{\sqrt{1 + \kappa^2} \sin z}{\sqrt{1 + \kappa^2 \sin^2 z}}, \tag{5}$$

and after some algebra, there results

$$I_\ell = \frac{1}{2\pi} \oint L \, d\ell = \frac{2\alpha_2 \sqrt{1 + \kappa^2}}{\pi \kappa \sqrt{\kappa^2 + \lambda^2}} ((\kappa^2 + \lambda^2)\Pi(\kappa^2, \lambda) - \lambda^2 K(\lambda)), \tag{6}$$

where K and Π are the complete elliptical integrals of the first and the third kind.

Since the Hamiltonian is $\mathcal{H} = \alpha_1$, we have that

$$\mathcal{H} = \frac{I_g^2}{2 I_1 I_3} \frac{I_3 \lambda^2 + I_1 \kappa^2}{\kappa^2 + \lambda^2}. \tag{7}$$

Let $\mathcal{W}$ be the generating function of the transformation from the Serret–Andoyer variables to the action and angle variables. It may be chosen as

$$\mathcal{W} = I_g g + I_h h + \mathcal{V}(\ell; I_\ell, I_g),$$

with the function $\mathcal{V}$ obtained directly from the quadrature

$$\mathcal{V}(\ell; I_\ell, I_g) = \int_{\ell_0}^{\ell} L(\ell; I_\ell, I_g) \, d\ell,$$

where we have to replace L by its value from expression (4).

To compute φ_ℓ and φ_g from the equations of the transformation, we previously need the value of the partial derivatives $\partial \lambda / \partial I_\ell$ and $\partial \lambda / \partial I_g$, since by the rule chain, and taking into account (4), we get

$$\begin{aligned}
\varphi_\ell &= \frac{\partial \mathcal{V}}{\partial I_\ell} = \frac{\partial \mathcal{V}}{\partial \lambda} \frac{\partial \lambda}{\partial I_\ell} = \frac{\partial \lambda}{\partial I_\ell} \int_{\ell_0}^{\ell} \frac{\partial L}{\partial \lambda} \, d\ell, \\
\varphi_g &= g + \frac{\partial}{\partial I_g} \int_{\ell_0}^{\ell} L \, d\ell = g + \frac{1}{I_g} \int_{\ell_0}^{\ell} L \, d\ell + \frac{\partial \lambda}{\partial I_g} \int_{\ell_0}^{\ell} \frac{\partial L}{\partial \lambda} \, d\ell.
\end{aligned} \tag{8}$$

With the help of the change (5), we compute the quadratures

$$\int_{\ell_0}^{\ell} \frac{\partial L}{\partial \lambda} \, d\ell = -\frac{I_g \kappa \lambda (1 + \kappa^2)^{1/2}}{(\kappa^2 + \lambda^2)^{3/2}} F(z, \lambda),$$

and

$$\int_{\ell_0}^{\ell} L \, d\ell = \frac{I_g}{\kappa} \sqrt{\frac{1 + \kappa^2}{\kappa^2 + \lambda^2}} [(\kappa^2 + \lambda^2)\Pi(z, \kappa^2, \lambda) - \lambda^2 K(\lambda)].$$

For the other partial derivatives, we put the expression (6) as $I_\ell = I_g f(\lambda)$, hence,

$$\frac{\partial I_\ell}{\partial \lambda} = I_g \frac{\partial f}{\partial \lambda} = \frac{1}{2\pi} \oint \frac{\partial L}{\partial \lambda} = -\frac{I_g 2\kappa \lambda (1 + \kappa^2)^{1/2}}{\pi (\kappa^2 + \lambda^2)^{3/2}} K(\lambda)$$

and

$$\frac{\partial I_g}{\partial \lambda} = -\frac{I_g}{f(\lambda)} \frac{\partial f}{\partial \lambda} = -\frac{I_g^2}{I_\ell} \frac{\partial f}{\partial \lambda}$$

$$= \frac{I_g \kappa^2 \lambda K(\lambda)}{(\kappa^2 + \lambda^2)[(\kappa^2 + \lambda^2)\Pi(\kappa^2, \lambda) - \lambda^2 K(\lambda)]}.$$

Replacing all these expressions into (8), one obtain

$$\varphi_\ell = \frac{\pi}{2} \frac{F(z, \lambda)}{K(\lambda)}, \varphi_g = g + \frac{1}{\kappa} \sqrt{(\kappa^2 + \lambda^2)(1 + \kappa^2)} \times$$

$$\times \left[\Pi(z, \kappa^2, \lambda) - \frac{\Pi(\kappa^2, \lambda) F(z, \lambda)}{K(\lambda)} \right]. \tag{9}$$

By inverting the first of these expressions, we get the argument z as $z = \mathrm{am}\,(2K(\lambda)/\pi\varphi_\ell, \lambda)$ and thus, we have the Serret–Andoyer angle g in terms of the action-angle variables

$$g = \varphi_g + \frac{1}{\kappa} \sqrt{(\kappa^2 + \lambda^2)(1 + \kappa^2)} \left[\frac{2}{\pi} \Pi(\kappa^2, \lambda) \varphi_\ell - \Pi(z, \kappa^2, \lambda) \right],$$

and the angle ℓ is obtained directly from the expression (5). Lastly, from Equation (4) we get L as

$$L = \frac{I_g \kappa}{\sqrt{\kappa^2 + \lambda^2}} \, \mathrm{dn}(2K(\lambda)/\pi \, \varphi_\ell, \lambda).$$

Note that dn is a cosine-like Jacobian elliptic function.

3. Motion Under a Newtonian Gravity Field

Let us consider a rigid body S, with center of mass C and mass m, that is moving under the gravitational influence of point a O of mass M. Besides the space $S(C, \mathbf{s}_1, \mathbf{s}_2, \mathbf{s}_3)$ and the body $\mathcal{B}(C, \mathbf{b}_1, \mathbf{b}_2, \mathbf{b}_3)$ reference systems, let us consider another inertial frame $S'(O, \mathbf{s}_1, \mathbf{s}_2, \mathbf{s}_3)$, parallel to the space system and with origin at the point O.

Under the assumption that the dimensions of the rigid body are small enough with respect to the dimensions of the relative orbit of C around O, and that the mass of S is small with respect to the mass of O, it is possible to assume that the orbital motion is independent of the rotation of S [10]. Besides, under these hypotheses, the gravitational potential V may be considered as a perturbation of the kinetic energy of rotation T. Hence, our problem will consist of finding the attitude of the rigid body, assuming that the orbital motion is a known function of the time.

When the rigid body has axial symmetry, $a_1 = a_2$, Serret–Andoyer variables are very convenient, since the angular variable ℓ does not appear in (1) and their Hamilton's equations are immediately integrated. In cases where the axial symmetry 'almost' exists, that is, when $a_1 \approx a_2$, some authors [11], by means of trigonometric relations, split the Hamiltonian (1) in two parts, one independent of the angle ℓ, and a second one depending on ℓ, but that have the small quantity a_1-a_2 as factor, and is considered as a perturbation. In this way, elliptic functions are avoided.

However, in general, the body is far from the axial symmetry, such as it happens in most of asteroids, and the previous assumption ($a_1 \approx a_2$) is no longer valid. Thus, it is necessary to cope with elliptic functions.

3.1. EXPANSION OF THE POTENTIAL FUNCTION

In order to apply a general perturbation method, it is desirable to have the unperturbed Hamiltonian (1) the simplest possible in order the Lie derivative $\mathcal{L}_0$, or the Poisson bracket, be easy to evaluate. In this section, we shall use Sadov's variables, although Kinoshita's of the ones given by Deprit and Elipe would serve for our purposes.

The potential acting on the rigid body S may be represented up to the second order by means of the well-known MacCullagh's formula

$$V = -\mathcal{G}\,\frac{mM}{r} - \mathcal{G}\,\frac{M}{2r^3}\,[(I_1 - I_2)(1 - 3\alpha^2) + (I_3 - I_2)(1 - 3\gamma^2)], \qquad (10)$$

where $\mathcal{G}$ is the gravitational constant, r is the distance between the centers of mass O and C, and (α, β, γ) the direction cosines of the unit position vector $\mathbf{u} = \mathbf{x}/r$ expressed in the body frame $\mathcal{B}$, that is to say,

$$\mathbf{u} = \mathbf{x}/r = \mathbf{b}_1\alpha + \mathbf{b}_2\beta + \mathbf{b}_3\gamma.$$

These direction cosines suggest to choose an orbital moving frame $\mathcal{O}(\mathbf{u}, \mathbf{v}, \mathbf{w})$, where $\mathbf{w}$ is in the direction of the orbital angular moment ($\mathbf{\Theta} = \mathbf{x} \times \dot{\mathbf{x}}$), and $\mathbf{v} = \mathbf{w} \times \mathbf{u}$. In this way, the two bases $\mathcal{O}$ and $\mathcal{B}$ are related by several compositions of rotations involving as arguments both the Serret–Andoyer (ℓ, g, h) and Polar-nodal (r, θ, ν) variables (see Figure 1). In matrix form, there results $(\alpha, \beta, \gamma) = R\,(1, 0, 0)$, with R the rotation obtained by the composition of seven rotations

$$R(-\ell, \mathbf{b}_3) \circ R(-\sigma, \mathbf{m}) \circ R(-g, \mathbf{n}) \circ R(-\epsilon, \ell) \circ$$
$$R(\nu - h, \mathbf{s}_3) \circ R(\mathbf{I}, \mathbf{N}) \circ R(\theta, \mathbf{w}),$$

where $\mathbf{N}$ and I are the node direction and inclination of the orbital plane ($\mathbf{uv}$) with respect to the space plane ($\mathbf{s}_1\mathbf{s}_2$).

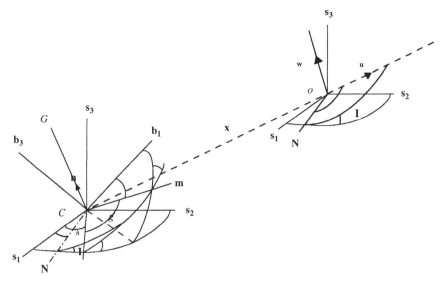

Figure 1. Orbital–rotational variables.

Once obtained the direction cosines α, β and γ, it is necessary to square their expressions and replace them in the Equation (10). By doing so, in the potential function will appear expressions of the type $\prod_{a,b} \sin^a \cos^b$, that is to say, combinations of products of powers of trigonometric sines and cosines of the involved variables. Note that the potential is given in terms of the Serret–Andoyer variables, whereas the kinetic energy was expressed in terms of the action and angle variables, namely, Sadov's variables. Thus, it is necessary to replace the solution obtained for the free rotation (§ 2), that is given in terms of elliptic functions. Hence, we deal with expressions of the type

$$\prod_{\alpha,\beta} \operatorname{sn}^a(z,k) \operatorname{cn}^b(z,k) \operatorname{dn}^c(z,k). \tag{11}$$

The computation of the potential (10) has been previously carried out by several authors (e.g. [4, 13]), although they avoid the powers in favor of linear combinations of angles. Here we maintain the powers of trigonometric functions in the potential because we will expand terms of the kind (11) as Fourier powers series of the amplitude which coefficients are rational functions of the Jacobian nome q. This type of expansions has proved to be very fast convergent, and it is possible to generate them automatically by computer by making use of the properties of the Theta functions and computing the coefficients by means of the Residues theorem. For details the reader is addressed to [1]. The complete expansion of the potential (10) is too large to be reproduced here, but it can be stored in the hard disk of our computer.

3.2. ANALYTICAL SOLUTION

With the above development, the Hamiltonian of the attitude dynamics of a triaxial rigid body under a Newtonian force field, may be put

$$\mathcal{H} = \mathcal{H}(\varphi_\ell, \varphi_g, \varphi_h, I_\ell, I_g, I_h) = \mathcal{H}_0 + V = \frac{I_g^2}{2\,I_1 I_3} \frac{I_3\lambda^2 + I_1\kappa^2}{\kappa^2 + \lambda^2} + \mathcal{H}_1,$$

in a set of canonical variables $(\varphi_\ell, \varphi_g, \varphi_h, I_\ell, I_g, I_h)$, in which the unperturbed Hamiltonian $\mathcal{H}_0$ depends on the momenta, and the perturbation is a truncated Fourier series, which arguments are the angular variables and the coefficients are functions on the momenta. Hence, it is possible to eliminate the angular variables $\varphi_\ell, \varphi_g, \varphi_h$ by means of a Lie transformation [8]. Such a transformation,

$$(\varphi_i', I_i') \xrightarrow{\quad \mathcal{W} \quad} (\varphi_i, I_i), \qquad (i = \ell, g, h),$$

we recall, makes the pullback of $\mathcal{H}$ belong to the kernel of the Lie derivative, that is to say, $\mathcal{L}_0(\mathcal{H}') = 0$. But since $\mathcal{H}_0$ is independent of the angles, up to the first order perturbation, we can select as the new Hamiltonian $\mathcal{H}_1'$ the average of $\mathcal{H}_1$ over the angles. The generator of the transformation $\mathcal{W}_1$ is obtained from the so called homologic equation, namely, the PDE

$$\{\mathcal{H}_0; \mathcal{W}_1\} = \mathcal{H}_1' - \mathcal{H}_1. \tag{12}$$

The procedure for both tasks, averaging the Hamiltonian and computing the generating function has been automated by means of an algebraic manipulator.

3.3. APPLICATION TO AN EROS-LIKE BODY

As an illustration, we apply the above exposed method to the attitude of a body, similar to the asteroid (433) Eros, the target of the NEAR mission, a natural body that is far away from the axial symmetry. Indeed, Eros is one of the most elongated asteroids, a 'potato-shaped' body with estimated dimensions of 40.5 by 14.5 by 14.1 km. Astronomers assign the asteroid a rotation period of 5.27 h. The value of the mass adopted by NASA is $5\,10^{15}$ kg. The following orbital elements of (433) Eros are extracted from Minor Planet Circular No. 24086, updated to a current osculating epoch (JDT 2450200.5):

$$a = 1.4583190\,AU, \qquad e = 0.2229864, \qquad I = 10°83085,$$
$$\Omega = 304°43847, \qquad \omega = 178°55858, \qquad P = 1.76\,\text{years}.$$

Besides these physical characteristics, we choose in rather arbitrary way the inclinations ϵ and σ to have the modulus of the elliptic functions involved in the problem very high, and close to the unit (the most unfavorable case). With these

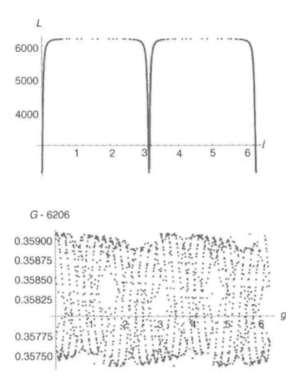

Figure 2. Evolution of the Serret angles ℓ and g (mod 2π) versus their respective moments after normalization. Note that normalization is carried out in Sadov's variables, then converted to Serret's ones. Since the graphics represent several revolutions, we observe ℓ is almost periodic, while g suffers a *precession*-like shift.

criteria, in terms of Serret–Andoyer variables, we choose as initial conditions the values

$\ell = 0.5235987755982988,$ $L = 6187.225633746185,$
$g = 0.5235987755982988,$ $G = 6206.357763463754,$
$h = 0.7853981633974483,$ $H = 5374.863488134382,$
$\varepsilon = 0.5235987755982988,$ $\sigma = 0.0785398163397448.$

Hence, there results $\lambda = 0.936057480558765$, $\kappa = 23.77968208793787$. In Figure 2 we present the evolution of the Serret–Andoyer variables after normalization that was carried out in Sadov's variables. Since the graphics represent several revolutions, we observe ℓ is almost periodic, while g suffers a *precession*-like shift.

Acknowledgements

The authors are indebted to Professor Cid, who motivated their interest for the rigid bodies, to Dr. Deprit for his geometric insights of the problem, to Dr. Barkin, who gave us the indications about the obtaining of the Fourier expansions in the Jacobian nome. This paper has been supported by the Spanish Ministry of Education and Science (Projects #PB98-1576 and #ESP99-1074-CO2-01).

References

1. Abad, A., Elipe, A. and Vallejo, M.: 1994, 'Automated Fourier series expansions for elliptic functions', *Mech. Res. Commun.* **21**, 361–366.
2. Abad, A., Elipe, A. and Vallejo, M.: 1997, 'Normalization of perturbed non-linear Hamiltonian oscillators', *Int. J. Non-lin. Mech.* **32**, 443–453.
3. Andoyer, H.: 1923, *Cours de mécanique céleste*, Paris, Gauthier-Villars et Cie.
4. Arribas, M. and Elipe, A.: 1993, 'Attitude dynamics of a rigid body on a Keplerian orbit: a simplification', *Celest. Mech. & Dyn. Astr.* **55**, 243–247.
5. Barkin, Y. A.: 1992, *Lecture notes* Universidad de Zaragoza (unpublished).
6. Brumberg, V. A.: 1995, *Analytical Techniques of Celestial Mechanics*. Springer, Berlin.
7. Deprit, A.: 1967, 'Free rotation of a rigid body studied in the phase plane', *Am. J. Phy.* **35**, 424–428.
8. Deprit, A.: 1969, 'Canonical transformation depending on a small parameter', *Celestial Mechanics*, **1**, 12–30.
9. Deprit, A. and Elipe, A.: 1993, 'Complete reduction of the Euler–Poinsot problem', *J. Astronaut. Sci.* **41**, 603–628.
10. Elipe, A., Abad, A. and Arribas, M.: 1992, 'Scaling Hamiltonians in attitude dynamics of two rigid bodies', *Rev. Acad. Ciencias de Zaragoza* **47**, 137–154.
11. Ferrándiz, J. M. and Sansaturio, M. E.: 1989, 'Elimination of the nodes when the satellite is a non-spherical rigid body', *Celest. Mech. & Dyn. Astr.* **46**, 307–320.
12. Galgani, L., Giorgilli, A. and Strelcyn, J. M.: 1981, 'Chaotic motion and transition to stochasticity in the classical problem of a heavy rigid body with a fixed point', *Il Nuovo Cimento B*, **61**, 1–20.
13. Kinoshita, H.: 1972, 'First-order perturbations of the two finite body problem', *Publ. Astron. Soc. Japan* **24**, 423–457.
14. Sadov, Yu. A.: 1970, 'The Action-Angles Variables in the Euler-Poinsot Problem', *Akad. Nauk SSSR*, Prep. **22**.
15. Serret, J. A.: 1866, 'Mémoire sur l'emploi de la méthode de la variation des arbitraires dans la théorie des mouvements de rotation', *Mémoi. Acad. Sci. Paris* **35**, 585–616.
16. Shuster, M.: 1993, 'A survey on attitude representations', *J. Astronaut. Sci.* **41**, 439–517.
17. Wisdom, J.: 1987, 'Rotational dynamics of irregularly shaped natural satellites', *Astron. J.* **94**, 1350–1360.

A SIMPLIFIED KINETIC ELEMENT FORMULATION FOR THE ROTATION OF A PERTURBED MASS-ASYMMETRIC RIGID BODY

JASON W. MITCHELL[1] and DAVID L. RICHARDSON[2]

[1]*Air Force Research Laboratory, AFRL/VACA, Wright-Patterson AFB, OH 45433-7521, U.S.A.*
[2]*Aerospace Engineering, University of Cincinnati, Cincinnati, OH 45221-0070, U.S.A.*

Abstract. Euler's equations, describing the rotation of an arbitrarily torqued mass asymmetric rigid body, are scaled using linear transformations that lead to a simplified set of first order ordinary differential equations without the explicit appearance of the principal moments of inertia. These scaled differential equations provide trivial access to an analytical solution and two constants of integration for the case of torque-free motion. Two additional representations for the third constant of integration are chosen to complete two new kinetic element sets that describe an osculating solution using the variation of parameters. The elements' physical representations are amplitudes and either angular displacement or initial time constant in the torque-free solution. These new kinetic elements lead to a considerably simplified variation of parameters solution to Euler's equations. The resulting variational equations are quite compact. To investigate error propagation behaviour of these new variational formulations in computer simulations, they are compared to the unmodified equations without kinematic coupling but under the influence of simulated gravity-gradient torques.

Key words: astronautics, satellite attitude dynamics, variation of parameters, rigid body motion, asymmetric rigid bodies

1. Introduction

Variational methods have played an important role in solving ordinary differential equations since their first application by Johann Bernoulli in 1697 to solve the general linear differential equation of first order. Perhaps the most famous application of a variational method, the variation of orbital elements or the variation of constants, was performed by Leonhard Euler between 1748 and 1752 to describe the mutual perturbations of Jupiter and Saturn. However, it was not until 1782 that Joseph-Louis Lagrange fully developed the method of the variation of parameters in an application involving cometary motion. His approach is widely used today particularly in astrodynamics applications where perturbed two-body motion is considered.

The first published work to attempt a variation of parameters solution to the problem of rotational motion was given by Sir Arthur Cayley and originally appeared in the *Memoirs of the Royal Astronomical Society* in 1861 [5]. In this work, Cayley chooses his rotational elements using a geometric analog with the two-body orbital motion's arbitrary planes of reference and departure [4]. Consequently, any disturbing torque appearing is the result of a disturbing function

Celestial Mechanics and Dynamical Astronomy **81:** 13–25, 2001.
© 2001 *Kluwer Academic Publishers.*

that must be written as the partial of a disturbing potential with respect to the rotational elements chosen. This limits the applicability of the resulting solution to only those torques that can result from a specified potential function. In addition, while Cayley discusses why he chose his rotational elements and fully develops the variational equations for those elements, he does not appear to specify a solution to the homogeneous problem, further limiting the usefulness of the work.

Since the development of artificial satellites, primarily during the mid to late 1970's, a variation of parameters solution for the rotational motion of rigid bodies has been investigated several times. Work by Kraige and Junkins [9] and Donaldson and Jezewski [6] used the body's kinetic energy and angular momentum as the primary parameters in a general variation of parameters scheme. The resulting equations were algebraically complex, involving complete and incomplete elliptic integrals of the first and second kinds, incomplete elliptic integrals of the third kind, and the Jacobian elliptic functions. The later work of Kraige and Skaar [10] and Kraige [8] did provide some simplification of the complexity in the form of the variational equations, however the choice of parameters remained the same. More recently, Bond [1] investigated a variation of parameters scheme for rigid body motion, choosing the case of mass symmetric, torque-free motion as the analytical basis for selecting the parameters.

With the previous work taken altogether, it seems that improvements may be made through a more judicious selection of parameters developed from the analytical solution to the general problem of torque-free rotation.

2. Re-examining Euler's Equations

The traditional method of developing an analytical solution for Euler's torque free motion equations is quite cumbersome and tedious to follow ([7], pp. 205–206). Moreover, immediately invoking the conservation laws for energy and angular momentum to obtain the first integrals of motion in the body frame complicates the solution development, since the resulting expressions fail to provide the easily recognized special function identities that could simplify the solution process. The following attempts to address these issues.

2.1. TRANSFORMING EULER'S EQUATIONS

The classical Euler's equations for rigid-body motion are given by the coupled first-order system

$$A\dot{\omega}_1 = (B - C)\,\omega_2\omega_3 + M_1, \tag{1a}$$
$$B\dot{\omega}_2 = (C - A)\,\omega_1\omega_3 + M_2, \tag{1b}$$
$$C\dot{\omega}_3 = (A - B)\,\omega_1\omega_2 + M_3. \tag{1c}$$

The ω_i, for $i = 1, 2, 3$, are the components of the angular velocity of the body expressed in a body-fixed, principal-axis coordinate frame. The M_i are the external torque components measured with respect to the principal axes. A, B, and C are the principal-axis moments of inertia which are ordered, without loss of generality, such that $A < B < C$.

Rewriting Equations (1), we have

$$\dot{\omega}_1 = -D_1\omega_2\omega_3 + M_1/A, \tag{2a}$$

$$\dot{\omega}_2 = D_2\omega_1\omega_3 + M_2/B, \tag{2b}$$

$$\dot{\omega}_3 = -D_3\omega_1\omega_2 + M_3/C, \tag{2c}$$

where

$$D_1 = (C - B)/A, \tag{3a}$$

$$D_2 = (C - A)/B, \tag{3b}$$

$$D_3 = (B - A)/C, \tag{3c}$$

with $D_i > 0$.

Upon performing a change of variables described by Richardson and Mitchell [14], and Livneh and Wei [11], given by

$$\Omega_i = \omega_i/\sqrt{D_i}, \tag{4a}$$

$$\frac{d\tau}{dt} = \sqrt{D_1 D_2 D_3}. \tag{4b}$$

Euler's equations become

$$\Omega_1' = -\Omega_2\Omega_3 + G_1, \tag{5a}$$

$$\Omega_2' = \Omega_1\Omega_3 + G_2, \tag{5b}$$

$$\Omega_3' = -\Omega_1\Omega_2 + G_3, \tag{5c}$$

where the prime indicates $d/d\tau$ and G_i are the rescaled external moments given by

$$G_1 = M_1/(AD_1\sqrt{D_2 D_3}), \tag{6a}$$

$$G_2 = M_2/(BD_2\sqrt{D_1 D_3}), \tag{6b}$$

$$G_3 = M_3/(CD_3\sqrt{D_1 D_2}). \tag{6c}$$

Finally, taking $G_i = 0$ in Equations (5) we have the transformed torque-free Euler equations,

$$\Omega_1' = -\Omega_2\Omega_3, \tag{7a}$$

$$\Omega_2' = \Omega_1\Omega_3, \tag{7b}$$

$$\Omega_3' = -\Omega_1\Omega_2. \tag{7c}$$

Two independent integrals of motion, c_1 and c_2, of the unperturbed system given by Equations (7) are immediately obtained by inspection and found to be

$$\Omega_1^2 + \Omega_2^2 = c_1^2, \tag{8a}$$

$$\Omega_2^2 + \Omega_3^2 = c_2^2. \tag{8b}$$

THEOREM 1. *The integrals of motion $c_1{}^2$ and $c_2{}^2$ are non-negative.*

Proof. Taking $\mathbf{\Omega} \in \mathbb{R}^3$ and forming $H^2 - 2TA$ and $2TC - H^2$, then using Equations (3), (4a), along with the strict ordering of the principal moments of inertia, we arrive at the desired result. $\qquad\square$

As consequence of proving Theorem 1, the constants of integration found in Equations (8) are related to the traditional constants of angular momentum H and kinetic energy T in the following.

COROLLARY 1. *The integrals of motion, $c_1{}^2 \equiv c_1{}^2\,(H, T)$ and $c_2{}^2 \equiv c_2{}^2\,(H, T)$ are given by*

$$c_1{}^2 = \frac{2TC - H^2}{(C - A)(C - B)}, \tag{9a}$$

$$c_2{}^2 = \frac{H^2 - 2TA}{(C - A)(B - A)}. \tag{9b}$$

2.2. TORQUE-FREE ANALYTICAL SOLUTION REVISITED

A solution for the unperturbed system, Equations (7), is developed by first considering Equation (7b). Combining Equation (7b) with Equations (8) gives

$$\Omega_2' = \sqrt{c_1{}^2 - \Omega_2{}^2}\sqrt{c_2{}^2 - \Omega_2{}^2}, \tag{10}$$

or separating variables

$$\int_0^{\Omega_2/c_1} \frac{ds}{\sqrt{1 - s^2}\sqrt{1 - k^2 s^2}} = c_2 \int_{\tau_0}^{\tau} ds. \tag{11}$$

Restricting the modulus $k = c_1/c_2$ such that $k \in \mathbb{R}(0, 1)$, we recognize Equation (11) as the inverse of a generalized sine function, Byrd and Friedman [3],

$$\operatorname{sn}^{-1}(\Omega_2/c_1) = c_2(\tau - \tau_0) \triangleq u, \tag{12}$$

where sn u is the Jacobian elliptic function sine amplitude u. Using this along with Equations (8) produces the familiar torque-free solution in terms of the Jacobian elliptic functions

$$\Omega_1 = c_1 \operatorname{cn} u, \tag{13a}$$
$$\Omega_2 = c_1 \operatorname{sn} u, \tag{13b}$$
$$\Omega_3 = c_2 \operatorname{dn} u, \tag{13c}$$
$$u = c_2(\tau - \tau_0), \tag{13d}$$

considering u to be the intrinsic angular variable or argument.

Returning to Equation (10) and following the same procedure but choosing $k = c_2/c_1$, we find

$$\text{sn}^{-1}(\Omega_2/c_2) = c_1\,(\tau - \tau_0) \triangleq u. \tag{14}$$

Again, using the above along with Equations (8), the unperturbed solution of Euler's equations is given by

$$\Omega_1 = c_1\,\text{dn}\,u, \tag{15a}$$
$$\Omega_2 = c_2\,\text{sn}\,u, \tag{15b}$$
$$\Omega_3 = c_2\,\text{cn}\,u, \tag{15c}$$
$$u = c_1\,(\tau - \tau_0). \tag{15d}$$

Since the the Jacobian elliptic functions found in Equations (13) and (15) depend on both the argument u and the modulus k, it is clear that there is a three-way nonlinear coupling through the amplitude, frequency, and modulus. While this is easily seen using the transformed differential equations, it is not as readily apparent from the analytical solutions to the unmodified torque-free equations.

3. Variational Equations

From Equations (13) and (15), the quantities c_1, c_2 and τ_0 form a natural set of parameters for the characterization of the solution to the torque free Euler equations. These quantities are analogous to a harmonic oscillator's amplitude, frequency, and initial time constant, respectively.

Directly obtaining a functional expression for τ_0 is difficult. Using either Lagrange's or Poisson's method to find the variation of τ_0 fails to provide a manageable solution due to the lack of sufficient elliptic function and integral identities necessary to render a suitable expression. As a result, it is more algebraically desirable to seek the variation of the argument u rather than τ_0 as in the following.

3.1. VARIATIONAL EQUATIONS FOR $c_1 < c_2$

The variations of the two independent integrals of motion, c_1 and c_2, are obtained by direct differentiation of Equations (8) along with substitutions from Equations (5) and (13), yielding

$$c_1' = \frac{1}{c_1}(\Omega_1 G_1 + \Omega_2 G_2) = G_1\,\text{cn}\,u + G_2\,\text{sn}\,u, \tag{16a}$$

$$c_2' = \frac{1}{c_2}(\Omega_2 G_2 + \Omega_3 G_3) = kG_2\,\text{sn}\,u + G_3\,\text{dn}\,u. \tag{16b}$$

To develop a variational equation for u, we begin by writing the derivatives, with respect to τ, of Equations (13a) and (13b) as

$$\Omega_i' = \frac{\partial \Omega_i}{\partial u}u' + \frac{\partial \Omega_i}{\partial c_1}c_1' + \frac{\partial \Omega_i}{\partial k}k', \tag{17}$$

for $i = 1, 2$. Forming $\Omega_2'\,\mathrm{cn}\,u - \Omega_1'\,\mathrm{sn}\,u$ and after extensive algebraic manipulation, we obtain

$$u' = c_2 + \frac{1}{c_1 k_c^2}\{[H(u)\,\mathrm{cn}\,u - \mathrm{sn}\,u\,\mathrm{dn}\,u]G_1 + [H(u)\,\mathrm{sn}\,u +$$
$$+ \mathrm{cn}\,u\,\mathrm{dn}\,u]k_c^2 G_2 - [H(u)\,\mathrm{dn}\,u - k^2\,\mathrm{sn}\,u\,\mathrm{cn}\,u]kG_3\}, \tag{18}$$

where $H(u) = E(u) - k_c^2 u$, $k_c^2 = 1 - k^2$ is the complementary modulus and $E(u)$ is the incomplete elliptic integral of the second kind.

3.2. VARIATIONAL EQUATIONS FOR $c_2 < c_1$

The variations of the two independent integrals of motion, c_1 and c_2, are obtained by direct differentiation of Equations (8) along with substitutions from Equations (5) and (15), yielding

$$c_1' = G_1\,\mathrm{dn}\,u + kG_2\,\mathrm{sn}\,u, \tag{19a}$$
$$c_2' = G_2\,\mathrm{sn}\,u + G_3\,\mathrm{cn}\,u. \tag{19b}$$

To obtain a variational equation for u, we write the derivatives, with respect to τ, of Equations (15b) and (15c) in the same fashion as Equations (17). Using this to form $\Omega_2'\,\mathrm{cn}\,u - \Omega_3'\,\mathrm{sn}\,u$, and again, after considerable algebraic manipulation, we find

$$u' = c_1 + \frac{1}{c_2 k_c^2}\{-[H(u)\,\mathrm{dn}\,u - k^2\,\mathrm{sn}\,u\,\mathrm{cn}\,u]kG_1 + [H(u)\,\mathrm{sn}\,u +$$
$$+ \mathrm{cn}\,u\,\mathrm{dn}\,u]k_c^2 G_2 + [H(u)\,\mathrm{cn}\,u - \mathrm{sn}\,u\,\mathrm{dn}\,u]G_3\}. \tag{20}$$

3.3. VARIATIONAL EQUATIONS FOR $c_1 = c_2$ AND $c_1 = 0$ OR $c_2 = 0$

The variational equations for the cases where $k = 1$ and $k = 0$, respectively, are of little interest since they represent two degenerate points in the continuous modulus k space. Consequently, they are omitted. However, a discussion of these cases can be found in Mitchell and Richardson [13] and Mitchell [12].

3.4. VARIATION OF τ_0

As previously mentioned, it is difficult to find the variation of τ_0 through the traditional applications of the variation of parameters. However, considering Equations (13d) and (18) for $c_1 < c_2$, the variation of τ_0 can be obtained with only a small additional effort in algebra. By directly differentiating Equation (13d) with respect to τ, we obtain

$$u' = c_2'(\tau - \tau_0) + c_2\left(1 - \tau_0'\right). \tag{21}$$

After substituting from Equations (16) and (18) into Equation (21) and rearranging, we find an expression for the variation of τ_0

$$\tau_0' = -\frac{1}{c_1 c_2 k_c^2}\Big\{ -[E(u)\,\mathrm{dn}\,u - k^2\,\mathrm{sn}\,u\,\mathrm{cn}\,u]kG_3 + [(E(u) - u)\,\mathrm{sn}\,u +$$
$$+ \mathrm{cn}\,u\,\mathrm{dn}\,u]k_c^2 G_2 + [(E(u) - k_c^2 u)\,\mathrm{cn}\,u - \mathrm{sn}\,u\,\mathrm{dn}\,u]G_1 \Big\}. \tag{22}$$

Similarly, the variation of τ_0 for the case of $c_2 < c_1$ is given by

$$\tau_0' = -\frac{1}{c_1 c_2 k_c^2}\Big\{ -[E(u)\,\mathrm{dn}\,u - k^2\,\mathrm{sn}\,u\,\mathrm{cn}\,u]kG_1 + [(E(u) - u)\,\mathrm{sn}\,u +$$
$$+ \mathrm{cn}\,u\,\mathrm{dn}\,u]k_c^2 G_2 + [(E(u) - k_c^2 u)\,\mathrm{cn}\,u - \mathrm{sn}\,u\,\mathrm{dn}\,u]G_3 \Big\}. \tag{23}$$

This provides the variation of the originally chosen parameters: c_1, c_2, and τ_0.

4. Numerical Testing

Avoiding the added complexity of a full kinematic model, a three-frequency torque model is used to simulate the gravity-gradient torques experienced by a satellite in a circular orbit about the Earth. Using the standard gravity-gradient torque description, appropriately scaled, we have

$$G_1 = \frac{3n^2}{\sqrt{D_2 D_3}}\cos\theta_1(\tau)\cos\theta_2(\tau), \tag{24a}$$

$$G_2 = \frac{3n^2}{\sqrt{D_1 D_3}}\cos\theta_1(\tau)\cos\theta_3(\tau), \tag{24b}$$

$$G_3 = \frac{3n^2}{\sqrt{D_1 D_2}}\cos\theta_2(\tau)\cos\theta_3(\tau), \tag{24c}$$

where $\cos\theta_i$ were chosen to simulate the gravity-gradient torques for a nominal 350 km orbital altitude and n corresponds to the mean motion. The mass separations, taken to be $D_1 \approx 0.4223$, $D_2 \approx 0.9403$, and $D_3 \approx 0.8592$, correspond to a gravity-gradient orientation for a early design of the International Space Station in its Man Tended Configuration.

The system specification is completed by choosing initial conditions. Table I shows the initial conditions and fixed step sizes used in the numerical simulations. These initial conditions were chosen to provide basic coverage of the range of the modulus since $k \in (0, 1)$.

Using the initial conditions from Table I, the variational equations, Equations (16) and (18), and Euler's equations, Equations (5), were integrated numerically with a Runge–Kutta 4/5 scheme using double-precision. Both systems of equations were integrated with the same fixed stepsize. The fundamental stepsize

TABLE I

Initial conditions and integrator stepsizes

Tag	Initial conditions	Step size
1	$\Omega_1 = 0.500, \Omega_2 = 0, \Omega_3 = 1$	$h = 0.010$
2	$\Omega_1 = 0.999, \Omega_2 = 0, \Omega_3 = 1$	$h = 0.010$
3	$\Omega_1 = 0.200, \Omega_2 = 0, \Omega_3 = 1$	$h = 0.010$
4	$\Omega_1 = 0.500, \Omega_2 = 0, \Omega_3 = 1$	$h = 0.001$
5	$\Omega_1 = 0.500, \Omega_2 = 0, \Omega_3 = 1$	$h = 0.100$

h was taken to be approximately 1/400th of the initial dn u period as specified by the baseline initial conditions 1, that is, $h = K/200$, where $K \equiv K(k)$ is the complete elliptic integral of the first kind. For this case, we have $K(0.5) \approx 1.686$.

Both sets of double-precision integrations were compared to an extended-precision integration of Equations (5). The `doubledouble` extended-precision library for C++ written by Keith Briggs [2] was used in this capacity. The extended-precision stepsize was approximately 1/10th of the of the step size of the corresponding double-precision integration. The extended-precision integration was performed with a 16th-degree first-order Chebyshev procedure [15].

For each integration, the l_2-norm of the body rate error was computed at each integration step, that is, $\|\Omega^* - \Omega\|_2$. Then, at each step, the l_2-norm of the errors for $[0, \tau]$ were computed, that is, $\| \|\Omega^* - \Omega\|_2 \|_2$. This error was graphed as a function of the scaled time τ. Figures 1 through 3 show the body rate error results for the initial conditions given in Table I. Each figure has an attached legend identifying each curve. Each curve is designated by two groups of three letter sequences where each sequence's digits represent:

1. **d** direct equations,
   v VOP form using u,
   t VOP form using τ_0.

2. **d** double-precision,
   q extended-precision.

3. **c** Chebyshev ($n = 16$) integrator,
   r Runge–Kutta (RK45) integrator.

The **b** appearing in the figure legends represents a form of the variation of parameters that is of little interest and not discussed here for brevity.

5. Conclusions

In this paper, a simplified variation of parameters solution to a normalized form of the arbitrarily-perturbed Euler's equations for attitude motion is developed. This

A SIMPLIFIED KINETIC ELEMENT FORMULATION FOR RIGID-BODY ROTATION

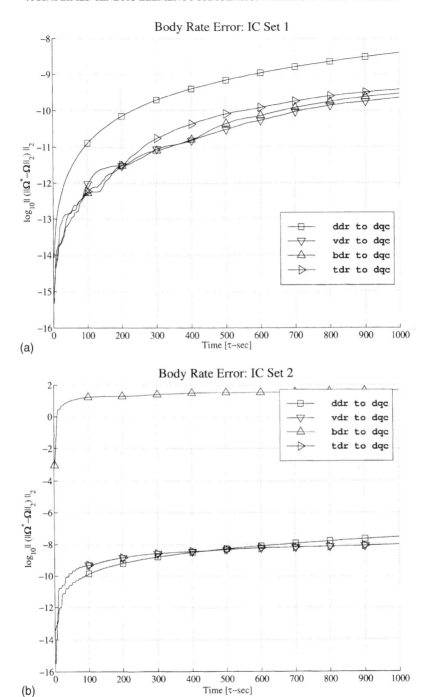

Figure 1. Body rate errors for initial conditions 1 and 2.

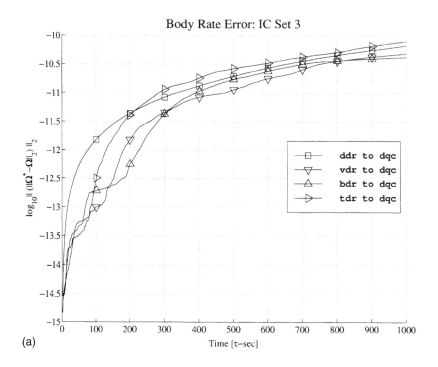

(a)

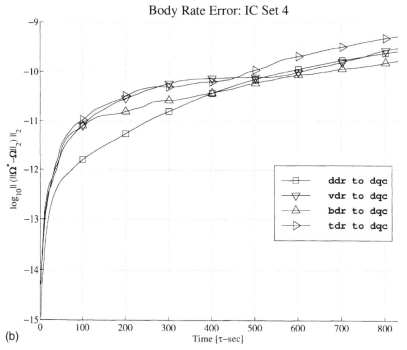

(b)

Figure 2. Body rate errors for initial conditions 3 and 4.

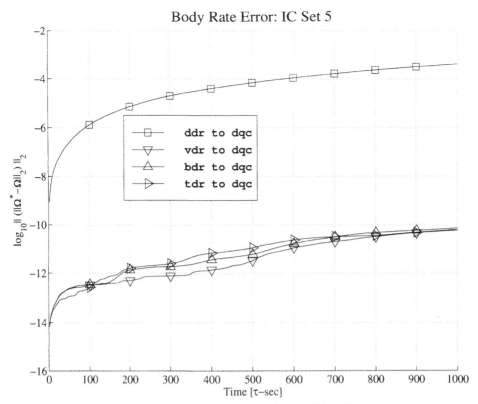

Figure 3. Body rate errors for initial conditions 5.

solution is valid over the entire modulus space, however a discussion of the solution form at the two degenerate points in the modulus space was omitted. The three parameters of the variation, two amplitudes and an angular displacement or initial time constant, lead to a very compact form of the equations. For the same fixed integrator stepsize, these equations generally yielded a higher numerical accuracy over longer simulation intervals as compared to a direct integration of Euler's equations. For the worst case performance, the variation of parameters performed as well as the direct Euler's equations. Moreover, since it is the equations describing the kinetics involved that have been recast, any improved accuracy is independent of the numerical integration scheme selected to perform the propagation.

While the accuracy of the integration did improve, one disadvantage of this method is additional computational overhead necessary to compute the required special functions. This overhead is consistent with the variation of parameters integrations of the orbital elements in astrodynamics applications and is expected. For applications involving algebraically-complex moment components, the variational equations will produce comparable execution timings, for example, a torque model that includes effects of zonal and tesseral harmonics of more than a few orders will exceed the computation time of the special functions used.

24 JASON W. MITCHELL AND DAVID L. RICHARDSON

Regarding the choice of stepsize for numerical integrations, the original differential equations and the variational equations both contain the same spectral content. Consequently, the variational equations cannot be integrated using stepsizes that are substantially greater than that needed for the original differential equations. In general, the suggested stepsize for any given numerical method is related to the highest frequency expected in the dynamics. Even so, the details in selecting a stepsize are still specific to a particular numerical method or class of methods.

Acknowledgements

This research was supported by the National Aeronautics and Space Administration under Graduate Student Researchers Program (GSRP) grant NGT-51639.

References

1. Bond, V. R.: 1996, 'A variation of parameters approach for the solution of the differential equations for the rotational motion of a rigid body', In: *Spaceflight Mechanics 1996: Proceedings of the AAS/AIAA Spaceflight Mechanics Conference*, Vol. 93 of *Advances in Astronautical sciences*, San Diego, California, p. 1113 (full text not published).
2. Briggs, K.: 1998, 'The doubledouble extended-precision software library for C++', http://epidem13.plantsci.cam.ac.uk/~kbriggs/, Department of Plant Sciences, University of Cambridge, U.K.
3. Byrd, P. F. and Friedman, M. D.: 1971, *Handbook of Elliptic Integrals for Engineers and Physicists*, 2nd edn, revised, Springer-Verlag, Berlin.
4. Cayley, A.: 1890a, 'A memoir on the problem of disturbed elliptical motion', In: *The Collected Mathematical Papers of Arthur Cayley*, Cambridge, Cambridge University Press, England, Vol. III, pp. 270–292.
5. Cayley, A.: 1890b, 'A memoir on the problem of the rotation of a solid body', In: *The Collected Mathematical Papers of Arthur Cayley*, Cambridge, Cambridge University Press, England, Vol. III, pp. 475–504.
6. Donaldson, J. D. and Jezewski, D. J.: 1977, 'An element formulation for perturbed motion about the center of mass', *Celest. Mech.* **16**(3), 367–387.
7. Goldstein, H.: 1980, *Classical Mechanics*, 2nd edn, Addison-Wesley, Philippines.
8. Kraige, L. G.: 1978, 'The development and numerical testing of a variation of parameters approach to the arbitrarily torqued, asymmetric rigid body problem', *Technical Report* VPI-E-78-9, Virginia Polytechnic Institute and State University, Blacksburg, Virginia.
9. Kraige, L. G. and Junkins, J. L.: 1976, 'Perturbation formulations for satellite attitude dynamics', *Celest. Mech.* **13**(1), 39–64.
10. Kraige, L. G. and Skaar, S. B.: 1977, 'A variation of parameters approach to the arbitrarily torqued, asymmetric rigid body problem', *J. Astronaut. Sci.* **25**(3), 207–226.
11. Livneh, R. and Wie, B.: 1997, 'New results for an asymmetric rigid body with constant body-fixed torques', *J. Guid. Contr. Dyn.* **20**(5), 873–881.
12. Mitchell, J. W.: 2000, 'A Simplified Variation of Parameters Solution for the Motion of an Arbitrarily Torqued Mass Asymmetric Rigid Body', PhD Thesis, University of Cincinnati.
13. Mitchell, J. W. and Richardson, D. L.: 2000, 'A simplified variation of parameters solution for the motion of an arbitrarily torqued mass asymmetric rigid body', In: *Astrodynamics 1999:*

Proceedings of the AAS/AIAA Astrodynamics Specialists Conference, Vol. 103 of *Advances in Astronautical Sciences*, San Diego, California, pp. 2489–2512.

14. Richardson, D. L. and Mitchell, J. W.: 1999, 'A simplified variation of parameters approach to Euler's equations', *ASME J. Appl. Mech.* **66**, 273–276.

15. Richardson, D. L., Schmidt, D. S. and Mitchell, J. W.: 1998, 'Improved Chebyshev methods for the numerical integration of first-order differential equations', In: *Spaceflight Mechanics 1998: Proceedings of the AAS/AIAA Spaceflight Mechanics Conference*, Vol. 99 of *Advances in Astronautical Sciences*, San Diego, California, pp. 1533–1544.

RESONANCE AND CAPTURE OF JUPITER COMETS

W. S. KOON[1], M. W. LO[2], J. E. MARSDEN[1] and S. D. ROSS[1]

[1]*Control and Dynamical Systems, California Institute of Technology, Pasadena, California, U.S.A.*
[2]*Navigation and Mission Design, Jet Propulsion Laboratory, Pasadena, California, U.S.A.*

Abstract. A number of Jupiter family comets such as *Oterma* and *Gehrels 3* make a rapid transition from heliocentric orbits outside the orbit of Jupiter to heliocentric orbits inside the orbit of Jupiter and vice versa. During this transition, the comet can be captured temporarily by Jupiter for one to several orbits around Jupiter. The interior heliocentric orbit is typically close to the 3:2 resonance while the exterior heliocentric orbit is near the 2:3 resonance. An important feature of the dynamics of these comets is that during the transition, the orbit passes close to the libration points L_1 and L_2, two of the equilibrium points for the restricted three-body problem for the Sun-Jupiter system. Studying the libration point invariant manifold structures for L_1 and L_2 is a starting point for understanding the capture and resonance transition of these comets. For example, the recently discovered heteroclinic connection between pairs of unstable periodic orbits (one around the L_1 and the other around L_2) implies a complicated dynamics for comets in a certain energy range. Furthermore, the stable and unstable invariant manifold 'tubes' associated to libration point periodic orbits, of which the heteroclinic connections are a part, are phase space conduits transporting material to and from Jupiter and between the interior and exterior of Jupiter's orbit.

Key words: comets, resonance, libration points, invariant manifolds, three-body problem

1. Introduction

A heteroclinic connection between periodic orbits about L_1 and L_2 was recently discovered by Koon et al. (2000). The existence of such a connection has important implications regarding the global dynamics of the three-body problem, and for the motion of bodies in the solar system experiencing such dynamics. The present paper summarizes earlier work on libration point orbits and their manifolds (see Conley, 1968; McGehee, 1969; Llibre et al., 1985; Koon et al., 2000) and applies the geometrical point of view to the comet resonance transition problem. The goal is to clearly state the qualitative dynamical picture that is forming, which any detailed investigation of transport between mean motion resonances must build upon. In Section 4, the particular case of transport between resonances interior and exterior to Jupiter's orbit is covered, following the example of the Jupiter family comet *Oterma*.

2. Jupiter Comets

2.1. RESONANCE TRANSITION IN COMET ORBITS

Some Jupiter comets such as Oterma and Gehrels 3 make a rapid transition from heliocentric orbits outside the orbit of Jupiter to orbits inside that of Jupiter and vice

Celestial Mechanics and Dynamical Astronomy **81:** 27–38, 2001.
© 2001 *Kluwer Academic Publishers.*

versa. During this transition, the comet may be captured temporarily by Jupiter for several orbits. The interior orbit is typically close to the 3:2 mean motion resonance while the exterior orbit is near the 2:3 resonance. See Figure 1(a). During the transition, the orbit passes close to the libration points L_1 and L_2, two of the equilibrium points (in a rotating frame) for the planar circular restricted three-body problem (PCR3BP) for the Sun–Jupiter system.

2.2. THE RELEVANCE OF INVARIANT MANIFOLDS

Lo and Ross (1997) used the two degree of freedom PCR3BP as the underlying model for resonance transition and related the transition to invariant manifolds, noticing that the orbits of Oterma and Gehrels 3 (in the Sun–Jupiter rotating frame) closely follow the computed invariant manifolds of L_1 and L_2.[1] Koon et al. (2000) developed this viewpoint along with another key ingredient, a heteroclinic connection between unstable periodic orbits around L_1 and L_2 with the same Jacobi constant (a multiple of the energy for the PCR3BP). The dynamical consequences of such an orbit are covered in great mathematical detail in that paper. Here, we focus on the study of exotic comet motion and resonance transition in terms of the libration point invariant manifolds.

2.3. HETEROCLINIC CONNECTIONS

A numerical demonstration is given in Koon et al. (2000) of a *heteroclinic connection* between pairs of equal energy periodic orbits, one around L_1, the other around L_2. This heteroclinic connection augments the previously known homoclinic orbits associated with the L_1 and L_2 periodic orbits (see McGehee, 1969). Linking these heteroclinic connections and homoclinic orbits leads to *dynamical homoclinic–heteroclinic chains* which form the backbone for temporary capture and rapid resonance transition of Jupiter comets. See Figure 1.

2.4. EXISTENCE AND CONSTRUCTION OF TRANSITION ORBITS

Koon et al. (2000) prove the existence of a large class of interesting orbits in the neighborhood of a chain which a comet can follow in its rapid transition between the inside and outside of Jupiter's orbit via a Jupiter encounter. One can label orbits near a chain with an itinerary giving their past and future whereabouts, making their classification and manipulation possible. Furthermore, a systematic procedure for the numerical construction of orbits with prescribed itineraries has been developed using the stable and unstable invariant manifold tubes of L_1 and L_2 periodic orbits.

---

[1] Belbruno and B. Marsden (1997) considered the comet transitions using a different approach, the 'fuzzy boundary' (or 'weak stability boundary') concept, which they said "can be viewed as a higher-dimensional analogue of the collinear Lagrange points L_1 and L_2 of Jupiter". During their investigation however, they suggested that resonance transition "does not seem to occur in the planar circular restricted problem of two degrees of freedom".

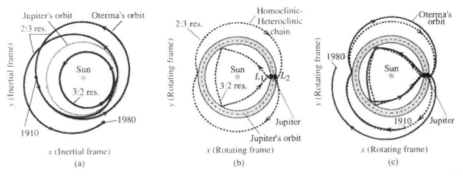

Figure 1. (a) Orbit of comet Oterma in Sun-centered inertial frame during time interval AD 1910–1980 (ecliptic projection). (b) The homoclinic-heteroclinic chain corresponding to the Jupiter comet Oterma. (c) The actual orbit of Oterma overlaying the chain.

3. A Few Key Features of the Three-Body Problem

3.1. PLANAR CIRCULAR RESTRICTED THREE-BODY PROBLEM

The comets of interest are mostly heliocentric, and the perturbations of their motion away from Keplerian ellipses are dominated by Jupiter's gravitation. Moreover, their motion is nearly in Jupiter's orbital plane, and Jupiter's small eccentricity (0.0483) plays little role during the fast resonance transition (less than or equal to one Jupiter period in duration). The PCR3BP is therefore an adequate starting model for illuminating the essence of the resonance transition process.

The PCR3BP describes the motion of a body moving in the gravitational field of two main bodies that are moving in circles. The two main bodies we consider are the Sun and Jupiter. The total mass is normalized to 1; they are denoted $m_S = 1 - \mu$ and $m_J = \mu$, where $\mu = 9.537 \times 10^{-4}$. The Sun and Jupiter rotate in the plane of their orbit in circles counterclockwise about their common center of mass and with angular velocity also normalized to 1.

3.2. EQUATIONS OF MOTION

Choosing a rotating coordinate system so that the origin is at the center of mass, the Sun and Jupiter are on the x-axis at the points $(-\mu, 0)$ and $(1 - \mu, 0)$ respectively, that is, the distance from the Sun to Jupiter is normalized to be 1. Let (x, y) be the position of the comet in the plane, then the equations of motion in this rotating frame are

$$\ddot{x} - 2\dot{y} = \Omega_x, \qquad \ddot{y} + 2\dot{x} = \Omega_y,$$

where $\Omega = (x^2 + y^2)/2 + (1 - \mu)/r_S + \mu/r_J$. Here, the subscripts of Ω denote partial differentiation in the variable. r_S, r_J are the distances from the comet to the Sun and the Jupiter, respectively. See Szebehely (1967) for the derivation.

3.3. ENERGY MANIFOLDS

These equations are autonomous and can be put into Hamiltonian form. They have an energy integral

$$E = \tfrac{1}{2}(\dot{x}^2 + \dot{y}^2) - \Omega(x, y),$$

which is related to the Jacobi constant C by $C = -2E$. Energy manifolds are three-dimensional surfaces foliating the four-dimensional phase space. For fixed energy, Poincaré sections are two-dimensional and therefore easily visualizable.

3.4. EQUILIBRIUM POINTS

The PCR3BP has three collinear equilibrium (Lagrange) points which are unstable, but for the comets of interest, we examine only L_1 and L_2. See Figure 2(a). Eigenvalues of the linearized equations at L_1 and L_2 have one real and one imaginary pair, having a saddle × center structure.

3.5. REGION OF POSSIBLE MOTION

The projection of the energy manifold onto the position space is the region in the xy-plane where the comet is energetically permitted to move around (known as the 'Hill's region'). The forbidden region is the region that is not accessible for a given energy. See Figure 2(b).

Our main concern is the behavior of orbits whose energy is just above that of L_2, for which the Hill's region is a connected region with an *interior region* (inside Jupiter's orbit), *exterior region* (outside Jupiter's orbit), and a *Jupiter (capture) region* (bubble surrounding Jupiter). These regions are connected by 'necks'

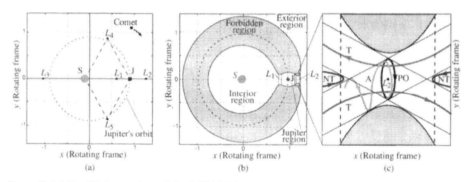

Figure 2. (a) Equilibrium points of the PCR3BP in the rotating frame. (b) Energetically forbidden region is gray 'C'. Hill's region (region in white), contains a 'neck' about L_1 and L_2. (c) The flow in the region near L_2, showing a periodic orbit around L_2 (labeled PO), a typical asymptotic orbit winding onto the periodic orbit (A), two transit orbits (T) and two non-transit orbits (NT). A similar figure holds for the region around L_1.

about L_1 and L_2 and the comet can make transitions between the regions only through these necks. This equilibrium neck region and its relation to the global orbit structure is critical and is discussed next.

3.6. FOUR TYPES OF ORBITS IN EQUILIBRIUM REGIONS

In each equilibrium region (one around L_1 and one around L_2), there exist four types of orbits (see Figure 2(c)) as given in Conley (1968): (1) an unstable *periodic* Lyapunov orbit; (2) four cylinders of *asymptotic* orbits that wind onto or off this period orbit, which form pieces of stable and unstable manifolds; (3) *transit* orbits which the comet must use to make a transition from one region to the other; and (4) *nontransit* orbits where the comet bounces back to its original region.

3.7. INVARIANT MANIFOLDS AS SEPARATRICES

McGehee (1969) first observed that the asymptotic orbits are pieces of the two-dimensional stable and unstable invariant manifold 'tubes' associated to the Lyapunov orbit and they form the boundary between transit and nontransit orbits. The transit orbits, passing from one region to another, are those inside the cylindrical manifold tube. The nontransit orbits, which bounce back to their region of origin, are those outside the tube. Most importantly, to transit from outside Jupiter's orbit to inside (or vice versa), or get temporarily captured, a comet *must* be inside a tube of transit orbits, as in Figure 3. The invariant manifold tubes are global objects – they extend far beyond the vicinity of the equilibrium region, partitioning the energy manifold into regions of qualitatively different orbit behavior.

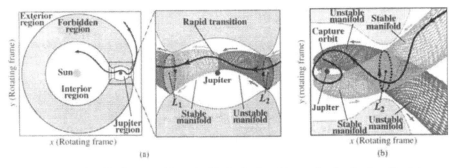

Figure 3. (a) Transit orbit from outside to inside Jupiter's orbit, passing by Jupiter. The tubes containing transit orbits (bounded by the cylindrical stable (lightly shaded) and unstable (darkly shaded) manifolds) intersect such that a transition is possible. (b) Orbit, beginning inside stable manifold tube in exterior region, is temporarily captured by Jupiter.

3.8. NUMERICAL COMPUTATION OF INVARIANT MANIFOLDS

Periodic Lyapunov orbits can be computed using a high order analytic expansion (see Llibre et al., 1985). Their stable and unstable manifolds can be approximated as given in Parker and Chua (1989). The basic idea is to linearize the equations of motion about the periodic orbit and then use the monodromy matrix provided by Floquet theory to generate a linear approximation of the periodic orbit's stable and unstable manifold. The linear approximation, in the form of a state vector, is numerically integrated in the nonlinear equations of motion to produce the approximation of the stable and unstable manifolds.

3.9. RAPID TRANSITION MECHANISM

The heart of the rapid transition mechanism from outside to inside Jupiter's orbit (or vice versa) is the intersection of transit orbit tubes. We can see the intersection clearly on a two-dimensional Poincaré section in the three-dimensional energy manifold. We take our section along a vertical line (parallel to the y-axis) through Jupiter as in Figure 4(a). Plotting $\dot{y}$ versus y along this line, we see that the tube cross-sections are distorted circles (see Figure 4(b)). Upon magnification, it is clear the tubes indeed intersect (see Figure 4(c)).

Any point within the region bounded by the curve corresponding to the stable tube cut is on an orbit that will go from the Jupiter region into the interior region. Similarly, a point within the unstable tube cut is on an orbit that came from the exterior region into the Jupiter region. A point inside the region bounded by the intersection of both curves (lightly shaded in Figure 4(c)) is on an orbit that makes the transition from the exterior region to the interior region, via the Jupiter region. The timescale for such a transition is short, less than one Jupiter period (Jupiter period $\approx$ 12 years).

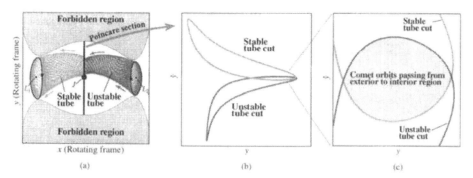

Figure 4. (a) Take a Poincaré section of the L_1 and L_2 periodic orbit invariant manifold tubes – a vertical line through Jupiter (J). (b) Look at unstable tube cut for L_2 and stable tube cut for L_1. (c) A small portion of the tubes intersect – this set in the phase space contains the comet orbits which pass from the exterior to the interior region.

4. Invariant Manifold Tubes and Resonance Transition

4.1. GENERIC TRANSPORT MECHANISM

This dynamical mechanism effecting transfer between the interior and exterior regions has not been previously recognized. It was previously believed that a third degree of freedom was necessary or that 'Arnold diffusion' was somehow involved. But clearly, only the planar CR3BP is necessary. The dynamics and phase space geometry involved in the heteroclinic connection now provide a language with which to discuss and further explore resonance transition.

4.2. TRANSPORT BETWEEN RESONANCES

The dynamical channel discussed in the previous section is a generic transport mechanism connecting the interior and exterior regions. We now focus on the case of transport between resonances, and in particular, the rapid transport mechanism connecting interior and exterior mean motion resonances (e.g. the 3:2 and 2:3 Jupiter resonances). By numerically computing the connection between the interior and exterior resonances, we will obtain a deeper understanding of the mean motion resonance transition of actual Jupiter comets, such as Oterma.

4.3. TUBE LOCATION

In Figure 5, the location of the tubes is shown schematically. To perform an Oterma-like transition from outside to inside Jupiter's orbit, a comet orbit would begin inside the stable manifold tube of L_2 on the outside, then pass through the L_2 equilibrium region to the L_2 unstable manifold in the Jupiter region (as in Figure 3(a)). Intersecting the L_1 stable manifold tube in the Jupiter region, the trajectory would pass by L_1 into the interior region. Note, we will occasionally refer to the interior, Jupiter, and exterior regions with the letters S, J, and X, respectively.

A comet orbit which circles the Sun once in the interior region of the rotating frame (as Oterma does as seen in Figure 1(c)) would have to be in a part of phase space where the L_1 stable and and unstable tubes intersect. We can see such an intersection along the U_1 section (see Figure 5).

4.4. INTERIOR AND EXTERIOR RESONANCES

In Figure 6(a), we see a cross-section of the stable and unstable tubes of the L_1 Lyapunov orbit, transformed into Delaunay variables (see Szebehely, 1967). The vertical axis is an angular variable, thus we can identify the top and bottom boundaries. The background points reveal the mixed character of the interior region phase space for this energy surface: stable periodic and quasiperiodic tori 'islands' embedded in a bounded chaotic 'sea'. The families of stable tori lie along strips of nearly constant semimajor axis, and correspond to mean motion resonances.

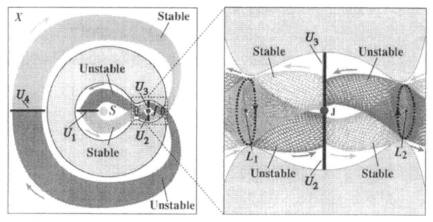

Figure 5. Location of L_1 and L_2 orbit invariant manifold tubes in position space (schematically). Stable manifolds are lightly shaded, unstable manifolds are darkly shaded. Location of Poincaré sections (U_1, U_2, U_3, and U_4) also shown. Magnification of Jupiter region at right.

The first cuts of the stable and unstable tubes intersect at the 3:2 resonance. Any point inside the unstable curve is on an orbit which came from J and any point inside the stable curve is on an orbit going toward J. Their intersection (the small diamond) contains all orbits that have come from J, have gone around the Sun once in the rotating frame, and will return to J. Because this intersection lies along the strip of 3:2 resonant orbits, we conclude that any comet which has an energy similar to Oterma's and which circles around the Sun once in the interior region *must* be in 3:2 resonance with Jupiter.

Similar to Figure 6(a) for the interior region, Figure 6(b) shows the first exterior region Poincaré cuts of the stable and unstable manifold tubes of an L_2 Lyapunov orbit with the U_4 section for the same energy, plotted using Delaunay variables. A similar mixed phase space structure is seen.

The stable and unstable tubes intersect at the region of the 2:3 resonance (the diamond). Any point inside the unstable curve is on an orbit which came from J and any point inside the stable curve is on an orbit going toward J. Although there is another intersection at the 1:2 resonance, the cross-section of the tube is widest near the 2:3 resonance. These are close to canonical coordinates, thus the vicinity of the 2:3 resonance is the more important. Therefore, we expect any comet with this energy which just came from J or is about to go to J, to be in 2:3 resonance with Jupiter. Their intersection (the diamond at the 2:3 resonance) contains the orbits that have come from J, circled the Sun once in the rotating frame, and will return to J.

4.5. CONNECTION BETWEEN RESONANCES

These two resonances (the 3:2 in the interior and the 2:3 in the exterior) are dynamically linked for this energy via the intersection between tubes in the Jupiter region.

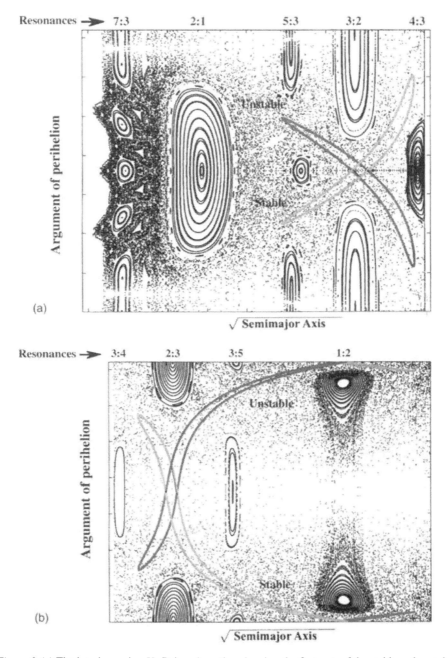

Figure 6. (a) The interior region U_1 Poincarè section showing the first cuts of the stable and unstable manifold tubes of an L_1 Lyapunov orbit. Notice their intersection at the 3:2 resonance. (b) The exterior region U_4 Poincarè section showing the first cuts of the stable and unstable manifold tubes of an L_2 Lyapunov orbit. Notice their intersections at the 2:3 and 1:2 resonances.

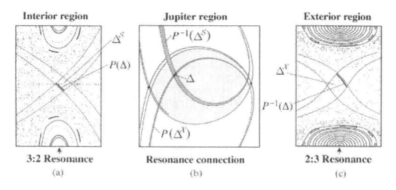

Figure 7. (a) Interior region tube intersection Δ^S. (b) The Jupiter region U_3 Poincarè section showing a portion of the image of Δ^X and the pre-image of Δ^S. Notice their intersections, the largest of which is labeled Δ. (c) Exterior region tube intersection Δ^X.

In Figure 7(b), we reproduce Figure 7(c), showing the collection of orbits passing from the exterior to interior region. Superimposed upon this large shaded region are pieces of the image and pre-image of the 2:3 and 3:2 intersection diamonds, respectively (Δ^X in the exterior region and Δ^S in the interior region, respectively). The diamonds are mapped to highly stretched and folded strips $P(\Delta^X)$ and $P^{-1}(\Delta^S)$ in the Jupiter region. Here, P denotes the Poincaré map connecting the sections U_1, U_2, U_3, and U_4.

Note that $P(\Delta^X)$ and $P^{-1}(\Delta^S)$ intersect; the largest of these intersections is labeled Δ. The image and pre-image of Δ are small strips in the interior and exterior regions respectively (see Figures 7(a) and (c)). This is an open set in the energy surface which dynamically links the 3:2 and 2:3 resonances via the Jupiter region. One can pick any point inside the strip Δ and integrate it forward and backward, generating an Oterma-like transition from the 2:3 to the 3:2 resonance. See Figure 8 and compare with Figure 1.

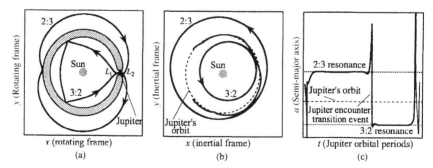

Figure 8. (a) An Oterma-like resonance transition in the rotating frame. (b) The same orbit in the heliocentric inertial frame. (c) Plot of semimajor axis versus time for the same orbit. Important mean motion resonances 3:2 and 2:3 are also shown for comparison.

We conclude that L_1 and L_2 invariant manifold tubes and their intersections lead to the resonance transition exhibited by comets like Oterma. We expect that Oterma executes a transition near the full model analogue of this dynamical channel. For example, the L_1 and L_2 invariant manifold structures in the three degree of freedom system are important for comets of similar energy like *Helin–Roman–Crockett* (see Howell et al., 2000).

4.6. OTHER RESONANCE CONNECTIONS

A similar resonance connection should exist for all nearby energies, as confirmed by numerical experiment. We have seen a link between only first order resonances ($p:q$, where $|p - q| = 1$) because we looked only at the first Poincaré cut of the tubes on our chosen surface in this study. Looking at cuts beyond the first reveals transitions between higher order resonances. In addition, higher energies have 'larger', more dispersive tubes, which have more intersections for a given cut number.

5. Conclusions

We have applied dynamical systems techniques developed in Koon et al. (2000) to the problem of resonance transitions and capture of Jupiter comets with energies near Oterma's. The fundamental mechanism is the rapid transport between the outside and inside of Jupiter's orbit via L_1 and L_2 periodic orbit invariant manifold tubes containing transit orbits. This mechanism provides a starting point for understanding the transport between mean motion resonances in more complicated models. Furthermore, the invariant manifold structures associated with L_1 and L_2 periodic orbits may prove valuable for understanding transport throughout the solar system.

References

Belbruno, E. and Marsden, B.: 1997, 'Resonance Hopping in comets', *Astr. J.* **113**(4), 1433–1444.
Conley, C.: 1968, 'Low energy transit orbits in the restricted three-body problem', *SIAM J. Appl. Math.* **16**, 732–746.
Howell, K. C., Marchand, B. G. and Lo, M. W.: 2000, 'Temporary Satellite Capture of Short-Period Jupiter Family Comets from the Perspective of Dynamical Systems', AAS/AIAA Space Flight Mechanics Meeting, *Paper No. 00-155*, Clearwater, Florida, U.S.A.
Koon, W. S., Lo, M. W., Marsden, J. E. and Ross, S. D.: 2000, 'Heteroclinic connections between periodic orbits and resonance transitions in celestial mechanics', *Chaos* **10**(2), 427–469.
Llibre, J., Martinez, R. and Simó, C.: 1985, 'Transversality of the invariant manifolds associated to the Lyapunov family of periodic orbits near L2 in the restricted three-body problem', *J. Differ. Eq.* **58**, 104–156.
Lo, M. and Ross, S.: 1997, 'SURFing the solar system: invariant manifolds and the dynamics of the solar system', *JPL IOM 312/97*, 2–4.

McGehee, R.: 1969, 'Some Homoclinic Orbits for the Restricted Three Body Problem', PhD Thesis, University of Wisconsin, Madison, Wisconsin, U.S.A.

Parker, T. S. and Chua, L. O.: 1989, *Practical Numerical Algorithms for Chaotic Systems*, Springer-Verlag, New York.

Szebehely, V.: 1967, *Theory of Orbits*, Academic Press, New York.

CHANGES IN ROTATIONAL ANGULAR MOMENTUM DUE TO GRAVITATIONAL INTERACTIONS BETWEEN TWO FINITE BODIES *

D.J. SCHEERES

Department of Aerospace Engineering, The University of Michigan, Ann Arbor, MI 48109-2140, U.S.A, e-mail: scheeres@umich.edu

Abstract. Interactions between an initially uniformly rotating body with a second degree and order gravity field and a sphere are analyzed. Explicit predictions of the change in rotational angular momentum of the non-spherical body are derived over one interaction (i.e. periapsis passage) between the bodies. The estimated changes are expressed in terms of trigonometric functions and generalized Hansen coefficients.

Key words: rotational dynamics, mutual gravitation, binary asteroids

1. Introduction

Interactions between a rotating second degree and order gravity field and a test particle can yield significant changes to the energy and angular momentum of the mutual orbit (Scheeres et al., 2000a). Previous work has developed estimates of the change in orbital energy and angular momentum following such an interaction (Scheeres, 1999). The current work generalizes this effect to the mutual orbit of a body with a second degree and order gravity field and a sphere with finite mass. Since no restricted approximation has been made the problem admits both the energy and angular momentum integrals, with these integrals involving contributions from both the translational and rotational dynamics of these bodies. Estimates of the change in rotational angular momentum of the non-spherical body are derived, applying a method used previously to estimate the change in mutual orbit of the bodies (Scheeres, 1999). The resulting estimates are a function of trigonometric terms and generalized Hansen coefficients. Applications of this result to the rotation of asteroids is given in Scheeres et al. (2000b).

2. Equations of Interaction

As shown in (Maciejewski, 1995) the equations of motion of two gravitationally interacting bodies can be expressed in terms of standard Lagrangian and Hamiltonian formulations. In the current application we allow one of the interacting bodies

---

* Research was supported by NASA's Planetary Geology and Geophysics Program via grant NAG5-9017 to The University of Michigan.

Celestial Mechanics and Dynamical Astronomy **81**: 39–44, 2001.
© 2001 *Kluwer Academic Publishers.*

to be a sphere of finite mass and the other body to be a non-spherical body with finite mass and a second degree and order gravity field. Since no restricted approximations are made in the analysis, the classical integrals of energy and angular momentum are retained for the system. For the current analysis we only need the energy and angular momentum integrals of this two-body problem

$$E = \frac{1}{2}\dot{\mathbf{r}} \cdot \dot{\mathbf{r}} + \frac{1}{2}\mathbf{W} \cdot \mathbf{I} \cdot \mathbf{W} - U(\mathbf{r}), \tag{1}$$

$$\mathbf{K} = \mathbf{H} + \frac{\mathcal{M}_c \mathcal{M}_s}{\mathcal{M}_c + \mathcal{M}_s} \mathbf{G}, \tag{2}$$

where E is the total energy of the system (minus the rotational energy of the sphere), $\mathbf{r}$ is the relative position vector between the body centers of mass, $\dot{\mathbf{r}}$ is the time derivative of the relative position vector (with respect to a non-rotating frame), $\mathbf{W}$ is the rotational velocity vector of the non-spherical body, $\mathbf{I}$ is its inertia tensor, $\mathcal{M}_s$ is the mass of the sphere, $\mathcal{M}_c$ is the mass of the non-spherical body, U is the mutual potential between these bodies, $\mathbf{K}$ is the total angular momentum of the system (minus the rotational angular momentum of the sphere), $\mathbf{H}$ is the rotational angular momentum vector of the non-spherical body in inertial space, and $\mathbf{G} = \mathbf{r} \times \dot{\mathbf{r}}$ is the specific angular momentum of the relative orbit of the bodies. Additional discussions of this specific system can be found in Scheeres et al. (2000b).

For the current system the mutual potential has the simplified form $U = -\mu/|\mathbf{r}| + U_{20} + U_{22}$, where

$$U_{20} = \frac{\mu C_{20}}{2|\mathbf{r}|^3}(1 - 3\sin^2\delta), \tag{3}$$

$$U_{22} = -\frac{3\mu C_{22}}{|\mathbf{r}|^3}\cos^2\delta\cos 2\lambda, \tag{4}$$

$$\mu = \mathcal{G}(\mathcal{M}_s + \mathcal{M}_c), \tag{5}$$

where $\mathcal{G}$ is the gravitational constant, and δ and λ are the latitude and longitude, respectively, of the sphere in the non-spherical body-fixed space. The parameters C_{20} and C_{22} are the second degree and order gravity coefficients, where we assume that their dimensions are given in units of length^2 and are defined with respect to a principal axis coordinate system.

In anticipation of future computations, the mutual perturbing potential $R = U_{20} + U_{22}$ can be rewritten as a function of the osculating elements of the relative orbit. In this formulation we assume that the non-spherical body is initially in uniform rotation about its maximum moment of inertia, its unperturbed rotational dynamics are then stated simply in terms of its rotational phase angle θ and its angular rotation rate $\dot{\theta} = W_z$, where W_z denotes the initial rotation rate about the non-spherical body's maximum moment of inertia. The perturbing potentials can be rewritten as

$$U_{20} = \frac{\mu C_{20}}{4r^3}[-1 + 3(\cos^2 i + \sin^2 i \cos 2(f + \omega))], \tag{6}$$

$$U_{22} = -\frac{3\mu C_{22}}{r^3}\left[\frac{1}{2}\sin^2 i \cos 2(\Omega - \theta) + \right.$$
$$+ \cos^4(i/2)\cos 2(f + \omega + \Omega - \theta) +$$
$$\left. + \sin^4(i/2)\cos 2(f + \omega - \Omega + \theta)\right], \tag{7}$$

where $r = p/(1 + e \cos f)$, and p, e, ω, Ω, and f are the usual osculating orbit elements of orbit parameter, eccentricity, argument of periapsis, longitude of the ascending node and true anomaly. Note that the body-fixed frame and the inertial frame are assumed to be aligned when $\theta = 0$.

3. Estimates of Change in the Rotational Angular Momentum

3.1. COMPUTING THE ESTIMATES

An equation for the change in rotational angular momentum of the non-spherical body can be found from the total angular momentum integral **K**. Differentiating Equation (2) and rewriting yields the relation

$$\dot{\mathbf{H}} = -\frac{\mathcal{M}_s \mathcal{M}_c}{\mathcal{M}_s + \mathcal{M}_c}\dot{\mathbf{G}}, \tag{8}$$

where the **G** vector can be decomposed in osculating elements as

$$\mathbf{G} = G\begin{bmatrix} \sin i \sin \Omega \\ -\sin i \cos \Omega \\ \cos i \end{bmatrix}. \tag{9}$$

Starting from the canonical form of the Lagrange planetary equations (Brouwer and Clemence, 1961) the following differential equations can be derived:

$$\frac{d(G \cos i)}{dt} = -\frac{\partial R}{\partial \Omega}, \tag{10}$$

$$\frac{d(G \sin i \sin \Omega)}{dt} = -\frac{\sin \Omega}{\sin i}\left[\frac{\partial R}{\partial \omega} - \cos i \frac{\partial R}{\partial \Omega}\right] - \cos \Omega \frac{\partial R}{\partial i}, \tag{11}$$

$$\frac{d(G \sin i \cos \Omega)}{dt} = -\frac{\cos \Omega}{\sin i}\left[\frac{\partial R}{\partial \omega} - \cos i \frac{\partial R}{\partial \Omega}\right] + \sin \Omega \frac{\partial R}{\partial i}, \tag{12}$$

where R is the perturbing potential. Combining these results yields a set of differential equations for the rotational angular momentum of the non-spherical body in terms of the osculating elements of the mutual orbit.

To generate a first-order estimate of the change in rotational angular momentum of the non-spherical body Equation (8) can be integrated over a single interaction. In performing the quadrature we assume that the basic constants of the orbit and body rotation do not change, generating an estimate of the total change in the angular momentum of the non-spherical body. This is discussed in more detail in Scheeres (1999) in the context of Picard's method of successive approximation (Moulton, 1958). Under these assumptions, the total change in the non-spherical body's rotational angular momentum vector is

$$\Delta \mathbf{H} = \int_{-T/2}^{T/2} \dot{\mathbf{H}} dt, \tag{13}$$

where T is the orbit period for an elliptic mutual orbit and is ∞ for a parabolic or hyperbolic mutual orbit. The only explicit functions of time under the integral is the true anomaly of the relative orbit and the rotational phase angle of the non-spherical body. We assume that the interaction is centered at periapsis (when $t = \theta = 0$) and thus the longitude of ascending node Ω and argument of periapsis ω in the resulting expressions are specified in the body-fixed frame.

3.2. EVALUATING THE INTERACTION FOR C_{20}

When the mutual orbit is elliptic or parabolic the quadrature in Equation (13) yields

$$\Delta \mathbf{H}_{20} = -\frac{3\pi}{2} C_{20} \frac{\mathcal{M}_c \mathcal{M}_s}{\mathcal{M}_c + \mathcal{M}_s} \sqrt{\frac{\mu}{p^3}} \sin(2i) \begin{bmatrix} \cos \Omega \\ \sin \Omega \\ 0 \end{bmatrix}. \tag{14}$$

For a mutual hyperbolic orbit between the bodies the total change in rotational angular momentum is

$$\Delta \mathbf{H}_{20} = -C_{20} \sqrt{e^2 - 1} \frac{\mathcal{M}_c \mathcal{M}_s}{\mathcal{M}_c + \mathcal{M}_s} \sqrt{\frac{\mu}{p^3}} \sin(i) \times$$

$$\times \begin{bmatrix} 3 \cos i \cos \Omega \left(1 + \frac{f_\infty}{\sqrt{e^2-1}} \right) + \left(\frac{e^2-1}{e^2} \right) \times \\ \times \{ \sin 2\omega \sin \Omega - \cos i \cos 2\omega \cos \Omega \} \\ \\ 3 \cos i \sin \Omega \left(1 + \frac{f_\infty}{\sqrt{e^2-1}} \right) - \left(\frac{e^2-1}{e^2} \right) \times \\ \times \{ \sin 2\omega \cos \Omega + \cos i \cos 2\omega \sin \Omega \} \\ \\ 0 \end{bmatrix}, \tag{15}$$

where $f_\infty = \arccos(-1/e)$ when $e > 1$.

CHANGES IN ROTATIONAL ANGULAR MOMENTUM

For interactions along an elliptic or parabolic orbit the total angular momentum of the rotation and orbit is conserved. For interaction during a hyperbolic flyby the projection of angular momentum along the z-axis is still conserved, but the total angular momentum magnitude is no longer conserved, and can have a change in its length. This implies that the mutual orbit can be given a change in inclination due to the interaction in this case.

3.3. EVALUATING THE INTERACTION FOR C_{22}

This case yields more complex results due to the interaction between the rotating gravity field and the mutual orbit. Performing the quadrature in Equation (13) yields

$$\Delta \mathbf{H}_{22} = 6C_{22} \frac{\mathcal{M}_c \mathcal{M}_s}{\mathcal{M}_c + \mathcal{M}_s} \sqrt{\frac{\mu}{p^3}} \times$$

$$\times \begin{bmatrix} \frac{1}{2}\sin i [\cos^2(i/2)\cos(2\omega + \Omega)I_2^1 + \\ + \sin^2(i/2)\cos(2\omega - \Omega)I_{-2}^1 - \\ - \cos i \cos \Omega I_0^1] \\[4pt] -\frac{1}{2}\sin i [\cos^2(i/2)\sin(2\omega + \Omega)I_2^1 + \\ + \sin^2(i/2)\sin(2\omega - \Omega)I_{-2}^1 - \\ - \cos i \sin \Omega I_0^1] \\[4pt] \cos^4(i/2)\sin 2(\omega + \Omega)I_2^1 - \\ - \sin^4(i/2)\sin 2(\omega - \Omega)I_{-2}^1 + \\ + \frac{1}{2}\sin^2 i \sin 2\Omega I_0^1 \end{bmatrix}, \tag{16}$$

where

$$I_m^n = \int_{-f_\infty}^{f_\infty} \left(\frac{p}{r}\right)^n e^{i(mf - 2W_z t)} \mathrm{d}f, \tag{17}$$

and the time t is specified in the quadrature as an explicit function of true anomaly using the elliptic, parabolic, or hyperbolic forms of Kepler's equation. Note that $f_\infty = \pi$ if $e \leqslant 1$, and $f_\infty = \arccos(-1/e)$ if $e > 1$. A similar definition for the I_m^n integrals was given in Scheeres (1999), and the mathematical properties and computations of these integrals were discussed there. In particular that paper established that the contributions from I_2^1 and I_0^1 are much larger than the contributions from I_{-2}^1, implying that retrograde interactions have a markedly decreased strength. In Scheeres et al. (2000b) the above results are validated using simplified estimates and numerical integrations of the dynamics.

The integrals I_m^n can be directly related to the Hansen coefficients when the mutual orbit is elliptic ($e < 1$)

$$I_m^n = 2(1 - e^2)^{n+1/2} \sum_{k=-\infty}^{\infty} X_k^{-(n+2),m} \frac{\sin \pi (k - 2\sigma)}{k - 2\sigma}, \tag{18}$$

$$\sigma = W_z \sqrt{\frac{a^3}{\mu}},\tag{19}$$

and the Hansen coefficients are defined as (Brumberg, 1995)

$$X_s^{q,r} = \frac{1}{2\pi} \int_{-\pi}^{\pi} \left(\frac{r}{a}\right)^q e^{i(rf-sM)} dM,\tag{20}$$

where M is the mean anomaly and s is generally assumed to be an integer. If we generalize the index s and allow it to be a real number, the integrals can be defined directly as

$$I_m^n = 2\pi (1 - e^2)^{n+1/2} X_{2\sigma}^{-(n+2),m}.\tag{21}$$

If the Hansen coefficients are further generalized for parabolic and hyperbolic orbits the relationships noted above can be extended to these cases as well. This specific generalization is not performed here.

4. Conclusions

A mathematical analysis of the effect of mutual gravitation on the rotational dynamics of a non-spherical body is presented. The model developed applies to the interaction of a non-spherical body with a sphere in a nominally conic orbit about the body. An estimate of the change in rotational angular momentum of the non-spherical body over one interaction is generated.

References

Brouwer, D. and Clemence, G. M.: 1961, *Methods of Celestial Mechanics*, Academic Press, pp. 283–291.

Brumberg, V. A.: 1995, *Analytical Techniques of Celestial Mechanics*, Springer, p. 39.

Maciejewski, A. J.: 1995, 'Reduction, relative equilibria and potential in the two rigid bodies problem', *Celest. Mech. & Dyn. Astr.* **63**, 1–28.

Moulton, F. R.: 1958, *Differential Equations*, Dover, pp. 179–186.

Scheeres, D. J.: 1999, 'The effect of C_{22} on orbit energy and angular momentum', *Celest. Mech. & Dyn. Astr.* **73**, 339–348.

Scheeres, D. J., Williams, B. G. and Miller, J. K.: 2000a, 'Evaluation of the dynamic environment of an asteroid: applications to 433 Eros', *J. Guid. Cont. Dyn.* **23**, 466–475.

Scheeres, D. J., Ostro, S. J., Werner, R. A, Asphaug, E. and Hudson, R. S.: 2000b, 'Effects of gravitational interactions on asteroid spin states', *Icarus* **147**, 106–118.

SYMMETRIES, REDUCTION AND RELATIVE EQUILIBRIA FOR A GYROSTAT IN THE THREE-BODY PROBLEM

F. MONDÉJAR[1], A. VIGUERAS[1]* and S. FERRER[2]
[1]*Departamento de Matemática Aplicada y Estadística, Universidad Politécnica de Cartagena, 30203 Cartagena, Spain*
[2]*Departamento de Matemática Aplicada, Universidad de Murcia, 30071 Espinardo, Spain*

Abstract. The problem of three bodies when one of them is a gyrostat is considered. Using the symmetries of the system we carry out two reductions. Global considerations about the conditions for relative equilibria are made. Finally, we restrict to an approximated model of the dynamics and a complete study of the relative equilibria is made.

Key words: gyrostat, reduction, relative equilibria, three-body problem

1. Introduction

In the last years new research about the problem of roto-translational motion of celestial bodies has appeared; some papers within differential geometry frame, others still with the classical approach. In particular, they show a new interest in the study of configurations of relative equilibria in different models. In the problem of three rigid bodies Vidiakin (1997) and Duboshin (1984) proved the existence of Euler and Lagrange configurations of equilibria when the bodies possess symmetries (for a review see Zhuravlev and Petrutskii, 1990). More recently Wang et al. (1991) consider the problem of a rigid body in a central Newtonian field and Maciejewski (1995) takes into account the problem of two rigid bodies in mutual Newtonian attraction. In the same way, these problems have been generalized to the case when the rigid bodies are gyrostats by Cid and Vigueras (1985), Wang et al. (1995) and Mondéjar and Vigueras (1999).

In order to study the configurations of equilibria of the general problem of three rigid bodies from a global geometrical point of view it is natural to consider first the problem when two bodies have spherical distribution of mass. Fanny and Badaoui (1997) study the configuration of the equilibria in terms of the global variables in the unreduced problem. There, simplifications such as considering spherical or axisymmetric bodies are made in order to get specific results. It is clear, as the papers of Maciejewski (1995) and Mondéjar and Vigueras (1999) show, that to work in the reduced system (if the problem has symmetries) produces natural

---

* Author for correspondence: Tel.: 34 968325582; Fax: 34 968325694;
e-mail: Antonio.Vigueras@upct.es

simplifications in the conditions of the equilibria, and then more general results can be obtained. This is the approach we will follow in this paper.

In the way just mentioned above, the problem of three rigid bodies when two are spherical and the other is a gyrostat is considered. Using the symmetries of the translational and rotational group possessed by the system, we perform a reduction process in two steps, giving explicitly at each step the Poisson structure of the reduced system. We make global consideration about the relative equilibria of the problem and give a general classification of the equilibria. Finally we restrict to the case when the gravitational potential function is approximated by zero order terms of its Taylor expansion. In this particular problem, we give a complete analysis of the relative equilibria. We enlarge some results obtained in Fanny and Badaoui (1998). We note also that the reduction procedure presented here applies immediately to rigid body case when we take the gyrostatic momentum to be zero.

2. Configuration and Phase Space

Let us denote by S_0 a gyrostat of mass m_0, by S_1 and S_2 two rigid bodies with spherical symmetry of masses m_1 and m_2 respectively. We remember that a gyrostat is a mechanical system G composed of a rigid body and other bodies (deformable or rigid) connected to it such that their relative motion do not change the distribution of mass of G. Let us consider an inertial reference frame $\mathcal{I} = \{O, u_1, u_2, u_3\}$ and a body frame $\mathcal{B} = \{C_0, b_1, b_2, b_3\}$ fixed at the center of mass C_0 of S_0. A particle in the body S_0 with coordinates Q in $\mathcal{B}$ is represented in the inertial frame $\mathcal{I}$ by the vector $q = R_0 + BQ$, where $B \in SO(3)$ and R_0 is the vector position of the center of mass of S_0 in $\mathcal{I}$. Let us denote by R_1 and R_2 the vector position of center of mass of the bodies S_1 and S_2 respectively in $\mathcal{I}$. Then, at any instant, the configuration of the system is uniquely determined by $((B, R_0), R_1, R_2)$. The configuration space of the problem is the Lie group $\mathbf{Q} = SE(3) \times \mathbb{R}^3 \times \mathbb{R}^3$, where $SE(3)$ is the known semidirect product of $SO(3)$ and $\mathbf{R}^3$.

The Kinetic energy of the system is

$$\mathcal{T} = 1/2 \int_{S_0} |\dot{q}|^2 dm(Q) + m_1 |\dot{R}_1|^2/2 + m_2 |\dot{R}_2|^2/2,$$

where $dm(.)$ denotes the mass measure of S_0, and $|\cdot|$ denotes the Euclidean norm in $\mathbb{R}^3$. The expression of the Kinetic energy simplifies (see Cid and Vigueras, 1985) to

$$\mathcal{T} = \frac{1}{2}\Omega \cdot \mathbf{I}\Omega + L_r \cdot \Omega + \frac{m_0}{2}|\dot{R}_0|^2 + \frac{m_1}{2}|\dot{R}_1|^2 + \frac{m_2}{2}|\dot{R}_2|^2 + \mathcal{T}_r,$$

where $\mathbf{I}$ is the tensor of inertia of S_0 in the body frame, Ω is the angular velocity of S_0 defined by $\dot{B} = B\widehat{\Omega}$, and L_r and $\mathcal{T}_r$ are the momentum and the Kinetic energy of the moving part of the gyrostat respectively. Here, $\widehat{\Omega}$ is the image by the standard isomorphism between the Lie algebras $so(3)$ and $\mathbb{R}^3$.

In what follows we assume that $\mathcal{T}_r$ is a known function of the time and L_r is constant.

The gravitational potential energy is the function $\mathcal{V} : \mathbf{Q} \to \mathbb{R}$

$$\mathcal{V} = -\frac{Gm_1 m_2}{|R_2 - R_1|} - \int_{S_0} \frac{Gm_1 \, dm(Q)}{|BQ + R_0 - R_1|} - \int_{S_0} \frac{Gm_2 \, dm(Q)}{|BQ + R_0 - R_2|}.$$

Then, the Lagrangian of the problem is $\mathcal{L} = \mathcal{T} - \mathcal{V} \circ \tau$ where $\tau \colon \mathbf{TQ} \to \mathbf{Q}$ is the canonical projection. The phase space is the cotangent bundle $\mathbf{T}^*\mathbf{Q}$. By the Legendre transformation we obtain the Hamiltonian of the problem

$$\mathcal{H} = \frac{1}{2}\Pi \cdot \mathbf{I}^{-1}\Pi - L_r \cdot \mathbf{I}^{-1}\Pi + \sum_{0 \leqslant i \leqslant 2} \frac{P_i^2}{2m_i} + \mathcal{V},$$

where $\Pi = I\Omega + L_r$ is the total angular momentum of the gyrostat in the body frame and $P_i = m_i \dot{R}_i$ are the linear momenta of the bodies in the fixed frame.

3. Symmetries and Reduction

The problem can be reduced by the action of the group $SE(3)$ (see Marsden et al., 1984a,b). However, in the case of semidirect products we can proceed by stages (see Marsden, 1992):

3.1. REGULAR REDUCTION BY THE TRANSLATION GROUP

In the first stage we use the symplectic reduction procedure by the action of the translation group $\mathbb{R}^3$ (see Marsden and Weinstein, 1974), we obtain a model for the reduced space $M_1 = SO(3) \times so(3)^* \times \mathbf{T}^*\mathbb{R}^3 \times \mathbf{T}^*\mathbb{R}^3$ and the reduced Hamiltonian on M_1 is the function

$$\mathcal{H}_1(z) = \frac{|\tilde{s}|^2}{2M} + \frac{|\tilde{r}|^2}{2\widetilde{M}} + \frac{1}{2}\Pi \cdot \mathbf{I}^{-1}\Pi - L_r \cdot \mathbf{I}^{-1}\Pi + \mathcal{V}(\Pi, r, s, \tilde{r}, \tilde{s}),$$

where

$$\mathcal{V}(z) = -\frac{Gm_1 m_2}{|r|} - \int_{S_0} \frac{Gm_1 \, dm(Q)}{\left|BQ + s + \frac{m_2}{M} r\right|} - \int_{S_0} \frac{Gm_2 \, dm(Q)}{\left|BQ + s - \frac{m_1}{M} r\right|}, \tag{1}$$

where $r = R_2 - R_1$, $s = R_0 - (m_1 R_1 + m_2 R_2)/\overline{M}$, $\tilde{r} = (m_1 P_2 - m_2 P_1)/\overline{M}$, $\tilde{s} = M_t(P_0 - m_0(P_1 + P_2))/\overline{M}$, $M_t = m_0 + m_1 + m_2$, $M = m_0\overline{M}/M_t$ and $\widetilde{M} = m_1 m_2/\overline{M}$.

3.2. REDUCTION BY THE ROTATION GROUP

In this second stage we use the Poisson reduction procedure by the action of the group $SO(3)$ on M_1 and we obtain the twice reduced Hamiltonian

$$\mathcal{H}_{\mathrm{II}}(z) = \frac{|P_\mu|^2}{2M} + \frac{|P_\lambda|^2}{2\widetilde{M}} + \frac{1}{2}\Pi \cdot \mathbf{I}^{-1}\Pi - L_r \cdot \mathbf{I}^{-1}\Pi + \mathcal{V}(z), \tag{2}$$

where

$$\mathcal{V}(z) = -\frac{Gm_1m_2}{|\lambda|} - \int_{S_0} \frac{Gm_1 dm(Q)}{\left|Q + \mu + \frac{m_2}{M}\lambda\right|} - \int_{S_0} \frac{Gm_2 dm(Q)}{\left|BQ + \mu - \frac{m_1}{M}\lambda\right|}, \tag{3}$$

with $\lambda = B^t r$, $P_\lambda = B^t \tilde{r}$, $\mu = B^t s$ and $P_\mu = B^t \tilde{s}$.

4. Relative Equilibria

The relative equilibria are the equilibria for the twice reduced space. We will denote the values at the equilibria with a subscript e. First, we begin writing explicitly the equations of the reduced dynamics:

$$\dot{\Pi} = \widehat{\Pi}\mathbf{I}^{-1}(\Pi - L_r) + \widehat{\lambda}\nabla_\lambda\mathcal{V} + \widehat{\mu}\nabla_\mu\mathcal{V}, \tag{4}$$

$$\dot{\lambda} = \widehat{\lambda}\mathbf{I}^{-1}(\Pi - L_r) + \frac{1}{\widetilde{M}}P_\lambda, \quad \dot{\mu} = \widehat{\mu}\mathbf{I}^{-1}(\Pi - L_r) + \frac{1}{M}P_\mu, \tag{5}$$

$$\dot{P}_\lambda = \widehat{P_\lambda}\mathbf{I}^{-1}(\Pi - L_r) - \nabla_\lambda\mathcal{V}, \quad \dot{P}_\mu = \widehat{P_\mu}\mathbf{I}^{-1}(\Pi - L_r) - \nabla_\mu\mathcal{V}, \tag{6}$$

where $\mathbf{I}^{-1}(\Pi - L_r) = \Omega$ is the angular velocity of S_0. Then, it is easy to prove that, in the equilibria, the angular velocity is parallel to the total angular momentum $\mathcal{M} = \Pi + \lambda \times P_\lambda + \mu \times P_\mu$. We observe that a similar result is obtained in the two rigid bodies and in the two gyrostats problems (see Maciejewski, 1995; Mondéjar and Vigueras, 1999). Moreover, the previous equations suggest the idea of considering two types of equilibria according to the vector Ω be orthogonal or not to one of the vectors λ or μ. Other way in order to study the previous equations is to impose the existence of equilibria of Euler or Lagrange type and then, to obtain necessary conditions about the angular velocity and the gyrostatic momentum of the gyrostat. For a complete treatment of these topics see Ferrer et al. (2001).

5. Zero Order Approximation Dynamics

In this section we study an approximate model of the twice reduced dynamics. More precisely we consider the zero order terms of the Taylor series expansion of the gravitational potential function $\mathcal{V}$ on the twice reduced space. Denoting by the same letter the approximate potential function, we write

$$\mathcal{V}(z) = -\frac{Gm_1 m_2}{|\lambda|} - \frac{Gm_0 m_1}{|\mu + (m_2/\overline{M})\lambda|} - \frac{Gm_0 m_2}{|\mu - (m_1/\overline{M})\lambda|}.$$

We can prove (for details see Ferrer et al., 2001) the following results about the configurations of equilibria.

THEOREM 1. *In the zero order approximation dynamics a necessary condition for equilibria is that the vector Ω_e is orthogonal to the vectors λ_e and μ_e.*

In what follows we assume that λ_e and Ω_e are orthogonal. First, we can prove the following proposition.

PROPOSITION 1. *One of the following conditions holds:*

(a) $\left| \mu_e + \frac{m_2}{M} \lambda_e \right| = \left| \mu_e - \frac{m_1}{M} \lambda_e \right|$;

(b) λ_e *and* μ_e *are parallel.*

Then, using the above Proposition, the following theorems follow:

THEOREM 2. *There exist three different equilibrium configurations of Euler type, that is where the vector λ_e and μ_e are parallel, and both orthogonal to Ω_e. For each equilibrium configuration the angular velocity Ω_e verifies*

$$|\Omega_e|^2 = G\overline{M}\left(1 + m_0/(\overline{M}v^2) - m_0/(\overline{M}(v-1)^2)\right)/|\lambda_e|^3,$$

where v is a unique positive root of the classical quintic algebraic equation of the three-body problem, which coefficients are functions of the masses. The cases 2 and 3 satisfy similar expressions, and the line defined by the bodies rotate around a fix direction.

We note that in the equilibrium configurations described in the above theorem, the three bodies are collinear, and the angular velocity of the gyrostat is orthogonal to the common direction determined by the bodies.

THEOREM 3. *There exists a Lagrangian equilibrium configuration when the angular velocity is orthogonal to the plane spanned by $< \lambda_e, \mu_e >$, and the conditions*

$$|\lambda_e| = \left| \mu_e - \frac{m_1}{M} \lambda_e \right| = \left| \mu_e + \frac{m_2}{M} \lambda_e \right|, \qquad |\Omega_e|^2 = GM_t \frac{1}{|\lambda_e|^3}, \qquad (7)$$

hold. Moreover, there are no other equilibrium configurations being Ω_e orthogonal to λ_e.

50 F. MONDÉJAR ET AL.

We observe that in these configurations the three bodies form a equilateral triangle in a plane orthogonal to the angular velocity. Finally, we note that Propositions 1 and 2 of Fanny and Badaoui (1997) are included in our configurations equilibria when the gyrostatic momentum is null, and the body is spherical. Thus, we enlarge some of the results obtained by Fanny and Badaoui.

Acknowledgement

This research was partially supported by the project PB98-1576 of the Ministerio de Educación y Cultura of Spain.

References

Cid, R. and Vigueras, A.: 1985, 'About the problem of motion of N gyrostats: I. the first integrals', *Celest. Mech. & Dyn. Astr.* **36**, 155–162.
Duboshin, G. N.: 1984, 'The problem of three rigid bodies', *Celest. Mech. & Dyn. Astr.* **33**, 31–47.
Fanny, C. and Badaoui, E.: 1997, 'Relative equilibrium in the three-body problem with a rigid body', *Celest. Mech. & Dyn. Astr.* **69**, 293–315.
Ferrer, S., Mondéjar, F. and Vigueras, A.: 2001 (in preparation).
Maciejewski, A.: 1995, 'Reduction, relative equilibria and potential in the two rigid bodies problem', *Celest. Mech. & Dyn. Astr.* **63**, 1–28.
Marsden, J. E.: 1992, *Lectures on Mechanics*, L. M. S., Lectures Note Series 174, Cambridge University Press.
Marsden, J. E., Ratiu, T. S. and Weinstein, A.: 1984a, 'Semidirect products and reductions in mechanics', *Trans. AMS* **281**, 147–177.
Marsden, J. E., Ratiu, T. S. and Weinstein, A.: 1984b, 'Reduction and Hamiltonian structures on duals of semidirect product Lie algebras', *Cont. Math. AMS* **28**, 55–100.
Marsden, J. E. and Weinstein, A.: 1974, 'Reduction of symplectic manifolds with symmetry', *Rep. Math. Phys.* **5**, 121–130.
Mondéjar, F. and Vigueras, A.: 1999, 'The Hamiltonian dynamics of the two gyrostats problem', *Celest. Mech. & Dyn. Astr.* **73**, 303–312.
Vidiakin, V. V.: 1977, 'Euler Solutions in the problem of translational-rotational motion of three-rigid bodies', *Celest. Mech. & Dyn. Astr.* **16**, 509–526.
Wang, L.-S., Lian, K.-Y. and Chen, P.-T.: 1995, 'Steady motions of gyrostat satellites and their stability', *IEEE Trans Automatic Control* **40**(10), 1732–1743.
Wang, L.-S., Krishnaprasad, P. S. and Maddocks, J. H.: 1991, 'Hamiltonian dynamics of a rigid body in a central gravitational field', *Celest. Mech. & Dyn. Astr.* **50**, 349–386.
Zhuravlev, S. G. and Petrutskii, A. A.: 1990, 'Current state of the problem of translational-rotational motion of three-rigid bodies', *Soviet Astron.* **34**, 299–304.

QUASI-STATIC MOTIONS OF VISCOELASTIC SATELLITES

BORIS S. BARDIN[1] and ANDRZEJ J. MACIEJEWSKI[2]

[1]*Department of Theoretical Mechanics, Faculty of Applied Mathematics, Moscow Aviation Institute, 4 Volokolamskoe Shosse, Moscow 125871, Russia, e-mail: boris.bardin@mailcity.com*
[2]*Institute of Astronomy, University of Zielona Góra, Lubuska 2, PL-65-265 Zielona Góra, Poland, e-mail: maciejka@astro.ca.wsp.zgora.pl*

Abstract. We consider the motion of a spacecraft which consists of a rigid body with a thin viscoelastic circular ring attached at some point of the body. Assuming that the stiffness of the ring is large and the dissipation is small enough, we study the quasi-static motion which is set in after free elastic oscillations have damped. In particular, the steady-state motions in the weakly elliptic orbit are found and their stability is investigated.

Key words: quasi-static motions, oscillations, rotations, stability

1. Introduction and Problem Statement

Fast progress in the development of space techniques leads to designing and construction of large space systems having a complex structure and working for a long period of time. That is why the study of the dynamics of such systems has attracted so much attention. In many theoretical works, flexible systems have been modeled in the form of connected rigid bodies or in the form of a single rigid body with flexible appendages. However, for large spacecrafts a continuum model seems to be more appropriate.

In this paper we study the dynamics of a large spacecraft modeled by a flexible solid body that is internally damped. The spacecraft moves on an elliptic orbit. It is supposed that the body is sufficiently rigid so that the following inequality $\omega_0 \ll \Omega_1$ is satisfied; ω_0 and Ω_1 are the mean motion of the center of mass and the least frequency of the free elastic vibrations of the spacecraft, respectively. We also assume that the damping time of the free elastic vibrations is smaller than the orbital period.

After dumping the natural elastic vibrations, the spacecraft moves under the influence of the gravitational and inertial torques. Such type of motion is called quasi-static motion.

In the presented paper we study quasi-static planar motions of a spacecraft about its center of mass. The spacecraft is modeled by a thin inextensible homogeneous circular viscoelastic ring rigidly connected with a rigid body. The elastic ring lies in a plane parallel to the orbital plane and is attached in a point C of the principal axis of the body perpendicular to the orbital plane. We investigate the problem within the framework of linear theory of elasticity.

Celestial Mechanics and Dynamical Astronomy **81**: 51–56, 2001.
© 2001 *Kluwer Academic Publishers.*

2. Equation of Motion

To describe the motion of the spacecraft about its center of mass we introduce the so-called *average frame of reference* $Ox_1x_2x_3$ (Canavin and Likins, 1977). The axes of this frame (in the absence of elastic strain) lie along the principal axes of inertia. In the process of deformation, the average frame moves in such a way that, at any moment of time, the following conditions are satisfied:

$$\int_V \mathbf{W} dm = 0, \quad \int_V \boldsymbol{\rho} \times \mathbf{W} dm = 0, \qquad (1)$$

where $\mathbf{W}$ is the elastic displacement of element dm, whose position in the undeformed state is specified by radius vector $\boldsymbol{\rho}$.

We can satisfy the first of conditions (1) taking the origin of $Ox_1x_2x_3$ in the centre of mass O of the spacecraft. Considering the planar motions we denote by φ the angle between the radius vector of the center of mass O_*O and axis Ox_1. In the average frame a displacement of an element dm of the spacecraft is an elastic deformation (Canavin and Likins 1977). This is why the angle φ describes the motion of the spacecraft as a whole in the orbital frame of reference, see Figure 1.

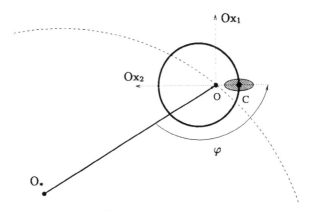

Figure 1. The rigid body with elastic ring.

Assuming that the angular velocity of the spacecraft is small, we can neglect the 'prestressing' by centrifugal forces, and we can expand an elastic displacement into a series

$$\mathbf{W} = \sum_{n=1}^{\infty} q_n(t) \mathbf{U}^{(n)}, \qquad (2)$$

of natural modes $\mathbf{U}^{(n)} = (U_1^{(n)}, U_2^{(n)}, U_3^{(n)})$ of spacecraft free vibrations. With the help of the technique developed by Bardin and Markeev (1993), we obtained functions $U_i^{(n)}$ solving the problem of free plane flexural vibrations of an elastic ring connected with a rigid body.

We describe the internal friction forces by means of the Rayleigh dissipative function (Landau and Lifshitz, 1959)

$$\Psi = \chi b \sum_{n=1}^{\infty} \Omega_n^2 \dot{q}_n^2,$$

where χ is a dimensionless parameter, b is a positive constant, Ω_n are frequencies of free elastic vibrations.

The motion of spacecraft in the above model is described by a system of second order ordinary differential equations, which includes one equation for the variable φ and an infinite set of equations for the coordinates q_n describing deformations of the spacecraft. For the general case these equations were obtained by Santiny (1976).

Such a system is rather complicated for analysis. However, in the case of quasi-static motion, applying the approach developed by Chernous'ko (1978) and Markeev (1989), we can achieve some simplifications. The main idea of this approach is based on the fact that under assumptions of Section 1 the parameter $\varepsilon = \omega_0/\Omega_1$ is small. That is why the approximate solution of the equations for coordinates q_n can be found by means of asymptotic methods of Vasil'eva Butuzov (1973). Substituting this solution into the equation for variable φ we obtain an equation of the planar quasi-static motion of the spacecraft as a whole, and the right-hand side of the obtained equation does not depend on variables q_n.

Assuming that $\chi \sim \varepsilon^\varkappa (1 < \varkappa < 2)$, and $e \sim \varepsilon^2$, where e is the orbital eccentricity, we can show similarly as in (Bardin and Markeev, 1993) that the equation of planar quasi-static motion of the spacecraft as a whole has the following form:

$$\ddot{\psi} + \alpha^2 \sin \psi = e(4 \sin \tau - 3\alpha^2 \cos \tau \sin \psi) + \varepsilon^2 F(\psi, \dot{\psi}) + O(\varepsilon^4), \qquad (3)$$

where $\psi = 2\varphi$, $\alpha^2 = 3(J_1 - J_2)/J_3$, J_i are the principal moments of inertia of undeformed spacecraft, and

$$F(\psi, \dot{\psi}) = \frac{1}{3}(3 - \alpha^2)[b_1 \sin \psi + b_2 \sin 2\psi + \chi b_3 \sin^2 \psi +$$
$$+ (b_4 \sin \psi + \chi b_5 \cos \psi + \chi b_6 \sin^2 \psi + \chi b_7 \cos^2 \psi)\dot{\psi} +$$
$$+ (b_8 \sin \psi + \chi b_9 \cos \psi)\dot{\psi}^2 + \chi b_{10} \cos \psi \dot{\psi}^3]. \qquad (4)$$

An explicit form of coefficients b_i is complicated and we do not give it here. The dot denotes differentiation with respect to $\tau = \omega_0 t$.

3. Steady-state Motions

Our assumption $e \sim \varepsilon^2 \ll 1$ allows us to apply the averaging method (Arnold et al., 1993) for investigation of Equation (3). Indeed, in the unperturbed motion $\varepsilon = e =$

0 (the ring is absolutely rigid and the orbit is circular) Equation (3) represents the equation of mathematical pendulum. We denote its general solution by $\psi = Q(\delta, k)$, $\delta = \omega(k)(\tau + \tau_0)$, where k, τ_0 are constants of integration and $\omega(k)$ is the frequency. Cases of oscillation and rotations have to be considered separately because the general solution $Q(\delta, k)$ for the respective cases has a different form.

Using the explicit form of the general solution of the unperturbed system we can carry out the change of variables $\psi, \dot{\psi} \rightarrow \delta, k$. In the perturbed system k and δ are the so-called fast and slow variables, respectively.

Calculations showed that if the frequency $\omega(k)$ is not close to $1/s$, $(s = 1, 2, ...)$ then in averaged equations the perturbing torque caused by the orbital ellipticity does not manifest itself in the approximation up to the second order in ε. That is why we will consider only the resonant case when $\omega(k) - 1/s$, $s = 1, 2, \ldots$ is small. We introduce resonance tuning $\vartheta = s\delta - p\tau$, which is slow near points of phase space where the resonance relation holds, and is fast at a distance from these points. Averaged with respect to the fast variable δ equations for k and ϑ are

$$\frac{dk}{d\tau} = \frac{1}{J'(k)}[e \sin \vartheta M_s(k) + \chi \varepsilon^2 P(k)],$$

$$\frac{d\vartheta}{d\tau} = s\omega(k) - \frac{1}{J'(k)}[eN_s(k) \cos \vartheta M_s(k) + \varepsilon^2 s R(k)], \tag{5}$$

where the coefficients $M_s(k)$, $N_s(k)$, $P(k)$, $R(k)$, $J'(k)$ are the following:

$$M_s(k) = -\frac{1}{4\pi} \int_{-\pi}^{\pi} \cos s\delta (8Q_\delta + 3s\omega^2 Q_\delta^2)\, d\delta, \qquad N_s(k) = -\frac{dM_s}{dk},$$

$$P(k) = \frac{3 - \alpha^2}{6\pi} \int_{-\pi}^{\pi} [\omega(b_5 \cos Q + b_6 \sin^2 Q + b_7 \cos^2 Q)Q_\delta^2 +$$

$$+ Q_\delta b_3 \sin^2 Q + Q_\delta^3 \omega^2 b_9 \cos Q + Q_\delta^4 \omega^3 b_{10} \cos Q]\, d\delta,$$

$$R(k) = \frac{3 - \alpha^2}{6\pi} \int_{-\pi}^{\pi} [b_1 + b_2 \cos Q + \omega b_4 Q_\delta + \omega^2 b_8 Q_\delta^2]Q_k \sin Q\, d\delta,$$

$$J'(k) = \frac{dJ}{dk}, \qquad J(k) = \frac{\omega(k)}{2\pi} \int_{-\pi}^{\pi} Q_\delta^2(\delta, k)\, d\delta.$$

Stationary points k_*, ϑ_* of system (5) correspond to steady-state modes of motion. In Figures 2–4 the so-called resonance curves $k_*(\alpha)$ are shown. A point on these curves corresponds to s stationary modes, whose phases ϑ_* differ by $\pi m (m = 1, 2, \ldots, s)$. In the case of oscillations, steady-state modes are $2\pi s$-periodic oscillations with respect to τ, while in the case of rotations they comprise rotations with a period of revolution about the center of mass equal to $2\pi s$.

For resonances $k = 1$ and $k = 2$ in the oscillatory motion the steady-state modes exist for all values of k. In other cases the steady-state modes disappear for small values of k and this phenomenon is caused by dissipation.

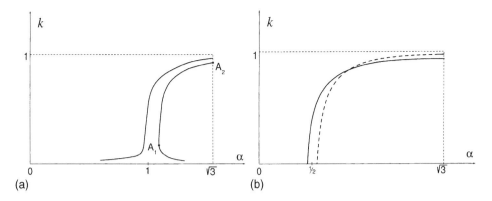

Figure 2. Resonance curves in the case of oscillations for resonances $s = 1$ (a) and $s = 2$ (b).

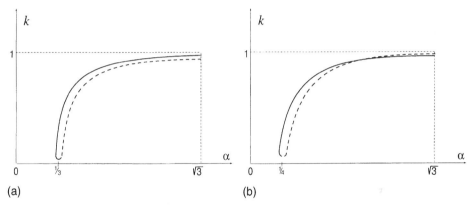

Figure 3. Resonance curves in the case of oscillations for resonances $s = 3$ (a) and $s = 4$ (b).

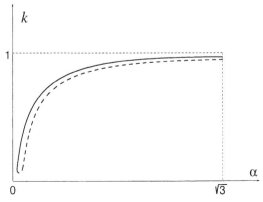

Figure 4. Resonance curves in the case of rotations.

We also solved the problem of stability of the steady-state modes analyzing the characteristic exponents of system (5), linearized in the neighborhood of ϑ_*, k_*. The result is also shown in Figures 2–4 where solid lines correspond to the stable steady-state modes, while dashed curves correspond to the unstable ones.

Acknowledgements

This work was supported by the RFFI (Russian Foundation of Fundamental Research) grant no. 99-01-00405. Boris S. Bardin acknowledges Toruń Centre for Astronomy, N. Copernicus University for excellent working conditions during his stay there.

References

Arnold, V. I., Kozlov, V. V. and Neishtadt, A. I.: 1993, *Dynamical Systems. III. Mathematical Aspects of Classical and Celestial Mechanics*, Springer-Verlag.

Bardin, B. S. and Markeev, A. P.: 1993, 'Plane resonant motions of a viscoelastic body', *Mech. Solids* **28**(3), 87–94.

Canavin, J. R. and Likins, P. W.: 1977, 'Floationg reference frames for flexible spacecraft', *J. Spacecrafts Rockets* **14**(2), 724 pp.

Chernous'ko, F. L.: 1979, 'On the motion of solid body with elastic and dissipative elements', *J. Appl. Math. Mech.* **42**(1), 32–41.

Landau, L. D. and Lifshitz, E. M.: 1959, *Theory of Elasticity*, Pergamon Press, London.

Markeev, A. P.: 1989, 'Dynamics of an elastic body in a gravitational field', *Cosmic Res.* **27**(2), 133–143.

Santiny, P.: 1976, 'Stability of flexible spacecraft', *Acta Astronaut.* **3**(9–10), 685 pp.

Vasil'eva, A. B. and Butuzov, V. F.: 1973, *Asymptotic Expansions of the Solutions of Singularly Perturbed Equations*, Nauka, Moscow.

FORMATION FLYING SATELLITES: CONTROL BY AN ASTRODYNAMICIST

KYLE T. ALFRIEND, SRINIVAS R. VADALI and HANSPETER SCHAUB

Texas A&M University, TX, U.S.A., e-mails: alfriend@aero.tamu.edu; svadali@aero.tamu.edu; hschaub@sandia.gov

Abstract. Satellites flying in formation is a concept being pursued by the Air U.S. Force and NASA. Potential periodic formation orbits have been identified using Hill's (or Clohessy Wiltshire) equations. Unfortunately the gravitational perturbations destroy the periodicity of the orbits and control will be required to maintain the desired orbits. Since fuel will be one of the major factors limiting the system lifetime it is imperative that fuel consumption be minimized. To maximize lifetime we not only need to find those orbits which require minimum fuel we also need for each satellite to have equal fuel consumption and this average amount needs to be minimized. Thus, control of the system has to be addressed, not just control of each satellite. In this paper control of the individual satellites as well as the constellation is addressed from an astrodynamics perspective.

Key words: formation flying, astrodynamics, satellite theory

1. Introduction

Previous studies on the relative motion of spacecraft in Earth orbit have typically used the Clohessy–Wiltshire (CW) equations (Carter, 1998; Kapila, 1999; Miller, 1999; Xing, 1999) to describe the relative equations of motion. With these linearized equations periodic motions in the relative motion reference frame have been identified. These periodic motions include in-plane, out-of-plane, and combinations of these two motion types. When one includes perturbations, some of these periodic orbits are no longer achievable without control to overcome the deviations. A simple example demonstrates this fact. Consider an out-of-plane relative motion caused by a difference in inclination angles. Due to the J_2 perturbation, the inclination difference will cause a differential nodal precession rate between the two satellites resulting in an oscillatory out-of-plane motion with increasing amplitude. However, the linear CW equations do not show this motion; they indicate an out-of-plane oscillatory motion with a constant amplitude. To maintain a relative orbit designed with the CW equations, periodic orbit corrections are necessary to cancel deviations caused by the J_2 perturbations. Further, a reference motion and the accompanying state transition matrix might result in an out-of-plane control that changes inclination because the state transition matrix does not indicate the increasing amplitude caused by the inclination difference. For these reasons it is necessary for the reference motion to include at least the J_2 gravitational perturb-

Celestial Mechanics and Dynamical Astronomy **81**: 57–62, 2001.
© 2001 *Kluwer Academic Publishers.*

ation effect. The satellites considered are assumed to be equal in size and shape. Therefore, compared to the J_2 effect, the differential drag and solar radiation effects are of lesser importance in this study and are neglected. In other formation flying scenarios they may be the dominant perturbations.

For two satellites to remain close together their periods, nodal precession rates, and perigee drift rates must be equal. For close satellits the only way for this to happen is for the two satellites to be in the same orbit. therefore, one of the conditions must be relaxed. For small eccentricities the differential perigee drift has the least effect and is the easiest to control, therefore we relax this condtion. The resulting orbits that match period and right ascension rate we call J_2 invariant orbits. First we derive the conditions for these orbits. Then, we consider control of the constellation, not just one satellite relative to the chief. For some relative motion orbits J_2 invariant orbits are not practical and the differential nodal precession rate must be controlled. Since this rate is proportional to the difference in inclination the fuel required to control this rate is likely to be different for each satellite. The system lifetime will be defined by the failure of the first satellite, therefore it is desirable that each satellite has the same fuel consumption. This is the focus of our constellation control.

2. J_2 Invariant Orbits

The orbit geometry is described through the Delaunay orbit elements l (mean anomaly), g (argument of perigee) and h (longitude of the ascending node) with the associated generalized momenta L, G and H defined as

$$L = \sqrt{\mu a}, \qquad G = \sqrt{\mu a(1 - e^2)} = L\eta, \qquad H = G \cos i, \tag{1}$$

where a is the semi-major axis, e is the eccentricity and i is the inclination angle.

From Brouwer's (1959) orbit theory the mean anomaly, argument of perigee and right ascension secular rates with $\epsilon = -J_2$ are

$$\dot{l} = \frac{1}{L^3} + \epsilon \frac{3}{4L^7} \left(\frac{L}{G}\right)^3 \left(1 - 3\frac{H^2}{G^2}\right) = \frac{1}{L^3} + \epsilon \frac{3}{4L^7 \eta^3} \left(1 - 3\cos^2 i\right), \tag{2}$$

$$\dot{g} = \epsilon \frac{3}{4L^7} \left(\frac{L}{G}\right)^4 \left(1 - 5\frac{H^2}{G^2}\right) = \epsilon \frac{3}{4L^7 \eta^4} \left(1 - 5\cos^2 i\right), \tag{3}$$

$$\dot{h} = \epsilon \frac{3}{2L^7} \left(\frac{L}{G}\right)^4 \left(\frac{H}{G}\right) = \epsilon \frac{3}{2L^7 \eta^4} \cos i. \tag{4}$$

For two satellites to remain close the secular growth of the three angles of the two satellites must be equal. When described by mean elements as above, these secular rates are a function of the momenta or semi-major axis, eccentricity and inclination. When described by osculating elements they are a function of all the orbital elements. Therefore, definition of the relative motion orbits is best described by mean elements. The transformation to osculating elements yields the initial conditions.

At any instant of time, the current inertial position and velocity vectors can be transformed into corresponding instantaneous orbit elements. In the absence of perturbations, these elements are constants. Adding the J_2 perturbation causes the elements to vary according to three types of motion, namely secular drift, short period motion and long period motion. The long period term is the period of the apsidal rotation. Over a short time this looks like a secular growth of order J_2^2. The short period growth manifests itself as oscillations of the orbit elements, but does not cause the orbits to drift apart. The relative secular growth is the type of growth that needs to be avoided for relative orbits to be J_2 invariant. This growth is best described through *mean orbit elements*. These are orbit averaged elements which do not show any of the short period oscillations. Mean elements can be obtained analytically or numerically. Highly accurate mean elements that must include atmospheric drag, tesseral harmonic and third body effects probably require numerical averaging. In this paper we use an analytical approach to help determine the accuracy that will be required. By studying the relative motion through the use of mean orbit elements, we are able to ignore the orbit period specific oscillations and address the secular drift directly. It is not possible to set the drift of each orbit to zero. However, instead we choose to set the difference in mean orbit element drifts to zero to avoid *relative secular growth*.

Using the fact that $\delta L = O(\epsilon)$ the resulting equations for the angle rate differences are

$$\delta\dot{\theta} = -\frac{3}{L_0^4}\delta L - \epsilon\frac{3}{4L_0^7\eta_0^5}\left[3\eta_0(1 - 3\cos^2 i_0) + 4(1 - 5\cos^2 i_0)\right]\delta\eta +$$

$$+\epsilon\frac{3}{2L_0^7\eta_0^4}(3\eta_0 + 5)\cos i_0 \sin i_0\delta i, \tag{5}$$

$$\delta\dot{g} = \epsilon\frac{3}{L_0^7\eta_0^5}[-2(1 - 5\cos^2 i_0)\delta\eta + 5\eta_0 \sin i_0 \cos i_0\delta i], \tag{6}$$

$$\delta\dot{h} = -\epsilon\frac{3}{2L_0^7\eta_0^5}[4\cos i_0)\delta\eta + \eta_0 \sin i_0\delta i]. \tag{7}$$

For the relative motion orbits to not drift apart the relative rates given by Equations (5)–(7) must be zero. Unfortunately, the only solution is the trivial solution $\delta L = \delta\eta = \delta i = 0$. Therefore, we can select only two conditions and the best decision is to control the perigee drift and let $\delta\dot{\theta} = \delta\dot{h} = 0$. These conditions lead to the orbits called J_2 invariant orbits. The conditions are

$$\delta\eta = -\frac{\eta_0}{4}\tan i_0\delta i$$

$$\delta L = -\frac{\epsilon}{4L_0^4\eta_0^5}(4 + 3\eta_0)(1 + 5\cos^2 i_0)L_0\delta\eta. \tag{8}$$

Two problems can arise when imposing these conditions. The first is the required large change in eccentricity (η) in near polar orbits resulting from the $\tan i_0$ term.

The second problem occurs in near circular orbits. Since $\eta^2 = (1 - e^2)$, $\delta\eta = -(\eta/e)\delta e$ and large changes in eccentricity are required to counter the differential nodal precession. If the required changes in eccentricity result in unacceptable relative motion orbits, it is best to invoke the condition that the projection of the deputy angular velocity vector along the chief orbit normal be zero. That is,

$$\delta\dot{\theta} + \delta\dot{h}\cos i_0 = 0. \tag{9}$$

This leads to the condition

$$\delta L = \frac{\epsilon}{4L_0^3\eta_0^5}(3\eta_0 + 4)[\eta_0 \sin 2i_0\delta i + (3\cos^2 i_0 - 1)\delta\eta]. \tag{10}$$

3. Constellation Control

As shown in the previous section the increase in the out of plane displacement caused by J_2^2 cannot be countered in some cases by small changes in the orbital element. From Gauss' variational equations it is easily shown that the Δv to counter this growth is

$$\Delta v = 3\pi J_2 \left(\frac{R_e}{R}\right)^2 v \sin^2 i\delta i. \tag{11}$$

Thus, the fuel consumption is a function of the difference in inclination. This means that the satellites in constellations that have out of plane motion will have different fuel maintenance requirements. From a lifetime and design viewpoint it is desirable that all satellites consume the same amount of fuel over the system lifetime. To demonstrate the concept for that results in equal fuel consumption for each satellite we will use as an example the relative motion orbit whose projection in the horizontal plane is a circle of radius ρ. Referring to Figure 1, which is a snapshot taken at the chief's equator crossing, the in track and out of plane displacements can be expressed as

$$\begin{aligned}
x &= 0.5\rho \sin(\psi_0 + \alpha), & \dot{x} &= 0.5\rho n_0 \cos(\psi_0 + \alpha), \\
y &= \rho \cos(\psi_0 + \alpha), & \dot{y} &= -\rho n_0 \sin(\psi_0 + \alpha), \\
z &= \rho \sin(\psi_0 + \alpha), & \dot{z} &= \rho n_0 \cos(\psi_0 + \alpha),
\end{aligned} \tag{12}$$

where ψ_0 is the chief argument of latitude and $\psi_0 = n_0 t$, the subscript 0 refers to the chief satellite and $t = 0$ occurs at the equator. Also

$$n_0 = \dot{\theta}_0 + \cos i_0\dot{\Omega}_0. \tag{13}$$

In the presence of J_2 and with the constraint in Equation (9) the modified CW equations (Carter, 1998) are

$$\begin{aligned}
\ddot{x} &- 2n_0\dot{y} - 3n - 0^2 x = u_x, \\
\ddot{y} &+ 2n_0\dot{x} = u_y, \\
\ddot{z} &+ n_0^2 z = u_z + 2An_0 \cos\alpha \sin\psi_0
\end{aligned} \tag{14}$$

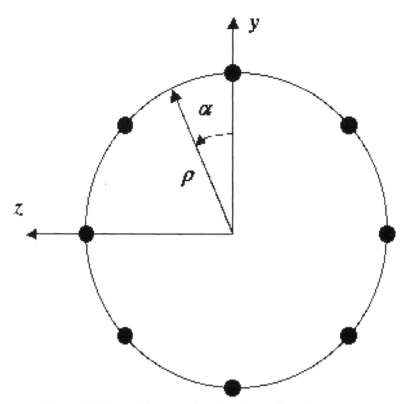

Figure 1. Horizontal plane snapshot when the chief is at the equator.

where

$$A = 1.5 J_2 n_0 (R_e/a_0)^2 \rho \sin^2 i_0. \tag{15}$$

The forcing frequency along the z-axis is very nearly equal to the natural frequency causing resonance. We detune the z-oscillator by introducing a small constant value for $\dot{\alpha}$ such that every satellite spends an equal amount of time with the same values of δi and $\delta \Omega$. In this way, the fuel consumption is balanced among all the satellites. The next question is what is the value of $\dot{\alpha}$ that minimizes the fuel consumption. Assuming $\dot{\alpha}$ is very small the controls necessary to rotate the formation and perfectly cancel the J_2 disturbance are

$$u_x = 2n_0 \dot{\alpha} x_r, \quad u_y = -n_0 \dot{\alpha} y_r, \quad u_z = -2n_0(\dot{\alpha} z_r + 2A \cos \alpha \sin \psi), \tag{16}$$

where x_r, y_r and z_r are given by Equations (12). The average quadratic cost per satellite, considering an infinite number of satellites, over one orbit of the chief can be represented by

$$J = \frac{n_0}{4\pi^2} \int_0^{2\pi/n_0} \int_0^{2\pi} (u_x^2 + u_y^2 + u_z^2) \, d\alpha \, dt. \tag{17}$$

Substituting for u_x, u_y and u_z and evaluating the integral gives

$$J = (3\dot{\alpha}^2\rho^2 + A^2 + 2A\rho\dot{\alpha})n_0^2. \tag{18}$$

Minimization of the above expression with respect to $\dot{\alpha}$ gives the optimal value of $\dot{\alpha}$

$$\dot{\alpha}_{\text{opt}} = -\frac{1}{2}J_2 n_0 \left(\frac{R_e}{a_0}\right)^2 \sin^2 i_0. \tag{19}$$

Since A is positive $\dot{\alpha}_{\text{opt}}$ is negative. If $\dot{\alpha}$ is zero, the average cost per satellite is $J = A^2 n_0^2$. Thus, this detuning strategy has resulted not only in an averaging of the fuel consumption over all the satellites, but also in a 33% reduction in the cost function and a substantial reduction in fuel consumption. For chief orbit parameters of $a_0 = 7100$ km and $i_0 = 70$ deg and a relative motion orbit with a radius of 0.5 km the period for the satellites to rotate around the circle is 179 days.

4. Conclusions

Two new concepts for minimizing fuel consumption for formation flying satellites have been presented. First, the changes in the orbital parameters for negating the out of plane drift and in track drift due to the J_2 perturbation were derived. Then, a strategy for equalizing the fuel consumption over all the satellites in the constellation was derived. This concept also resulted in a substantial reduction in the total fuel consumption for the constellation.

Acknowledgement

This research was supported by the Air Force Office of Scientific Research under grant F49620-99-1-0075.

References

Brouwer, D.: 1959, 'Solution of the problem of artificial satellite theory without drag', *Astronaut. J.* **64** (1274), 378–397.

Carter, T. E.: 1998, 'State transition matrix for terminal rendezvous studies: brief survey and new example', *J. Guid. Navigat. Cont.* 148–155.

Kapila, V., Sparks, A. G., Buffington, J. M. and Yan, Q.: 1999, 'Spacecraft formation flying: dynamics and control', In: *Proceedings of the American Control Conference*, San Diego, California, pp. 4137–4141.

Sedwick, R., Miller, D. and Kong, E.: 1999, 'Mitigation of differential perturbations in clusters of formation flying satellites', In: *AAS/AIAA Space Flight Mechanics Meeting*, Paper No. AAS 99-124.

Xing, G. Q., Parvez, S. A. and Folta, D.: 1999, 'Implementation of autonomous GPS guidance and control for the spacecraft formation flying', In: *Proceedings of the American Control Conference*, San Diego, California, pp. 4163–4167.

LOW ENERGY TRANSFER TO THE MOON

W. S. KOON[1], M. W. LO[2], J. E. MARSDEN[1] and S. D. ROSS[1]

[1]*Control and Dynamical Systems, Caltech, Pasadena, California, U.S.A.*
[2]*Navigation and Mission Design, Jet Propulsion Laboratory, Pasadena, California, U.S.A.*

Abstract. In 1991, the Japanese Hiten mission used a low energy transfer with a ballistic capture at the Moon which required less ΔV than a standard Hohmann transfer. In this paper, we apply the dynamical systems techniques developed in our earlier work to reproduce systematically a Hiten-like mission. We approximate the Sun–Earth–Moon-spacecraft 4-body system as two 3-body systems. Using the invariant manifold structures of the Lagrange points of the 3-body systems, we are able to construct low energy transfer trajectories from the Earth which execute ballistic capture at the Moon. The techniques used in the design and construction of this trajectory may be applied in many situations.

Key words: earth–moon transfer, invariant manifolds, three-body problem, space mission design

1. How to Get to the Moon Cheaply

1.1. HITEN MISSION

The traditional approach to construct a spacecraft transfer trajectory to the moon from the Earth is by Hohmann transfer. This type of transfer uses only 2-body dynamics. It is constructed by determining a two-body Keplerian ellipse from an Earth parking orbit to the orbit of the moon. See Figure 1(a). The two bodies involved are the Earth and a spacecraft. Such a transfer requires a large ΔV for the spacecraft to catch up and get captured by the moon.

In 1991, the Japanese mission, Muses-A, whose propellant budget did not permit it to transfer to the moon via the usual method was given a new life with an innovative trajectory design, based on the work of Belbruno and Miller (1993). Its re-incarnation, renamed Hiten, used a low energy transfer with a ballistic capture at the moon. An Earth-to-Moon trajectory of this type, which utilizes the perturbation by the Sun, requires less fuel than the usual Hohmann transfer. See Figures 1(b) and (c).

1.2. COUPLED THREE-BODY MODEL

In this paper, we present an approach to the problem of the orbital dynamics of this interesting trajectory by implementing *in a systematic way* the view that the Sun–Earth–Moon-spacecraft 4-body system can be approximated as two coupled 3-body systems. Below is a schematic of this trajectory in the Sun–Earth rotating

Celestial Mechanics and Dynamical Astronomy **81:** 63–73, 2001.
© 2001 *Kluwer Academic Publishers.*

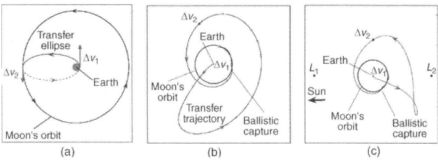

Figure 1. (a) Hohmann transfer. (b) Low energy transfer trajectory in the geocentric inertial frame. (c) Same trajectory in the Sun–Earth rotating frame.

frame, showing the two legs of the trajectory: (1) the Sun–Earth Lagrange point portion and (2) the lunar capture portion. See Figure 2(a).

Within each 3-body system, using our understanding of the invariant manifold structures associated with the Lagrange points L_1 and L_2, we transfer from a 200 km altitude Earth orbit into the region where the invariant manifold structure of the Sun–Earth Lagrange points interact with the invariant manifold structure of the Earth–Moon Lagrange points. See Figure 2(b). We utilize the sensitivity of the 'twisting' of trajectories near the invariant manifold tubes in the Lagrange point region to find a fuel efficient transfer from the Sun–Earth system to the Earth–Moon system. The invariant manifold tubes of the Earth–Moon system provide the dynamical channels in phase space that enable ballistic captures of the spacecraft by the Moon.

The final Earth-to-Moon trajectory is integrated in the bi-circular 4-body model where both the Moon and the Earth are assumed to move in circular orbits about the Earth and the Sun respectively in the ecliptic, and the spacecraft is an infinitesimal mass point. This bi-circular solution has been differentially corrected to

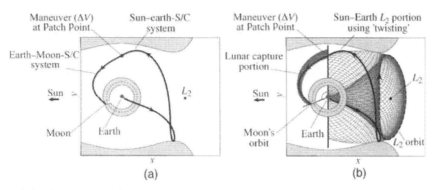

Figure 2. (a) Two legs of a Hiten-like trajectory in the Sun–Earth rotating frame. (b) The interaction of invariant manifold tubes of the Sun–Earth and the Earth–Moon systems permits a fuel efficient Earth-to-Moon transfer with the perturbation of the Sun.

a fully integrated trajectory with the JPL ephemeris using JPL's LTool (libration point mission design tool). LTool is JPL's new mission design tool currently under development which is based on dynamical systems theory. This will be described in a subsequent paper.

2. Planar Circular Restricted Three-Body Problem–PCR3BP

We start with the PCR3BP as our first model of the mission design space. Let us recall some basic features of this model summarized in Koon et al. (2000).

2.1. THREE-BODY MODEL

The PCR3BP describes the motion of a body moving in the gravitational field of two main bodies (the primaries) that are moving in circles. The two main bodies could be the Sun and Earth, or the Earth and Moon, etc. The total mass is normalized to 1; they are denoted $m_S = 1 - \mu$ and $m_E = \mu$. The two main bodies rotate in the plane in circles counterclockwise about their common center of mass and with angular velocity also normalized to 1. In the actual solar system: the Moon's eccentricity is 0.055 and Earth's is 0.017 which are quite close to circular. The third body, the spacecraft, has an infinitesimal mass and is free to move in the plane. The planar restricted 3-body problem is used for simplicity. Generalization to the 3-dimensional problem is of course important, but many of the essential dynamics can be captured well with the planar model.

2.2. EQUATIONS OF MOTION

Choose a rotating coordinate system so that the origin is at the center of mass, the Sun and Earth are on the x-axis at the points $(-\mu, 0)$ and $(1 - \mu, 0)$ respectively, that is, the distance from the Sun to Earth is normalized to be 1. See Figure 3(a). Let (x, y) be the position of the spacecraft in the plane, then the equations of motion are

$$\ddot{x} - 2\dot{y} = \Omega_x, \qquad \ddot{y} + 2\dot{x} = \Omega_y,$$

where $\Omega = (x^2 + y^2)/2 + (1 - \mu)/r_s + \mu/r_e$. Here, the subscripts of Ω denote partial differentiation in the variable. r_s, r_e are the distances from the spacecraft to the Sun and the Earth respectively. See Szebehely (1967) for the derivation.

2.3. ENERGY MANIFOLDS

The system of equations is autonomous and can be put into Hamiltonian form with 2 degrees of freedom. It has an energy integral

$$E = \tfrac{1}{2}(\dot{x}^2 + \dot{y}^2) - \Omega(x, y),$$

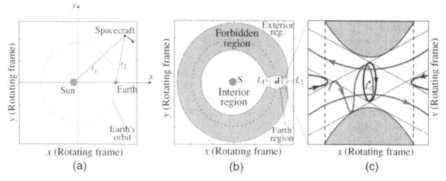

Figure 3. (a) Equilibrium points of the PCR3BP in the rotating frame. (b) Hill's region (schematic, the region in white), which contains a 'neck' about L_1 and L_2. (c) The flow in the region near L_2, showing a periodic orbit around L_2, a typical asymptotic orbit winding onto the periodic orbit, two transit orbits and two non-transit orbits. A similar figure holds for the region around L_1.

which is related to the Jacobi constant C by $C = -2E$. Energy manifolds are 3-dimensional surfaces foliating the 4-dimensional phase space. For fixed energy, Poincaré sections are then 2-dimensional, making visualization of intersections between sets in the phase space particularly simple.

Our main concern is the behavior of orbits whose energy is just above that of L_2. Roughly, we refer to this small energy range as the temporary capture energy range. Fortunately, the temporary capture energy manifolds for the Sun–Earth and Earth–Moon systems intersect in phase space, making a fuel efficient transfer possible.

2.4. EQUILIBRIUM POINTS

The PCR3BP has three collinear equilibrium (Lagrange) points which are unstable, but for the cases of interest to mission design, we examine only L_1 and L_2 in this paper. Eigenvalues of the linearized equations at L_1 and L_2 have one real and one imaginary pair. Roughly speaking, the equilibrium region has the dynamics of a saddle × harmonic oscillator.

2.5. HILL'S REGIONS

The Hill's region which is the projection of the energy manifold onto the position space is the region in the xy-plane where the spacecraft is energetically permitted to move around. The forbidden region is the region which is not accessible for the given energy. For an energy value just above that of L_2, the Hill's region contains a 'neck' about L_1 and L_2 and the spacecraft can make transition through these necks. See Figure 3(b). This equilibrium neck region and its relation to the global orbit structure is critical: it was studied in detail by Conley (1968), McGehee (1969), and Llibre et al. (1985).

2.6. THE FLOW NEAR L_1 AND L_2

More precisely, in each region around L_1 and L_2, there exist four types of orbits (see Figure 3(c)): (1) an unstable periodic Lyapunov orbit (black oval); (2) four cylinders of asymptotic orbits that wind on to or off this period orbit; they form pieces of stable and unstable manifolds; (3) transit orbits which the spacecraft can use to make a transition from one region to the other; for example, passing from the exterior region (outside Earth's orbit) into the Earth temporary capture region (bubble surrounding Earth) via the neck region; (4) non-transit orbits where the spacecraft bounces back to its original region.

2.7. INVARIANT MANIFOLDS AS SEPARATRICES

Furthermore, invariant manifold tubes are global objects – they extend far beyond the vicinity of L_1 and L_2. These tubes partition the energy manifold and act as separatrices for the flow through the equilibrium region: those inside the tubes are transit orbits and those outside the tubes are non-transit orbits. For example in the Earth–Moon system, for a spacecraft to transit from outside the Moon's orbit to the Moon capture region, it is possible *only* through the L_2 periodic orbit stable manifold tube. Hence, stable and unstable manifold tubes control the transport of material to and from the capture region. These tubes can be utilized also for ballistic capture. See Figure 4.

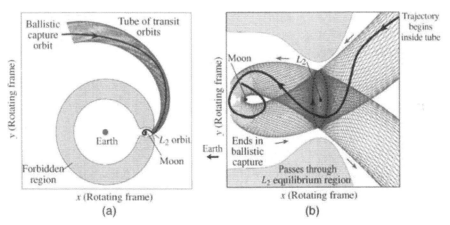

Figure 4. (a) Trajectories inside the stable manifold tube will transit from outside Moon's orbit to Moon capture region. (b) Trajectory that ends in ballistic capture at the Moon.

3. Low Energy Transfer to the Moon with Ballistic Capture

3.1. TWO COUPLED THREE-BODY MODEL

By taking full advantage of the dynamics of the 4-body system (Earth, Moon, Sun, and spacecraft), the fuel necessary to transfer from the Earth to the Moon can be significantly less than that required by a Hohmann transfer. However, since the structure of the phase space of the 4-body system is poorly understood in comparison with the 3-body system, we initially model it as two coupled planar circular restricted 3-body systems. By doing this, we can utilize the Lagrange point dynamics of both the Earth–Moon-spacecraft and Sun–Earth-spacecraft systems. In this simplified model, the Moon is on a circular orbit about the Earth, the Earth (or rather the Earth–Moon center of mass) is on a circular orbit about the Sun, and the systems are coplanar. In the actual solar system: the Moon's eccentricity is 0.055, Earth's is 0.017. and the Moon's orbit is inclined to Earth's orbit by 5°. These values are low, so the coupled planar circular 3-body problem is considered a good starting model. An orbit which becomes a real mission is typically obtained first in such an approximate system and then later refined through more precise models which include effects such as out-of-plane motion, eccentricity, the other planets, solar wind, etc. However, tremendous insight is gained by considering a simpler model which reveals the essence of the transfer dynamics.

This is similar to the more standard approach in mission design where the solar system is viewed as a series of 2-body problems where Keplerian theory applies. JPL's spectacular multiple fly-by missions such as Voyager and Galileo are based on this Keplerian decomposition of the solar system. But when one needs to deal with the ballistic capture regime of motion, a 3-body decomposition of the solar system is absolutely necessary.

However, the success of this approach depends greatly on the configuration of the specific four bodies of interest. In order for low energy transfers to take place, the invariant manifold structures of the two 3-body systems must intersect within a reasonable time. Otherwise, the transfer may require an impractically long time of flight. For the Sun–Earth–Moon-spacecraft case, this is not a problem. The overlap of these invariant manifold structures provide the low energy transfers between the Earth and the Moon.

3.2. CONSTRUCTION OF EARTH-TO-MOON TRANSFER

The construction is done mainly in the Sun–Earth rotating frame using the Poincaré section Γ (along a line of constant x-position passing through the Earth). This Poincaré section helps to glue the Sun–Earth Lagrange point portion of the trajectory with the lunar ballistic capture portion. See Figures 7(c) and (d), p. 71.

The basic strategy is to find an initial condition (position and velocity) for a spacecraft on the Poincaré section such that when integrating forward, the spacecraft will be guided by the L_2 Earth–Moon manifold and get ballistically captured by the Moon; when integrating backward, the spacecraft will hug the Sun–Earth manifolds and return to Earth.

We utilize two important properties of the Lagrange point dynamics of the 3-body problem. The stable manifold tube is key in targeting a capture orbit for the Earth–Moon portion of the design. The twisting of orbits in the equilibrium region

3.3. LUNAR BALLISTIC CAPTURE PORTION

Recall that by targeting the region enclosed by the stable manifold tube of the L_2 Lyapunov orbit in the Earth–Moon system, we can construct an orbit which will get ballistically captured by the Moon. When we transform this Poincaré cut of the stable manifold of an Earth–Moon L_2 Lyapunov orbit into the Poincaré section of the Sun–Earth system, we obtain a closed curve. A point interior to this curve will approach the Moon when integrated forward. See Figure 5. Assuming the Sun is a negligible perturbation to the Earth–Moon-spacecraft 3-body dynamics during this leg of the trajectory, any spacecraft with initial conditions within this closed curve will be ballistically captured by the Moon. 'Ballistic capture by the Moon' means an orbit which under natural dynamics gets within the sphere of influence of the Moon (20,000 km) and performs at least one revolution around the Moon. In such a state, a slight ΔV will result in a stable capture (closing off the necks at L_1 and L_2).

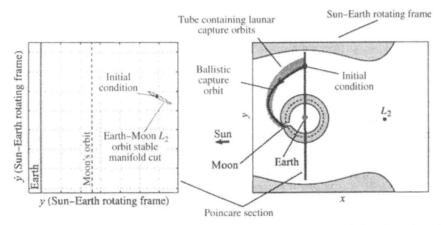

Figure 5. (a) The stable manifold cut of an Earth–Moon L_2 orbit in the Poincaré section of the Sun–Earth system. (b) A point interior to this cut, with the correct phasing of the Moon, will reach the Moon's ballistic capture region when integrated forward.

3.4. TWISTING OF ORBITS AND SUN–EARTH LAGRANGE POINT PORTION

Since the twisting of orbits in the equilibrium region is key in finding the Sun–Earth Lagrange point portion of the design, we would like to review this property briefly. In Koon et al. (2000), we learn that orbits twist in the equilibrium region following roughly the Lyapunov orbit. The amount of twist of an orbit depends sensitively on its distance from the manifold tube. The closer to the manifold tube an orbit begins on its approach to the equilibrium region, the more it will be twisted when it exits the equilibrium region. Hence, with small change in the initial condition (such as a small change in velocity at a fixed point), we can change the destination of an orbit dramatically. In fact, we can use this sensitivity to target the spacecraft back to a 200 km Earth parking orbit.

Look at the Poincaré section Γ in Figure 6(a). Notice that how a minute line strip q_2q_1 of orbits just outside of the unstable manifold cut, when integrated backward, gets stretched into a long strip $P^{-1}(q_2)P^{-1}(q_1)$ of orbits that wraps around the whole stable manifold cut. Recall that points on q_2q_1 represent orbits which have the same position but slightly different velocity. But their pre-image $P^{-1}(q_2)P^{-1}(q_1)$ can reach any position on the lower line where the stable manifold tube intersects (see Figure 6(b)).

Pick an energy in the temporary capture range of the Sun–Earth system which has L_2 orbit manifolds that come near a 200 km altitude Earth parking orbit. Compute the Poincaré section Γ (see Figure 6(a)). The curve on the right is the Poincaré cut of the unstable manifold of the Lyapunov orbit around the Sun–Earth L_2. Picking an appropriate initial condition just outside this curve, we can backward integrate to produce a trajectory coming back to the Earth parking orbit.

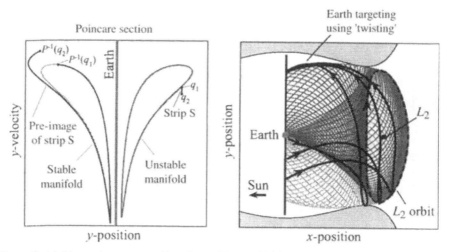

Figure 6. (a) Line strip q_2q_1 outside of unstable manifold cut gets stretched into a long strip $P^{-1}(q_2)P^{-1}(q_1)$ that wraps around stable manifold cut. (b) With infinitesimal changes in velocity, any point near lower tube cross-section can be targeted (integrating backward).

3.5. CONNECTING THE TWO PORTIONS

We can vary the phase of the Moon until the Earth–Moon L_2 manifold cut intersects the Sun–Earth L_2 manifold cut. See Figures 7(a) and (b). In the region which is in the interior of the gray curve but in the exterior of the black curve, an orbit will get ballistically captured by the Moon when integrated forward; when integrated

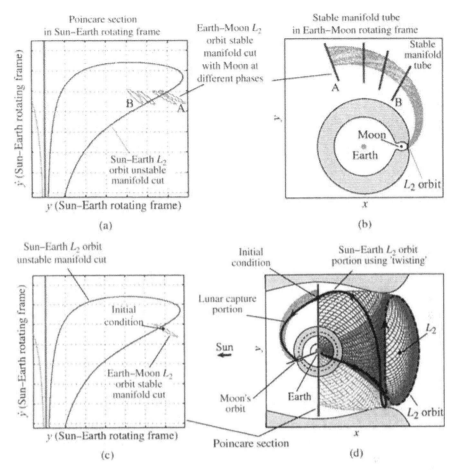

Figure 7. (a) and (b) Vary the phase of the Moon until Earth–Moon L_2 manifold cut intersects Sun–Earth L_2 manifold cut. (c) Pick a point in the interior of gray curve but in the exterior of black curve. (d) An orbit will get ballistically captured by the Moon when integrated foreword; when integrated backward, orbit will hug the invariant manifolds back to the Earth.

backward, the orbit will hug the unstable manifold back to the Sun–Earth L_2 equilibrium region with a twist, and then hug the stable manifold back towards the position of the Earth parking orbit. See Figures 7(c) and (d).

With only a slight modification (a small mid-course ΔV of 34 m/s at the patch point), this procedure produces a genuine solution integrated in the bi-circular 4-body problem. Since the capture at the Moon is natural (zero ΔV), the amount of on-board fuel necessary is lowered by about 20% compared to a traditional Hohmann transfer. This bi-circular solution has been differentially corrected to a fully integrated trajectory with JPL ephemeris using JPL's LTool. This is the subject of a future paper.

3.6. WHY DOES IT WORK?

What follows is a couple of heuristic arguments for using the coupled 3-body model. When outside the Moon's small sphere of influence (20,000 km), which is most of the pre-capture flight, we can consider the Moon's perturbation on the Sun–Earth-spacecraft 3-body system to be negligible. Thus, we can utilize Sun–Earth Lagrange point invariant manifold structures. The mid-course ΔV is performed at a point where the spacecraft is re-entering the Earth's sphere of influence (900,000 km), where we can consider the Sun's perturbation on the Earth–Moon-spacecraft 3-body system to be negligible. Thus, Earth–Moon Lagrange point structures can be utilized for the lunar portion of the trajectory.

Moreover, the fact that the patch point ΔV is so small and may even be eliminated can be understood by considering the following. From a 200 km circular orbit around the Earth, it requires approximately 3150 m/s (provided by the launch vehicle) to reach the Earth–Moon L_1 and L_2. For another 50 m/s, you can reach the Sun–Earth L_1 and L_2! In other words, a spacecraft needs roughly the same amount of energy to reach the Sun–Earth and the Earth–Moon L_1 and L_2. This fortuitous coincidence is what enables these low energy lunar transfer and capture orbits.

4. Conclusion

We have laid bare the dynamical mechanism for a Hiten-like mission. The theory of Lagrange point dynamics of the 3-body system developed in Koon et al. (2000) is crucial in understanding this problem. In many previous applications of dynamical systems theory to mission design, the focus has been on using the trajectory arcs on the invariant manifolds as initial guesses for the desired end-to-end trajectory. In this paper, we have shown that the tubular regions enclosed by the manifolds, the regions exterior to the manifolds, as well as the manifolds themselves all may be used to advantage depending on the desired characteristics of the final trajectory. Mission designers with this knowledge can pick and choose to their hearts' content, an infinite variety of trajectories to suit almost any purpose at hand.

References

Belbruno, E. A. and Miller, J. K.: 1993, 'Sun-perturbed Earth-to-Moon transfers with ballistic capture', *J. Guid. Cont. Dyn.* **16**, 770–775.

Conley, C.: 1968, 'Low energy transit orbits in the restricted three-body problem', *SIAM J. Appl. Math.* **16**, 732–746.

Koon W. S., Lo, M. W., Marsden, J. E. and Ross, S. D.: 2000, 'Heteroclinic connections between Lyapunov orbits and resonance transitions in celestial mechanics', *Chaos* **10**, 427–469.

Llibre, J., Martinez, R. and Simó, C.: 1985, 'Transversality of the invariant manifolds associated to the Lyapunov family of periodic orbits near *L2* in the restricted three-body problem', *J. Differ. Eq.* **58**, 104–156.

McGehee, R. P.: 1969, 'Some Homoclinic Orbits for the Restricted Three Body Problem', PhD Thesis, University of Wisconsin, Madison, Wisconsin.

Szebehely, V.: 1967, *Theory of Orbits, The Restricted Problem of Three Bodies*, Academic Press, New York and London.

DISSIPATIVE FORCES AND EXTERNAL RESONANCES

SYLVIE JANCART and ANNE LEMAITRE

Département de mathématique FUNDP, 8, Rempart de la Vierge, B-5000 Namur, Belgique,
e-mail: sylvie.jancart@fundp.ac.be

Abstract. We analyze the process of resonance trapping due to Poynting–Robertson drag and Stokes drag in the frame of the restricted 3-body problem and in the case of external mean motion resonances. The numerical simulations presented are computed by using the 3-dimensional extended Schubart averaging (ESA) integrator developed by Moons (1994) for all mean motion resonances. We complete it by adding the contributions of the dissipative forces. To follow the philosophy of the initial integrator, we average the drag terms, but we do not make any expansion in series of eccentricity or inclination. We show our results, especially capture around asymmetric equilibria, and compare them to those found by Beaué and Ferraz-Mello (1993, 1994) and Liou et al. (1979).

Key words: Schubart averaging, dissipative forces, resonance trapping

1. Introduction

In the beginning of the nineties, the discoveries of exoplanets around different stars spurred researches on models combining the effects of mean motion resonances and dissipative forces on particles. To study the dynamical effects of these forces, we develop an averaged integrator by adding the contributions of the averaged dissipative forces to the equations of motion in the case of an inclined elliptic restricted 3-body problem.

2. Extended Schubart Integrator and Dissipative Contributions

In her 3-dimensional averaged integrator, Moons (1994) computed the disturbing function (and its derivatives) which appears in the resonant Hamiltonian of the inclined elliptic restricted 3-body problem, in closed form and for any given value of the eccentricity and inclination of the two small bodies. This integrator is known as the 'extended Schubart averaging' (ESA) and gives the averaged motion of an asteroid in mean motion resonance $p + q/p$ with a perturbing planet. The adopted procedure uses regular variables appropriate to these resonances and related to the eccentricity and inclination.

$$x = \sqrt{2J}\cos\sigma, \qquad y = \sqrt{2J}\sin\sigma,$$
$$X = \sqrt{2J_z}\cos\sigma_z, \qquad Y = \sqrt{2J_z}\sin\sigma_z,$$

Celestial Mechanics and Dynamical Astronomy **81:** 75–80, 2001.
© 2001 *Kluwer Academic Publishers.*

$$N = L + G - \frac{p}{q}L, \qquad v = -\frac{p+q}{q}\lambda' + \frac{p}{q}\lambda,$$

where σ, σ_z are the resonant angles, $J = L - G$ and $J_z = G - H$ and L, G, H are the Delaunay's variables. λ, λ' represent the mean longitudes of the asteroid and the planet, respectively. The expression of the resonant angles are given by

$$\sigma = \frac{(p+q)\lambda' - p\lambda - q\tilde{\omega}}{q}, \qquad \sigma_z = \frac{(p+q)\lambda' - p\lambda - q\Omega}{q}.$$

Following the same idea, we calculate the closed forms, without any expansion in series of the eccentricity or the inclination, for the drag terms and we add the averaged components of the dissipative force to the equations of motion given by the ESA. This work was done for the Poynting–Robertson drag and the Stokes drag. We compare our results with the ones obtained by Beaugé and Ferraz-Mello in the planar restricted problem of 3 bodies and give some results for the inclined problem.

2.1. THE POYNTING–ROBERTSON DRAG

The Poynting–Robertson drag (PR) is caused by the nonuniform reemission of the sunlight that a particle absorbs. It causes the orbit of the particle to decay at a rate dependent on its size. Usually, we express this force by a parameter γ which compares it to the gravitational force. So, the PR drag force has the form (see Burns, 1979, Murray and Dermott, 1999)

$$\mathbf{F}_R = \frac{\gamma\mu}{r^2}\left(\hat{r} - \frac{\dot{r}}{c}\hat{r} - \frac{1}{c}\mathbf{v}\right),$$

where, $\mathbf{r}$ and $\mathbf{v}$ are respectively the position vector and the velocity vector of the particle, $\hat{r} = \mathbf{r}/r$, c the speed of light, $\mu = k^2 M_c$ with M_c the primary mass and k^2 the universal gravitation constant and

$$\gamma = \frac{\Phi S Q_{\mathrm{pr}}|r|^2}{c\mu},$$

where S is the geometrical cross section of the particle, Φ is the energy flux density at distance r, Q_{pr} the radiation pressure efficiency factor. The components of this force in Delaunay's variables, after averaging, are

$$F_M = 0, \qquad F_L = \frac{-\mu\gamma n(1 + (3/2)e^2)}{(c\beta^3)},$$

$$F_\omega = 0, \qquad F_G = \frac{-\mu\gamma n}{c},$$

$$F_\Omega = 0, \qquad F_H = \frac{-(\mu\gamma n \cos i)}{c},$$

DISSIPATIVE FORCES AND EXTERNAL RESONANCES

where $\beta = \sqrt{1 - e^2}$, and M, ω, Ω and n are the mean anomaly, the pericenter, the node and the mean motion of the particle, respectively. The symbol F_z is the non conservative part to add to the right-hand side of the differential equation $\dot{z}$ of ESA.

2.2. THE STOKES DRAG

What is generally called Stokes drag effect is the pression of gas molecules on particles of different sizes. This force is a linear function of the relative velocity of the particle with respect to the gas (see Paterson 1987).

$$\mathbf{F} = -C\mathbf{v}_r = -C(\mathbf{y} - \alpha(\boldsymbol{\omega} \times \mathbf{x})),$$

with C a nondynamical parameter which depends on the physical properties of the gas and on the density and the size of the particles. $\mathbf{x}$, $\mathbf{y}$ are the Cartesian heliocentric coordinates and the Cartesian velocity vector, respectively. In this case, the gas has a circular velocity a bit less than the Keplerian velocity at the same point. α is the ratio of gas velocity over Keplerian velocity and ω is the circular angular velocity at that point. The averaged components of this force are given by

$$
\begin{aligned}
F_M &= 0, \\
F_\omega &= -F_\Omega \cos i, \\
F_\Omega &= -C\alpha/\beta.(d_5 - d_6 - e^2(d_5 + d_1) + 2ed_3) \sin \omega \cos \omega, \\
F_L &= CL(-\beta \cos i + \alpha A \cos^2 i + \alpha B \sin^2 i), \\
F_G &= CL(-\beta + \alpha A \cos i), \\
F_H &= CL(-1 + \alpha\beta d_1 \cos i),
\end{aligned}
\tag{1}
$$

where $B = \cos^2 \omega A - \beta^2(\cos^2 \omega - \sin^2 \omega)d_5$, $\beta = \sqrt{1 - e^2}$, d_1, d_3, d_5, d_6 and A are simple combinations of complete elliptic integrals of the first and second kind $\mathbf{K}$ and $\mathbf{E}$

$$
\begin{aligned}
d_1 &= \frac{4}{\sqrt{1 + e}}\mathbf{K}(a_k), \\
d_3 &= \frac{4}{\sqrt{1 + e}}\left(\frac{2 - a_k^2}{a_k^2}\mathbf{K}(a_k) - \frac{2}{a_k^2}\mathbf{E}(a_k)\right), \\
d_5 &= \frac{-16}{\sqrt{1 + e}}\left(\frac{a_k^2 - 1}{3a_k^2}\mathbf{K}(a_k) - \frac{2 + a_k^2}{a_k^4}\mathbf{E}(a_k)\right), \\
d_6 &= d_1 - d_5, \\
A &= \frac{-4}{3}\sqrt{1 + e}(4\mathbf{E}(a_k) - (1 - e)\mathbf{K}(a_k)),
\end{aligned}
\tag{2}
$$

where $a_k = \sqrt{2e/(1 + e)}$ and u is the eccentric anomaly of the particle.

The averaging of these expressions (1) with respect to the mean anomaly was done by using the tables of Gradshteyn and Ryzhik (1963) after numerical checking. These results were obtained without any expansion in eccentricity and inclination, we just had to compute the elliptic or trigonometric functions.

Numerical results are shown in Section 3 and 4. To obtain the dissipative contributions in resonant variables in both cases, PR drag and Stokes drag, we use

$$F_\sigma = -Fv - F_\omega, \qquad\qquad F_J = F_H - F_G,$$
$$F_{\sigma_z} = -F\sigma - F_M - F_\omega, \qquad F_{J_z} = F_G - F_L,$$
$$F_v = \frac{p}{q}(F_M + F_\omega), \qquad\qquad F_N = 2LF_H,$$

and we add them to the right-hand side of the differential equations of ESA in σ, σ_z, v, J, J_z and N.

3. Results: Planar Problem

In order to check the validity of the results, we performed a series of numerical integrations in the case of the planar problem in presence of PR drag and Stokes drag and compared our results with those of Beaugé and Ferraz-Mello in the case of the 1/2 resonance. In all cases, the initial conditions are chosen well outside the resonance region.

3.1. THE POYNTING–ROBERTSON DRAG

As Beaugé and Ferraz-Mello (1994), we chose a drag coefficient $\gamma = 0.2$ and a perturbing planet of mass $m' \geqslant 5 \times 10^{-4}$ the mass of the central body on a circular orbit. In fact, the authors presented a minimum masse of 5×10^{-4} for which captures are possible. All our simulations are in accordance with this result. Using the same variables and units, we obtain temporary captures around asymmetric equilibria with the semi-major axis and universal eccentricity $e = 0.48$ as proposed in the above article (1994). Going further, we computed orbits over 8×10^6 years and always obtained transition from the asymmetric equilibrium region to the horseshoe region.

3.2. THE STOKES DRAG

For this perturbation, the authors have chosen Jupiter as perturbing body with $e' = 0.05$ and $C = 1.4 \times 10^{-4}$/day for the drag coefficient which corresponds to a perturbation on a particle with radius equal to 886 cm. For many values of the initial eccentricity, we obtain temporary capture around asymmetric equilibria. In the case of the circular problem ($e' = 0$), the captures are effective and whatever the initial eccentricity is, we always obtain captures around asymmetric equilibria with oscillations around the universal eccentricity $e = 0.054$.

4. Results: Inclined Problem

All the simulations above consider only planar orbits. Since our integrator is three dimensional, we performed a series of integrations in the case of the inclined problem for particles under the effect of PR drag and Stokes drag for the 1/2 resonance. In all cases, the initial conditions are chosen well outside the resonance region and we chose the same values for the drag parameters as for the planar problem. We give here some results. More complete results will be presented in a following paper.

4.1. THE POYNTING–ROBERTSON DRAG

In a 3-dimensional space, the captures are always temporary as announced in the article of Liou et al. (1994) for the case of external resonances. But to obtain these captures, the perturbing body must have a mass of a least 10^{-3} in units of the mass of the central body. We performed simulations for inclinations up to 50 degrees. For initial inclinations under 40 degrees, as for the planar case, these captures appear always around asymmetric equilibria and occur in less than 4000 years. After 10^5 years, the particles usually enter in a Kozai resonance and the eccentricity has regular variations around the universal eccentricity found for the planar case. These variations are in phase with those of the inclination. For initial inclinations over 40 degrees, the possible captures are very short in time, the particles stay in the resonance region 5×10^3 years and then, their semi-major axis falls rapidly.

4.2. THE STOKES DRAG

We first present results in the case of a circular orbit for Jupiter, the perturbing body.

As for the Poynting–Robertson drag, the perturbing body must have a mass of at least 10^{-3} in units of the mass of the central body to obtain captures. These effective captures appear always around asymmetric equilibria and occur in less than 4000 years as long as the initial inclination is not higher than 14 degrees. Whatever the initial condition is, the inclination falls to a value close to 10^{-4} degrees and the eccentricity oscillates around the universal eccentricity found for the planar case.

For the elliptic problem, we find many temporary captures but never if the initial inclination is over 14 degrees. The particles stay in the resonance region for at least 10^4 years.

5. Conclusion

The presented integrator is a very fast tool to test the captures of particles under a drag effect. Interested by these asymmetric equilibria and the transitions from one resonance region to another one, we developed a simple model in the case of the circular planar restricted three-body problem. This model is based on the Andoyer Hamiltonian as suggested in the paper of Beaugé (1994). Like in the second fundamental model of resonance developed by Henrard and Lemaitre (1983), we calculate the equilibria, their stability and capture probability. The results of this research are presented in Jancart, et al. (2001).

Acknowledgements

The authors wish to express their gratitude to J. Henrard for helpful comments and discussions and to C. Beaugé for data and explanations.

References

Beaugé, C. and Ferraz-Mello, S.: 1993, 'Resonance trapping in the primordial solar nebula: the case of a Stockes drag dissipation', *Icarus* **103**, 301–318.

Beaugé, C. and Ferraz-Mello, S.: 1994, 'Capture in exterior mean motion resonances due to Poynting-Robertson drag', *Icarus* **110**, 239–260.

Beaugé, C.: 1994, 'Asymmetric librations in exterior resonances', *Celest. Mech. & Dyn. Astr.* **60**, 225–248.

Burns, J. A., Lamy, P. L. and Soter, S: 1979, 'Radiation forces on small particles in the solar system', *Icarus* **40**, 1–48.

Gradshteyn, I. and Ryzhik, I. M.: 1963, *Tables of Integrals*, Academic Press.

Jancart, S., Lemaitre, A. and Istace, A.: 2001 'Second fundamental model of resonance with asymmetric equilibria', *Celest. Mech. & Dyn. Astr.* (submitted).

Henrard, J. and Lemaitre, A.: 1983, 'Second fundamental model of resonance', *Celest. Mech. & Dyn. Astr.* **30**, 197–218.

Liou, J-C., Zook, H. A. and Jackson, A. A.: 1979, 'Radiation pressure, Poynting-Robertson drag and solar wind drag in the restricted three-body problem', *Icarus* **116**, 186–201.

Murray, C. D. and Dermott, S. F.: 1999, *Solar System Dynamics*, Cambridge University Press.

Moons, M.: 1994, 'Extended Schubart averaging', *Celest. Mech. & Dyn. Astr.* **60**, 173–186.

Patterson, C.: 1987, 'Resonance capture and the evolution of the planet', *Icarus* **70**, 319–333.

LUNISOLAR RESONANCES REVISITED

SŁAWOMIR BREITER

A. Mickiewicz University, Astronomical Observatory, Słoneczna 36,
Poznań PL 60-286, Poland, e-mail: breiter@amu.edu.pl

Abstract. Lunisolar resonances arise in the artificial satellite problem without short-periodic terms. The basic model including the Earth's J_2 and a Hill-type model for the Sun or the Moon admits 20 different periodic terms which may lead to a resonance involving the satellite's perigee, node and the longitude of the perturbing body. Some of the resonances have been studied separately since 1960s. The present paper reviews all single resonances, attaching an appropriate fundamental model to each case. Only a part of resonances match known fundamental models. An extended fundamental model is proposed to account for some complicated phenomena. Most of the double resonance cases still remain unexplored.

Key words: artificial satellites, lunisolar perturbations, resonances, analytical methods, fundamental models

1. Introduction

Forty years ago the artificial satellite theory boosted the progress of celestial mechanics. The main types of resonances which may occur in this problem were immediately recognized in that pioneering era: the critical inclination in the main problem, tesseral/sectorial resonances, and lunisolar resonances. The first two categories have attracted a lot of attention and each of them possess a vast bibliography. The more surprising is the fact, that the problem of lunisolar resonances has remained seriously underrated. Quite possibly it was due to the lack of interest in the effects whose typical time scale may exceed decades or even centuries, but this situation have changed once we realized the dangers imposed by space debris. The present paper aims at the first concise synthesis of the problem.

By a strange caprice of history, the modern celestial mechanics was founded by an alchemist – Isaac Newton (James, 1993). Not pushing the analogy too far, we can say that analytical procedures of celestial mechanics still resemble the efforts of alchemists. We distil Hamiltonians in order to achieve the integrable 'essence' and since the times of Poincaré we know that the ultimate goal, a transmutation of a non-integrable system into an integrable one, remains out of our reach. Nevertheless, the principal objective of this paper is to classify the lunisolar resonances with respect to the 'fundamental models' they match. The notion of a fundamental model stems from the theory of asteroids where it represents the 'fifth essence' of motion, that is, the most simple Hamiltonian capable of properly representing the qualitative features of the problem.

Celestial Mechanics and Dynamical Astronomy **81:** 81–91, 2001.
© 2001 *Kluwer Academic Publishers.*

2. Basic Model

Attacking a problem in its full complexity is a bold but seldom fruitful strategy. Following the well-established tradition (Musen, 1960; Cook, 1962; Kudielka, 1997) let us assume the following *basic model*.

1. The geopotential is restricted to the J_2 part like in the main problem of the satellite theory.
2. Sun and Moon move on circular orbits in the fixed Ecliptic plane; their potentials are truncated at the second Legendre polynomial term, like in the Hill problem.
3. All the remaining perturbing forces are neglected.
4. The variables have been transformed so that the 'distilled' Hamiltonian does not depend on the mean anomaly ℓ; the contribution of J_2^2 is neglected.

Our basic model is a two-and-a-half degrees of freedom Hamiltonian with a general form (Breiter, 1999)

$$\mathcal{H} = \mathcal{H}_0(J_2) + \varepsilon \sum_{k=0}^{22} (S_k \cos s_k + M_k \cos m_k), \tag{1}$$

where $\mathcal{H}_0$ and the amplitudes S_k (solar terms) or M_k (lunar terms) are functions of the Delaunay momenta $L = \sqrt{\mu a}$, $G = L\eta$, $H = G \cos I$, with μ denoting the geocentric gravity parameter and $\eta = \sqrt{1 - e^2}$. The arguments are functions of the angles $g = \omega$, $h = \Omega$ and of time t via the mean longitudes of the Sun λ_s (in s_k) and Moon λ_m (in m_k). All angles are measured with respect to the equator-equinox reference frame, which implies that both perturbing bodies have the right ascensions of their ascending nodes equal to zero and thus their mean longitudes are equal to the mean arguments of latitude. A symbol λ' will refer to either λ_s or λ_m, depending on the context. To simplify further discussion, let us similarly write C_k' to stand for S_k or M_k and ϕ_k' to stand for m_k or s_k in the situations where the distinction is not necessary. Most of the arguments are provided explicitly in next sections (see Breiter, 1999 for the full list). Let us recall, that L is a constant of motion in the basic model.

The expression of $\mathcal{H}_0$ is classical (Brouwer, 1959)

$$\mathcal{H}_0 = J_2 \frac{a_e^2 \mu^4 (G^2 - 3H^2)}{4 L^3 G^5}, \tag{2}$$

where a_e is the Earth equatorial radius. Dividing the amplitudes C_k' (Breiter, 1999) by the presumably greater $\mathcal{H}_0$, we can define the small parameter ε as the ratio

$$\varepsilon = \frac{a^5 N^2}{J_2 a_e^2 \mu}.$$

For the Sun-related terms $N = \dot{\lambda}_s$, whereas for the lunar terms $N = \dot{\lambda}_m \sqrt{\nu}$, where $\nu \approx 0.0123$ is the Moon/Earth mass ratio. For both bodies the small parameter takes approximately the same value ranging from 10^{-5} close to the Earth surface, up to 10^{-1} at $a \approx 5\,a_e$. The latter value of a fixes the limit of 'low orbits' in the context of lunisolar perturbations. Only the low orbits are to be discussed in this paper.

3. Lunisolar Resonances

Let us define the lunisolar resonance as a phenomenon occurring when for at least one of the angles ϕ'_k its unperturbed frequency may vanish

$$\frac{d\phi'_k}{dt} = \frac{\partial \phi'_k}{\partial t} + \left\{ \phi'_k, \mathcal{H}_0 \right\} = 0. \tag{3}$$

Condition (3) admits a simple solution in terms of $c = \cos I$ and of a variable $w = (a/a_e)\,\eta^{8/7}$ (Cook, 1962; Hughes, 1980, 1981). At this stage a first distinction can be made.

1. If $\partial \phi'_k/\partial t = 0$, the 'Kozai–Lidov[1] resonances' occur at a specific inclination regardless of w.
2. If $\partial \phi'_k/\partial t \neq 0$ and $\partial \phi'_k/\partial g = 0$, the 'semi-secular inclination resonances' occur along the curves $w = (\text{const.}) \times c^{2/7}$.
3. If $\partial \phi'_k/\partial t \neq 0$ and $\partial \phi'_k/\partial g \neq 0$, the 'semi-secular eccentricity resonances' occur along the curves $w = (\text{const.}) \times (-5c^2 + (\partial \phi'_k/\partial h)c + 1)^{2/7}$.

Figure 1 presents all the resonance lines $w = w(c(I))$ of the basic model; it is similar to the one published by Cook (1962) but supplemented with the inclination resonances. Observing, that the solar and lunar terms generating the Kozai–Lidov resonances are actually additive, we find 26 different resonances (note that some resonance curves consist of two separate branches). The numbers labeling resonances in Figure 1 are due to (Cook, 1962), save for 16–20 which refer to inclination resonances not considered by Cook; a resonance labeled k involves ϕ'_k as a critical argument. The marks S and M refer to the resonances with Sun and Moon respectively, whereas the absence of this mark indicates that the contribution of both bodies is merged.

Lunisolar resonances seem to form a separate class somewhere in between of the traditional resonance types. Not depending on the satellite's mean anomaly they share with the mean motion (or tesseral/sectorial) resonances an important property: their Hamiltonian has a 'secular part' $\mathcal{H}_0$ one order of magnitude greater than the periodic term, which limits the amplitude of the resonance to $O(\sqrt{\varepsilon})$. On the other hand, they are partially similar to the secular resonances of the planetary type or to the Kozai resonances: they occur in the averaged or doubly averaged

---

[1] Proposing this name, the author wants to pay tribute to the contribution of Lidov (1961), but also to indicate certain similarity with the asteroids' Kozai resonances (Kozai, 1962).

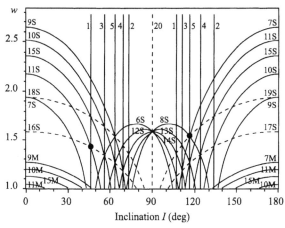

Figure 1. Location of the lunisolar resonances in the (I, w) plane. Dashed lines indicate inclination resonances.

system and thus their time-scale is very slow with respect to the orbital period (ranging from decades or centuries in typical cases to millennia in almost degenerate situations). There are also two limit cases of the lunisolar resonances: if $J_2 \to 0$ we obtain the Kozai resonances problem (Kozai, 1962) whereas for $\dot{\lambda}' \to 0$ we recover a classical frozen orbits case (Coffey et al., 1994).

The double nature of the lunisolar resonances seems to be responsible for the variety of interesting and unusual phenomena associated with the problem.

4. Single Resonances

Let us consider the case when a lunisolar resonance is not coupled with another one. In such a single resonance case we can easily (at least for low orbits) transform variables

$$(g, h, G, H) \leftrightarrows (\varphi, \psi, \Phi, \Psi), \tag{4}$$

to obtain a one degree of freedom, 'redestilled' resonant Hamiltonian

$$\mathcal{K} = \mathcal{A}(\Phi, \Psi) + \mathcal{B}_1(\Phi, \Psi) \cos \varphi + \mathcal{B}_2(\Phi, \Psi) \cos 2\varphi, \tag{5}$$

where in most of the cases either $\mathcal{B}_1$ or $\mathcal{B}_2$ is absent. The momentum Ψ is a constant of motion. Instead of using the critical angle φ and its conjugate Φ, one should rather adopt some Poincaré type variables like

$$x = \sqrt{2\Phi} \sin \varphi, \qquad X = \sqrt{2\Phi} \cos \varphi, \tag{6}$$

to avoid virtual singularities at $\Phi = 0$. The study of each single resonance actually amounts to plotting the level curves of $\mathcal{K}$ on the (x, X) plane. Let us group all the single resonances according to the fundamental models $\mathcal{M}$ which may reasonably approximate the Hamiltonian $\mathcal{K}$.

4.1. THE SFM GROUP

Let us begin with two inclination resonances: 16S and 17S. They are both the first order ($\mathcal{B}_2 = 0$) resonances, leading to the resonant Hamiltonian $\mathcal{K}$ which expanded locally around $x = X = 0$ admits a reduction to the one-parameter second fundamental model of Henrard and Lemaitre (1983)

$$\mathcal{M}_{\text{SFM}} = R^2 + \beta R - 2\sqrt{2R} \cos r. \tag{7}$$

The transformation $(\varphi, \Phi) \rightarrow (r, R)$ is a simple scaling operation described in (Henrard and Lemaitre, 1983) and it will not be discussed here. The critical arguments φ, remaining variables and the remainder of each transformation (4) are listed in Table I. In all tables $\langle \mathcal{H} \rangle$ means the average of $\mathcal{H}$ as defined in equation (1) with all angles except the critical argument φ removed. The types of the phase flow generated by $\mathcal{M}_{\text{SFM}}$ are presented in Figure 2, where $y = \sqrt{2R} \sin r$, $Y = \sqrt{2R} \cos r$.

4.2. THE SFM2 GROUP

First-order resonances are rather exceptional in the whole family. All eccentricity resonances and two of the inclination ones are of the second order, with $\mathcal{B}_1 = 0$. Table I enumerates all of the second-order cases whose Hamiltonians can be successfully reduced to the one-parameter model of Lemaitre (1984)

$$\mathcal{M}_{\text{SFM2}} = 2R^2 + \beta R + R \cos 2r. \tag{8}$$

TABLE I

Simple cases of single resonances

No.	φ	Φ	ψ	Ψ	$\mathcal{K}$
SFM					
16S	$-(h + 2\lambda_s)$	$G - H$	$g + h$	G	$\langle \mathcal{H} \rangle - 2\dot{\lambda}_s (G - H)$
17S	$h - 2\lambda_s$	$G + H$	$g - h$	G	$\langle \mathcal{H} \rangle - 2\dot{\lambda}_s (G + H)$
SFM2					
3	$-(g + \frac{1}{2}h)$	$L - G$	$-\frac{1}{2}h$	$G - 2H$	$\langle \mathcal{H} \rangle$
4	$-(g - \frac{1}{2}h)$	$L - G$	$\frac{1}{2}h$	$G + 2H$	$\langle \mathcal{H} \rangle$
10S,M	$-(g - \frac{1}{2}h - \lambda')$	$L - G$	$\frac{1}{2}h$	$G + 2H$	$\langle \mathcal{H} \rangle - \dot{\lambda}' G$
11S,M	$-(g + \frac{1}{2}h - \lambda')$	$L - G$	$-\frac{1}{2}h$	$G - 2H$	$\langle \mathcal{H} \rangle - \dot{\lambda}' G$
15S,M	$-(g - \lambda')$	$L - G$	h	H	$\langle \mathcal{H} \rangle - \dot{\lambda}' G$
18S	$-(h + \lambda_s)$	$G - H$	$g + h$	G	$\langle \mathcal{H} \rangle - \dot{\lambda}_s (G - H)$
19S	$h - \lambda_s$	$G + H$	$g - h$	G	$\langle \mathcal{H} \rangle - \dot{\lambda}_s (G + H)$

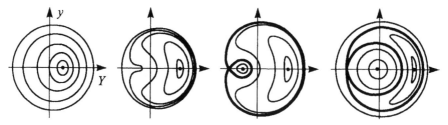

Figure 2. Phase flow types in the second fundamental model (SFM).

Stemming from the same family of Andoyer Hamiltonians (Andoyer, 1903), $\mathcal{M}_{SFM2}$ is closely related with $\mathcal{M}_{SFM}$, although there is an important difference between them: SFM admits only a saddle-node bifurcation, whereas the phase flow of SFM2 changes due to a double pitchfork bifurcation. Nevertheless, both models merit the name 'simple' when compared with remaining resonant cases.

Although the resonance 5 ($\dot{g} = 0$) formally belongs to the SFM2 class, it is not enumerated in Table I. The reason is that the basic model assumed in Section 2 is too simplistic for the frozen orbits problem with g as a critical angle (Hough, 1981a; Coffey et al., 1994).

4.3. THE POLAR GROUP (EFM2)

Even a short glimpse at the results presented in (Breiter, 1999) indicates that some of the second-order eccentricity resonances admit too many of critical points to be properly accounted for by the SFM2. A good example of this problem is the group of five resonances whose inclination curves form a family intersecting at $I = 90°$ (Figure 1).

Observing the behavior of this 'polar group' (Table II) we can derive from Equation (5) an 'extended fundamental model' (EFM2) capable of reproducing qualitative patterns of the group:

$$\mathcal{M}_{EFM2} = \gamma R^3 + 2R^2 + \beta R + R \cos 2r. \tag{9}$$

The details concerning derivation and properties of this two-parameter model will be the subject of a separate study. Here let us only remark, that EFM2 reproduces a rich family of saddle-node, pitchfork and saddle connection (separatrix)

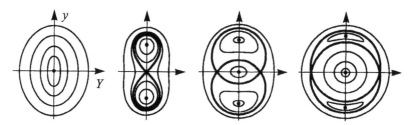

Figure 3. Phase flow types in the second fundamental model for second-order resonances (SFM2).

TABLE II
Resonances of the polar group

No.	φ	Φ	ψ	Ψ	$\mathcal{K}$
EFM2					
6S	$-(g+h+\lambda_S)$	$L-G$	$-h$	$G-H$	$\langle\mathcal{H}\rangle - \dot\lambda_S(L-G)$
8S	$-(g-h+\lambda_S)$	$L-G$	h	$G+H$	$\langle\mathcal{H}\rangle - \dot\lambda_S(L-G)$
12S	$-(g+\frac{1}{2}h+\lambda_S)$	$L-G$	$-\frac{1}{2}h$	$G-2H$	$\langle\mathcal{H}\rangle - \dot\lambda_S(L-G)$
13S	$-(g-\frac{1}{2}h+\lambda_S)$	$L-G$	$\frac{1}{2}h$	$G+2H$	$\langle\mathcal{H}\rangle - \dot\lambda_S(L-G)$
14S	$-(g+\lambda_S)$	$L-G$	h	H	$\langle\mathcal{H}\rangle - \dot\lambda_S(L-G)$

type bifurcations (see Figure 4). Notice that SFM2 model of Lemaitre (1984) is a particular case of EFM2 proposed in Equation (9).

4.4. MIXED MODEL RESONANCES

Even the introduction of EFM2 does not account for all cases arising in the basic model. There still remain four of the second-order eccentricity resonances whose phase portraits resemble a composition of two SFM2 or SFM2 and EFM2. Each of these resonances, listed in Table III, consist of two separate branches in Figure 1 – one for prograde and one for retrograde orbits. In effect, the libration zones associated with prograde and retrograde orbits never approach each other. Thus, for example, the prograde orbits in the resonance No. 7S can be analyzed independently on the retrograde ones as in (Breiter, 2000). The dashed line in Figure 5 has been traced to emphasize this separation.

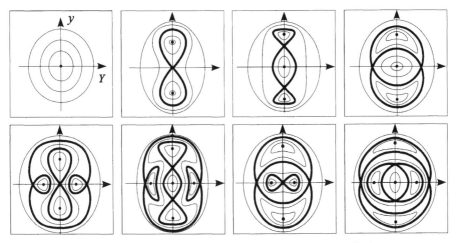

Figure 4. Phase flow in the extended fundamental model (EFM2).

TABLE III

Mixed model resonances

	No.	φ	Φ	ψ	Ψ	$\mathcal{K}$
SFM2/SFM2						
	1	$-(g+h+\lambda_s)$	$L-G$	$-h$	$G-H$	$\langle \mathcal{H} \rangle - \dot{\lambda}_s(L-G)$
	2	$-(g-h+\lambda_s)$	$L-G$	h	$G+H$	$\langle \mathcal{H} \rangle - \dot{\lambda}_s(L-G)$
SFM2/EFM2						
	7S,M	$-(g+\frac{1}{2}h+\lambda')$	$L-G$	$-\frac{1}{2}h$	$G-2H$	$\langle \mathcal{H} \rangle - \dot{\lambda}'(L-G)$
	9S,M	$-(g-\frac{1}{2}h+\lambda')$	$L-G$	$\frac{1}{2}h$	$G+2H$	$\langle \mathcal{H} \rangle - \dot{\lambda}'(L-G)$

Figure 5 presents only the most developed types of the phase flow in both cases. Gradually changing the value of Ψ we first meet the bifurcation sequence of SFM2; this pattern is common to both types of the mixed model resonances. Then, when the SFM2 phase portrait is already well formed (the dark grey zone in Figure 5 – compare with the rightmost case in Figure 3), new bifurcations appear at the origin of axes leading to the occurrence of either a new SFM2 or EFM2 (the light grey patterns in Figure 5).

The absence of interaction between the libration zones suggests, that there is no point in designing a new fundamental model. It seems more reasonable to use either SFM2 or EFM2 locally in appropriately restricted domains of Φ.

4.5. FROZEN NODE RESONANCE

All lunisolar resonances considered until now had a nice property, that either $\mathcal{B}_1$ or $\mathcal{B}_2$ vanished. The only exception from this rule is the problem of frozen nodes,

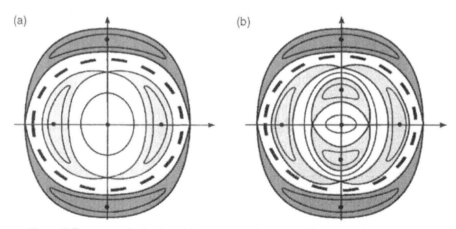

Figure 5. Two types of mixed model resonances: SFM2/SFM2 (a) and SFM2/EFM2 (b).

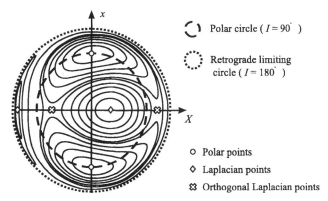

Figure 6. Phase flow in the frozen nodes problem.

where the Hamiltonian (5) contains two harmonics of the critical angle $\varphi = -h$ (resonance 20). Recent results concerning this problem can be found in (Kudielka, 1997).

Up to an arbitrary constant and a constant factor, the resonant Hamiltonian $\mathcal{K} = \langle \mathcal{H} \rangle$ assumes a form

$$\mathcal{K} = (2\Psi - \Phi)(\Phi + 2\beta\Phi \cos 2\varphi) + \gamma(\Psi - \Phi)\sqrt{2\Phi(2\Psi - \Phi)} \cos \varphi, \quad (10)$$

where $\Phi = G - H$ and the second pair of variables is $\psi = g + h$, $\Psi = G$. As usually, β and γ stand for unspecified parameters. The phase flow resulting from (10) is shown in Figure 6; regardless of the value of the constants Ψ, β and γ, there are two centers of libration ('orthogonal equatorial points' according to Kudielka (1997), here simply named 'polar points') located *always* at $h = 90°, 270°$ with $\Phi = 0$ (i.e. $I = 90°$). There is also one of the two stable 'Laplacian points' close to $\Phi = 2\Psi$ (i.e. $I = 180°$) that we do not find in the general Andoyer Hamiltonian AFM2 (Beaugé, 1994)

$$\mathcal{M}_{\text{AFM2}} = \Phi^2 + \beta_1 \Phi + \beta_2\sqrt{2\Phi} \cos \varphi + \beta_3 \Phi \cos 2\varphi, \quad (11)$$

which was a first candidate for a fundamental model of this problem. We may conclude the presentation of the frozen nodes resonance with two remarks

(i) The Hamiltonian (10) is too complicated and probably too particular to call it a fundamental model, whereas known fundamental models do not represent it properly.
(ii) Poincaré variables on a plane are not well suited to this problem whose spherical topology[2] calls for some CDM-type variables (Coffey et al., 1994).

---

[2] Indeed, looking at Figure 6 one may think of a map of the Earth with the north pole at its centre and the Antarctic smeared along the map's circular boundary.

5. Double Resonances

Figure 1 contains about 60 points where at least two resonance curves intersect. These points indicate the location of double resonances, where two critical arguments have vanishing frequencies. Let it be recalled that the number of resonances curves intersecting at one point is not important: a double resonance occurs equally for two or six crossing curves.

Only two of these numerous cases have been discussed until now; they are marked as black dots in Figure 1. Hough (1981b) analyzed the coupling between the Sun-synchronous motion of a satellite's node (No. 19S) and a critical inclination (No. 5). More recently Breiter (2001) examined the coupling of resonances No. 16S and 1. The Poincaré sections presented in the latter contribution indicate the presence of a long-term chaos in the problem.

For obvious reasons (nonintegrability) we do not know much about double resonances in general. The more interesting should be to study couplings between various fundamental models. Such a general work still remains to be performed.

6. Conclusions

The lunisolar resonances are interesting phenomena not only for their practical significance in space mission design, but also due to the breakdown of classical fundamental models derived from Andoyer Hamiltonians. It seems, that question of lunisolar resonances calls for the joint effort of experts in solar system dynamics and artificial satellites theory. In contrast to a widespread *cliché*, the satellite problems still require the research on the level more fundamental than just tracing the microscopic influence of yet another tesseral harmonic.

7. Acknowledgement

Thanks are due to Alessandro Morbidelli whose careful review helped to improve the final version of the article.

References

Andoyer, H.: 1903, 'Contribution à la théorie des petites planétes dont le moyen mouvement est sensiblement double de celui de Jupiter', *Bull. Astron.* **20**, 321–356.

Beaugé, C.: 1994, 'Asymmetric librations in exterior resonances', *Celest. Mech. & Dyn. Astr.* **60**, 225–248.

Breiter, S.: 1999, 'Lunisolar apsidal resonances at low satellite orbits', *Celest. Mech. & Dyn. Astr.* **74**, 253–274.

Breiter, S.: 2000, 'The prograde C7 resonance for Earth and Mars satellite orbits', *Celest. Mech. & Dyn. Astr.* **77**, 201–214.

Breiter, S.: 2001, 'On the coupling of lunisolar resonances for Earth satellite orbits', *Celest. Mech. & Dyn. Astr.* **80**, 1–20.

Brouwer, D.: 1959, 'Solution of the problem of artificial satellite theory without drag', *Astron. J.* **64**, 378–397.

Coffey, S. L., Deprit, A. and Deprit, E.: 1994, 'Frozen orbits for satellites close to an Earth-like planet', *Celest. Mech. & Dyn. Astr.* **59**, 37–72.

Cook, G. E.: 1962, 'Luni-solar perturbations of the orbit of an Earth satellite', *Geophys. J.* **6**, 271–291.

Henrard, J. and Lemaitre, A.: 1983, 'A second fundamental model for resonance', *Celest. Mech.* **30**, 197–218.

Hough, M. E.: 1981a, 'Orbits near critical inclination, including lunisolar perturbations', *Celest. Mech.* **25**, 111–136.

Hough, M. E.: 1981b, 'Sun-synchronous orbits near critical inclination', *Celest. Mech.* **25**, 137–157.

Hughes, S.: 1980, 'Earth satellite orbits with resonant lunisolar perturbations, I. Resonances dependent only on inclination', *Proc. R. Soc. Lond.* A **372**, 243–264.

Hughes, S.: 1981, 'Earth satellite orbits with resonant lunisolar perturbations, II. Some resonances dependent on the semi-major axis, eccentricity and inclination', *Proc. R. Soc. Lond.* A **375**, 379–396.

James, J.: 1993, *The Music of the Spheres*, Grove Press, New York.

Kozai, Y.: 1962, 'Secular perturbations of asteroids with high inclination and eccentricity', *Astron. J.* **67**, 591–598.

Kudielka, V.: 1997, 'Equilibria bifurcations of satellite orbits', In: R. Dvorak and J. Henrard (eds), *The Dynamical Behaviour of our Planetary System*, Kluwer, pp. 243–255.

Lidov, M. L.: 1961, 'Evolution of artificial planetary satellites under the action of gravitational perturbations due to external bodies', *Isskustvennye Sputniki Zemli* (in Russian) **8**, 5–45.

Lemaitre, A.: 1984, 'High-order resonances in the restricted three-body problem', *Celest. Mech. & Dyn. Astr.* **32**, 109–126.

Musen, P.: 1960, 'Contributions to the theory of satellite orbits', In: H. K. Bijl (ed.), *Space Research*, North-Holland, New York, pp. 434–447.

COLLISION PROBABILITY FOR EARTH-CROSSING ASTEROIDS USING ORBITAL RANGING

KARRI MUINONEN[1], JENNI VIRTANEN[1] and EDWARD BOWELL[2]

[1]*Observatory, Kopernikuksentie 1 (P.O. Box 14), FIN-00014 University of Helsinki, Finland*
[2]*Lowell Observatory, 1400 West Mars Hill Road, Flagstaff, AZ 86001, U.S.A.*

Abstract. We introduce new techniques for the computation of the collision probability for Earth-crossing asteroids in the case of short observational arcs and/or small numbers of observations. The techniques rely on the orbital element probability density computed using statistical orbital ranging. We apply the techniques to the Earth-crossing asteroid 1998 OX$_4$ with non-vanishing collision probability in numerous close approaches after the year 2012 (inclusive). We study the invariance of the collision probability in transformations between different orbital element sets, and develop a Spearman rank correlation measure for the validity of the linear approximation. We introduce an optimized, fast version of the statistical ranging method.

Key words: collision probability, orbit determination, Earth-crossing asteroids

1. Introduction

There has been a rapid evolution of theoretical and computational tools for short-term collision probability assessment beginning from the collision analysis for the asteroid 1997 XF$_{11}$ in spring 1998. However, the foundations for the current work were laid decades ago by Öpik (1951, 1976), Wetherill (1967), and Kessler (1981), who carried out theoretical studies on the average, long-term collision probability.

Milani and Valsecchi (1999) and Milani et al. (1999, 2000a, b, c) assessed the collision probability by developing linear and semilinear methods for the propagation of orbital uncertainty (Milani, 1999; Chesley and Milani, 1999). They apply the principle line of variation of the orbital element covariance matrix to sample the spatial confidence region for the potential impactor, which is then propagated nonlinearly to the close approach date. They illustrate the probability of a collision with 'virtual impactors' (see 2000b) and discuss the means by which to exclude the impact possibility by performing negative observations (an idea by, e.g. Bowell et al., 1993). Milani et al. (1999) also study the relevance of resonant returns in the close approach analysis, an impact mechanism brought up by B. Marsden for asteroid 1997 XF$_{11}$.

Valsecchi et al. (2001) show how the variety of pre-encounter conditions can be restricted by the semimajor axis of the perturbed orbit of the small body. They derive an analytic expression for the distribution of energy perturbations at close encounters that can be explicitly illustrated in the so-called *b*-plane of the encounter.

Celestial Mechanics and Dynamical Astronomy **81:** 93–101, 2001.
© 2001 *Kluwer Academic Publishers.*

Chodas (1993) and Yeomans and Chodas (1994) approximate the impact probability by examining the spatial error ellipsoid at the predicted time of the closest Earth approach. Chodas and Yeomans (1996, 1999) discuss a more general method that uses the Monte-Carlo approach but is nonetheless restricted by the use of a least-squares covariance matrix.

In the collision probability computations here, we assume that the Bayesian *a posteriori* probability density of the orbital elements is known (see Muinonen and Bowell, 1993a). We use the method of statistical orbital ranging (Virtanen et al., 2001) to map the element probability density rigorously. The present analysis is a continuation of the work described in Muinonen and Bowell (1993b), Muinonen (1996, 1999), and Virtanen et al. Whereas Milani et al. (2000a) and Chodas and Yeomans (1999) provide so-called target-plane tools for rapid initial screening of collision probability for Earth-crossing asteroids, we aim at maximum rigor of the probabilistic analysis, in spite of the sometimes considerable computational load.

We apply the techniques to the Earth-crossing asteroid 1998 OX$_4$ (tentative size $\sim$300 m), discovered by J. Scotti with the 0.9-m Spacewatch Telescope on July 26, 1998. Unfortunately, the asteroid 1998 OX$_4$ is currently lost.

In Section 2, we present the theoretical methods and computational tools for assessing the collision probability. Section 3 provides their application to the asteroid 1998 OX$_4$ and, in Section 4, we close the article with discussion.

2. Collision Probability

2.1. DEFINITIONS

The planetary collision probability of a small body on a given time interval $\tau = \{t \mid t \in [t_1, t_2]\}$ equals the integral (Muinonen and Bowell, 1993a)

$$P_c(\tau) = \int_{C(\tau)} dP\, p_p(P), \tag{1}$$

where p_p is the *a posteriori* probability density of the orbital elements P and the phase space domain $C(\tau)$ contains the elements leading to a collision on the interval τ. Because collisions are irreversible in time, the epoch of the orbital elements t_0 must precede τ. In what follows, the basic orbital elements are Keplerian, $P = (a, e, i, \Omega, \omega, M_0)^T$, where a is the semimajor axis, e the eccentricity, i the inclination, Ω the longitude of ascending node, ω the argument of perihelion, and M_0 the mean anomaly at the epoch t_0. The three angular elements ω, Ω, and i are referred to the ecliptic at a specified equinox (presently J2000.0).

The orbital element probability density p_p is proportional to the *a priori* (p_{pr}) and observational error (p_ϵ) probability densities, the latter being evaluated for the sky-plane ('O-C') residuals $\Delta\psi(P)$,

$$p_p(P) \propto p_{pr}(P) p_\epsilon(\Delta\psi(P)), \tag{2}$$

where p_ϵ can usually be assumed to be Gaussian. For the mathematical form of p_p to be invariant in transformations from one orbital element set to another (e.g. from Keplerian to equinoctial or Cartesian), we regularize the statistical analysis by

$$p_{pr}(P) \propto \sqrt{\det \Sigma^{-1}(P)}, \qquad \Sigma^{-1}(P) = \Phi(P)^T \Lambda^{-1} \Phi(P), \qquad (3)$$

where Σ^{-1} is the information matrix (or the inverse covariance matrix) evaluated for the orbital elements P, Φ contains the partial derivatives of right ascension (R.A.) and declination (Dec.) with respect to the orbital elements at the observation dates, and Λ is the covariance matrix for the observational errors. The transformation of rigorous probability densities thus becomes analogous to that of Gaussian densities.

The final *a posteriori* probability density of the orbital elements is

$$p_p(P) \propto \sqrt{\det \Sigma^{-1}(P)} \, \exp\left[-\frac{1}{2}\chi^2(P)\right],$$

$$\chi^2(P) = \Delta\psi^T(P)\Lambda^{-1}\Delta\psi(P). \qquad (4)$$

As a consequence of securing the invariance in orbital element transformations, the collision probability is independent of the orbital element set. Assuming constant p_{pr} is acceptable, when the exponent part of Equation (4) confines the probability density into a phase space regime, where the determinant part is essentially constant.

In the validity regime of the linear approximation,

$$\chi^2(P) \approx \chi^2(P_{ls}) + \Delta P^T \Sigma^{-1}(P_{ls})\Delta P, \qquad \Delta P = P - P_{ls}$$

$$\det \Sigma^{-1}(P) \approx \det \Sigma^{-1}(P_{ls}), \qquad (5)$$

where P_{ls} denotes the least-squares orbital elements, and the orbital element probability density is Gaussian,

$$p_p(P) \propto \exp\left[-\frac{1}{2}\Delta P^T \Sigma^{-1}(P_{ls})\Delta P\right]. \qquad (6)$$

Note that the present p_p in Equation (4) allows a properly weighted local linearization in the vicinity of arbitrary orbital elements P.

We can study the validity of the linear approximation quantitatively by comparing the probability density values in Equations (4) and (6) for a large number of sample orbital elements. To measure the validity, we compute the Spearman rank correlation coefficient (Press et al., 1994) between the two sets of probability density values, avoiding detailed normalizations of the probability densities. In essence, for sample orbital elements drawn from the Gaussian probability density (Eq. 6), a rank correlation coefficient close to unity is a *necessary* condition for the approximation. For orbital elements drawn from the rigorous probability density (Eq. 4), a rank correlation coefficient close to unity is a *sufficient* condition for the linear approximation.

2.2. UPPER BOUNDS USING COLLISION ORBITS

Once collision orbits are found in the neighborhood of the maximum likelihood orbit P_{ml} (Muinonen, 1999), one can further determine the maximum likelihood collision orbit; that is, the collision orbit that lies probabilistically closest to P_{ml}, and compute the confidence boundary

$$\Delta \chi^2(\tau) = \min_{P \in C(\tau)} \left\{ \chi^2(P) - \chi^2(P_{ml}) + \log_e \left[\frac{\det \Sigma(P)}{\det \Sigma(P_{ml})} \right] \right\}. \tag{7}$$

Thereafter, an upper bound for the collision probability follows from the computation of the probability outside the confidence boundary of the maximum likelihood collision orbit.

In the linear approximation, the maximum likelihood collision orbit lies closest to the least–squares elements P_{ls} in terms of the covariance matrix Σ,

$$\Delta \chi^2(\tau) \approx \min_{P \in C(\tau)} \Delta P^T \Sigma^{-1}(P_{ls}) \Delta P. \tag{8}$$

The upper bound follows from the probability outside the hyper-ellipsoid defined by the confidence boundary $\Delta \chi^2(\tau)$ of the closest collision orbit,

$$P_c(\tau) \leqslant Y_6 = 10^{-10^{X_6}}, \tag{9}$$

where

$$X_6 = \log_{10} \left(-\log_{10} Y_6 \right), \qquad Y_6 = \frac{1}{8} [(\Delta \chi^2(\tau) + 2)^2 + 4] \times$$
$$\times \exp \left[-\frac{1}{2} \Delta \chi^2(\tau) \right]. \tag{10}$$

The super-exponent X_6 is a useful measure of primary close approaches.

For the rigorous non–Gaussian probability density, Equations (9) and (10) can be taken to be approximately true. Extensive analyses of the probability density (using, e.g. orbital ranging) would be required to improve the accuracy of the upper bound. The confidence boundary $\Delta \chi^2(\tau)$ is computed numerically with the help of the downhill simplex minimization method for both Gaussian and rigorous probability densities.

2.3. COLLISION PROBABILITY USING STATISTICAL RANGING

We compute the orbital element probability density using the technique of statistical ranging (Virtanen et al., 2001). We have optimized the computational algorithm significantly so that, for certain examples, the optimized technique is several orders of magnitude faster than the original one. We no longer select the two pairs of observations randomly from the entire observation set but, instead, make use of a single pair that efficiently produces sample orbital elements. Moreover, we have improved the iteration of the topocentric range intervals; in particular, a careful refinement of the second interval has proved invaluable in reducing the computational time.

To compute the collision probability, we first map the extent of the orbital element probability density by generating a large, unbiased set of sample orbital elements. Second, in the vicinity of each element set, presently along the principal eigenvector of the least squares covariance matrix, we compute a one-dimensional interval of statistically acceptable elements. Third, within each one-dimensional interval, a sub-interval is computed: it includes the elements that result in a collision within the given time interval.

To summarize, the collision probability is

$$
P_{\mathrm{c}}(\tau) = \frac{\sum_{i=1}^{N_1} \frac{\Delta_i^{(\mathrm{c})}}{\Delta_i^2} \left[\sum_{j=1}^{N_2} \sqrt{\det \Sigma^{-1}(\boldsymbol{P}_{ij}^{(\mathrm{c})})} \, \exp \left(-\frac{1}{2} \chi^2(\boldsymbol{P}_{ij}^{(\mathrm{c})}) \right) \right]}{\sum_{i=1}^{N_1} \frac{1}{\Delta_i} \left[\sum_{j=1}^{N_2} \sqrt{\det \Sigma^{-1}(\boldsymbol{P}_{ij})} \, \exp \left(-\frac{1}{2} \chi^2(\boldsymbol{P}_{ij}) \right) \right]}, \tag{11}
$$

where N_1 is the number of initial orbital intervals Δ_i and N_2 is the number of sample orbits per orbital interval. Moreover, $\Delta_i^{(\mathrm{c})}$ denotes the orbital interval i allowing collisions, χ^2 is given in Equation (4), and $\boldsymbol{P}_{ij}$ and $\boldsymbol{P}_{ij}^{(\mathrm{c})}$ are the sample orbits j on the orbital interval i and collision orbital interval i, respectively.

A rough approximation can be obtained from Equation (11) by omitting the probability density values,

$$
P_{\mathrm{c}}(\tau) \approx \left(\sum_{i=1}^{N_1} \frac{1}{\Delta_i} \right)^{-1} \sum_{i=1}^{N_1} \frac{\Delta_i^{(\mathrm{c})}}{\Delta_i^2}, \tag{12}
$$

which further reduces to $\Delta_I^{(\mathrm{c})}/\Delta_I$ for a single orbital interval I. These approximations are computationally faster but can result in completely erroneous collision probabilities.

3. Results

In the orbital analysis, DE–405 positions and velocities (Standish et al., 1995) are used for the planets with the Earth and Moon as separate bodies, while the perturbations due to the largest asteroids (1) Ceres, (2) Pallas, and (4) Vesta are excluded. Parts of the OrbFit free orbit determination software package are used (ftp://copernico.dm.unipi.it/pub/orbfit/). The software derives from the work by Milani (1999), Carpino and Knězević (1996), and Muinonen and Bowell (1993a).

For the Earth-crossing asteroid 1998 OX$_4$, follow-up observations were secured over only a 9.1-day arc, with altogether 21 observations at four different observing sites. Initial orbit determination using Gauss's method succeeded, and the subsequent differential correction of the orbital elements converged rapidly. There was a single outlier observation, and the standard deviations for R.A. and Dec. observational errors were 0.57 and 0.38 arcsec, respectively, including the slippage factor of 1.24. To specify the probability density, some 2500 orbits were computed using the technique of statistical ranging.

The linear approximation was studied for a large sample of orbital elements drawn from the Gaussian probability density (Eq. 6) specified by the least-squares orbital elements and covariance matrix. The Spearman rank correlation coefficient between the rigorous and approximate probability density values of Equations (4) and (6), respectively, was only 0.43, signalling of major differences in the probability densities and the unequivocal failure of the linear approximation, thus showing the definite need to rely on the rigorous technique of statistical ranging.

Throughout the present work for 1998 OX$_4$, we have assumed a constant *a priori* probability density; studies of the orbital element probability density showed that, using the constant *a priori* density instead of that in Equation (3) can here produce a maximum error of less than an order of magnitude. Thus, all our present collision probability estimates are order-of-magnitude estimates.

We start the collision analyses by noting that the collision probabilities in the possible primary close approaches of 2001 and 2003 are negligible: the upper-bound metric is $X_6 = 7.1$ for the former and $X_6 = 9.7$ for the latter. Indeed, we saw no evidence of Earth collision for 1998 OX$_4$ during the January 2001 close approach opportunity.

The computation of the orbital intervals (Eq. 11) in the vicinity of the sample orbital elements and the subsequent computation of the collision intervals was the most severe limitation of the analysis; for the results that follow, the task required several weeks of computing time on a modern workstation.

The comparison of the present computations to those by Milani et al. (2000b) is shown in Table I. The collision probabilities agree within two orders of magnitude. Recalling the compromises between computational speed and mathematical rigor in the work by Milani et al., the differences in the two dynamical models, and the differences in assigning observational weights, the agreement is satisfactory. Our rigorous and approximate collision probabilities are strikingly different; the latter approach appears to overestimate the collision probability systematically by several orders of magnitude. The collision probabilities by Milani et al., are in between the rigorous and approximate probabilities. The upper bounds, giving only loose constraints on the probability, are not particularly useful for the analysis of the secondary close approaches.

Table II shows the new collision probabilities, together with the 2014 probability, discovered in the present work for primary close approaches in 2001. Generally, the new collision probabilities are smaller than those in Table I. The earliest year

TABLE I

The collision probability and its upper bounds in Jan. 19.0-22.0 2014, 2038, 2044, and 2046

	2014	2038	2044	2046
$\Delta\chi^2$	22	13	13	24
X_6	0.5	0.1	0.1	0.5
Y_6	1(−3)	5(−2)	5(−2)	4(−4)
P_c (Eq. 11)	2(−10)	4(−9)	1(−7)	1(−8)
P_c (Eq. 12)	5(−7)	2(−5)	9(−6)	4(−5)
NeoDys	–	2.7(−7)	5.1(−7)	6.6(−7)
MCV	3(−8)	1(−7)	2(−7)	2(−7)

Also shown are the NeoDys (http://newton.dm.unipi.it/cgi-bin/neodys/neoibo) probabilities, and those by Milani et al. (2000b). Note the abbreviation $1(−3) = 1 \times 10^{-3}$.

TABLE II

Collision probabilities for secondary close approaches in Jan. 19.0-22.0 2012, 2014, 2017, 2022, and 2023 after a primary close approach in 2001

	2012	2014	2017	2022	2023
P_c (Eq. 11)	4(−11)	2(−10)	4(−11)	5(−17)	1(−24)
P_c (Eq. 12)	1(−6)	5(−7)	4(−6)	2(−6)	3(−9)

of non-vanishing collision probability is 2012, and the present study is complete to the year 2026 (inclusive).

For curiosity, collision is further possible with an extremely small probability in a tertiary close approach in 2015 after a primary approach in 2001 and a grazing secondary one in 2014. Such a collision opportunity showed up in the systematic search for collision orbits in 2015, beginning from a collision orbit in 2014.

Another discovery is that the 2005 and 2007 primary close approaches contribute to the collision probabilities in 2044 and 2046 (Table III), respectively. Indeed, for 2046, the contribution of the 2007 primary approach dominates over that of the 2003 approach.

TABLE III

Contributions of the 2003, 2005, and 2007 primary approaches to the collision probabilities in 2038, 2044, and 2046

	2038	2044	2046
2003	2(−9)	6(−8)	8(−10)
2005		1(−11)	
2007			5(−9)

4. Discussion

We have offered mathematical definitions for the collision probability between a small body and a planet (or the Sun), the orbital element probability density, and for the linear approximation. We have outlined an efficient, Spearman rank correlation measure for assessing the validity of the linear approximation. The validity depends on the orbital element set used in the computation, a topic that we will be returning to in future studies.

We have offered upper bounds for the collision probability in the linear approximation and using the rigorous probability density. We have optimized the statistical ranging technique for orbit determination, and established a Monte-Carlo technique for the computation of the collision probability for short-arc Earth-crossing asteroids.

While collision orbits are not seen to be useful in giving upper bounds for the collision probability in secondary close approaches, they are a useful tool for assessing the primary approaches; cf., the primary approach of 1997 XF_{11} to the Earth in 2028 (Muinonen, 1999).

We confirm the 1998 OX_4 collision probabilities in 2014, 2038, 2044, and 2046 by Milani et al. (2000b) within two orders of magnitude. We report new collision probabilities for 2012, 2015, 2017, 2022, and 2023 but emphasize that our analyses of the secondary approaches after the 2001 primary approach still continue. Furthermore, secondary approaches after primary ones in 2003, 2005, and 2007 remain mostly unexplored except that we have discovered contributions of the 2005 and 2007 close approaches to the 2044 and 2046 collision probabilities, respectively. We note that computational noise can still lurk in the present results and, in general, that the collision probability is rather sensitive to the hypotheses of the observational errors. We will optimize the numerical integration (Kaasalainen and Laakso, 2001), will continue to study the case of 1998 OX_4, and will participate in the ongoing negative observation campaign (A. Boattini, private communication).

Acknowledgements

The authors are grateful to an anonymous reviewer. Research funded, in part, by the Academy of Finland (K. Muinonen, J. Virtanen) and by NASA (E. Bowell).

References

Bowell, E., Wasserman, L. H., Muinonen, K., McNaught, R. H. and West, R. M.: 1993, 'A search for the lost asteroid (719) Albert', *Bull. Am. Astron. Soc.* **25**, 1118.

Carpino, M. and Knežević, Z.: 1996, 'Determination of the mass of (1) ceres from close approaches of other asteroids', In: S. Ferraz–Mello, B. Morando and J.–E. Arlot (eds), *IAU Symposium 172, Dynamics, Ephemerides, and Astrometry of Solar System Bodies*, Kluwer Academic Publishers, Dordrecht, pp. 203–206.

Chesley, S. R. and Milani, A.: 1999, 'Nonlinear Methods for the Propagation of Orbital Uncertainty', *Paper 99-148 in AAS/AIAA Space Flight Mechanics Meeting*, 16–19 August 1999, Girkwood, Alaska.

Chodas, P. W.: 1993, 'Estimating the impact probability of a minor planet with the Earth', *Bull. Amer. Astron. Soc.* **25**, 1226.

Chodas, P. W. and Yeomans, D. K.: 1996, 'The orbital motion and impact circumstances of comet Shoemaker-Levy 9', In: K. S. Noll, H. A. Weaver and P. D. Feldman (eds), *The Collision of Comet Shoemaker-Levy 9 and Jupiter*, Cambridge University Press.

Chodas, P. W. and Yeomans, D. K.: 1999, 'Orbit Determination and Estimation of Impact Probability for Near Earth Objects', *Paper 99-002 in 21st Annual AAS Guidance and Control Conference.*

Kaasalainen, M. and Laakso, T.: 2001, 'Near-integrability as a numerical tool in solar system dynamics', *Astron. Astrophys.* **368**, 706–711.

Kessler, D. J.: 1981, 'Derivation of the collision probability between orbiting objects: the lifetimes of Jupiter's outer Moons', *Icarus* **48**, 39–48.

Milani, A.: 1999, 'The asteroid identification problem I. Recovery of lost asteroids', *Icarus* **137**, 269–292.

Milani, A. and Valsecchi, G. B.: 1999, 'The asteroid identification problem II. Target plane confidence boundaries', *Icarus* **140**, 408–423.

Milani, A., Chesley, S. R. and Valsecchi, G. B.: 1999, 'Close approaches of asteroid 1999 AN$_{10}$: resonant and non-resonant returns', *Astron. Astrophys.* **346**, L65–L68.

Milani, A., Chesley, S. R. and Valsecchi, G. B.: 2000a, 'Asteroid close encounters with earth: risk assessment', *Planet. Space Sci. ACM'99 Spl Issue* (in press).

Milani, A., Chesley, S. R., Boattini, A., Valsecchi, G. B. and Giovanni, B.: 2000b, 'Virtual impactors: search and destroy', *Icarus* **144**, 12–24.

Milani, A., LaSpina, A., Sansaturio, M. E. and Chesley, S. R.: 2000c, 'The asteroid identification problem III. Proposing identifications', *Icarus* **144**, 39–53.

Muinonen, K.: 1996, 'Spaceguard survey for near–Earth objects: collision probability assessment', *Mem. S. A. It.* **67**, 999–1004.

Muinonen, K.: 1999, 'Asteroid and comet impacts with the Earth. Impact hazard and collision probability', In: Bonnie A. Steves and Archie E. Roy (eds), *Dynamics of Small Bodies in the Solar System*, Kluwer Academic Publishers, Dordrecht, pp. 127–158.

Muinonen, K. and Bowell, E.: 1993a, 'Asteroid orbit determination using Bayesian probabilities', *Icarus* **104**, 255–279.

Muinonen. K. and Bowell, E.: 1993b, 'Collision probability for Earth–crossing asteroids on stochastic orbits', *Bull. Amer. Astron. Soc.* **25**, 1116.

Öpik, E. J.: 1951, 'Collisional probabilities with the planets and the distribution of interplanetary matter', *Proc. Roy. Irish Acad. A.* **54**, 165–199.

Öpik, E. J.: 1976, *Interplanetary Encounters, Close–Range Gravitational Interactions*, Elsevier, Amsterdam.

Press, W. H., Teukolsky, S. A., Vetterling, W. T. and Flannery, B. P.: 1994, *Numerical Recipes in FORTRAN, the Art of Scientific Computing*, Cambridge University Press.

Standish, E. M., Newhall, X. X., Williams, J. G. and Folkner, W. F.: 1995, 'JPL planetary and lunar ephemerides, DE403/LE403', JPL IOM 314.10–127.

Valsecchi, G .B., Milani, A., Gronchi, G. F. and Chesley, S. R.: 2001, 'The distribution of energy perturbations at planetary close encounters' (in press).

Wetherill, G. W.: 1967, 'Collisions in the asteroid belt', *J. Geophys. Res.* **72**, 2429–2444.

Virtanen, J., Muinonen, K. and Bowell, E.: 2001, 'Statistical ranging of asteroid orbits', *Icarus* (in press).

Yeomans, D.K. and Chodas, P.W.: 1994, 'Predicting close approaches of asteroids and comets to earth', In: T. Gehrels (ed), *Hazards Due to Comets and Asteroids*, University of Arizona Press, Tucson, pp. 241–258.

THE DYNAMICS OF METEOROID STREAMS

I. P. WILLIAMS
Astronomy Unit, Queen Mary and Westfield College, Mile End Road, London E1 4NS, U.K.

Abstract. Meteors are streaks of light seen in the upper atmosphere when particles from the interplanetary dust complex collide with the Earth. Meteor showers originate from the impact of a coherent stream of such dust particles, generally assumed to have been recently ejected from a parent comet. The parent comets of these dust particles, or meteoroids, fortunately, for us tend not to collide with the Earth. Hence there has been orbital changes from one to the other so as to cause a relative movement of the nodes of the meteor orbits and that of the comet, implying changes in the energy and/or angular momentum. In this review, we will discuss these changes and their causes and through this place limits on the ejection process. Other forces also come into play in the longer term, for example perturbations from the planets, and the effects of radiation pressure and Poynting–Robertson drag. The effect of these will also be discussed with a view to understanding both the observed evolution in some meteor streams. Finally we will consider the final fate of meteor streams as contributors to the interplanetary dust complex.

Key words: meteors, meteor streams, orbital dynamics, solar system

1. Introduction

Unless we live in very special and unique times, meteors must have been witnessed by the human race since antiquity. Records of their appearances described in terms like *many falling stars* date back for at least two millenia (see Hasegawa, 1993), but serious scientific analysis only dates back for a tenth of this time-span, or about 200 years. The principal reason for this must be the christian view prevalent at that time that the heavens were perfect and so bits of solid material moving on orbits that intersect that of the Earth could not exist. Hence, the observed displays could not actually be *falling stars*, or even falling dust grains and so they were regarded as atmospheric phenomena akin, for example, to lightning. Indeed, the name *meteors* implies an atmospheric phenomenon. The first indication that this interpretation of meteors was not correct had come when Benzenberg and Brandes (1800) had simultaneously observed the same meteors from two different locations and through parallax determined their height to be about 90 km, well above all usual visible atmospheric phenomenon. (This, as it turned out was a remarkably accurate determination of the typical height of meteors.)

The spectacular Leonid displays of 1799 and 1833 had prompted observations of meteors which, in turn, lead to establishing facts and formulating theories. Both Olmstead (1834) and Twining (1834) noticed that shower meteors radiated from a fixed point, since called the *radiant*. A few years later, Herrick (1837, 1838) pointed out that the annual showers were periodic on a siderial rather than a tropical year.

Celestial Mechanics and Dynamical Astronomy **81:** 103–113, 2001.
© 2001 *Kluwer Academic Publishers.*

104   I. P. WILLIAMS

These observations naturally led to the correct idea that meteor streams existed in interplanetary space and that a meteor shower was observed when the Earth passed through such a stream. The Leonid storms of 1799, 1833, and 1866 helped Adams (1867), LeVerrier (1867) and Schiaparelli(1867) to conclude that the orbit of the Leonid meteors were very similar to that of comet 55P/Tempel–Tuttle and that 33 years were very close to the orbital period of this comet, so that spectacular streams are seen only when the parent comet is close to the Earth. Since then comet–meteor stream pairs have been identified for virtually all recognizable significant stream, the identification being made on the basis of the similarity in the orbital elements, usually using some objective measure of the differences such as that proposed by Southworth and Hawkins (1963) or Drummond (1981). A more physical criterion has also recently been proposed by Jopek (1993). A list of meteoroid stream-comet pairings was produced by Cook (1973) and most of the gaps in Cook's list have now been filled. As observational data has improved and more meteor orbits being determined and catalogued (e.g. Lindblad and Steel, 1994; Betlem et al., 1997), so the agreement between comet and meteor orbits have improved, a good indication that the pairings are correct.

A physical understanding of the relationship between meteors and comets became possible about half a century ago, after Whipple (1950) had proposed a new model for comets in which a single cometary nucleus existed. The nucleus was composed of an icy matrix within which dust grains of varying sizes were embedded. As the comet approached the Sun, the ice would sublime, releasing the grains and accelerating them away from the nucleus through drag from the outflowing gas. The very small grains would also react to radiation pressure and flow nearly radially outwards to form the well-known dust tail, while larger grains would remain bound and not have a large velocity relative to the nucleus and so move on fairly similar orbits to the comet–hence forming a meteor stream. Whipple (1951) also modelled this process obtained a formula the ejection speed of the meteoroids relative to the nucleus. A number of authors (e.g. Gustafson, 1989a; Harris and Hughes, 1995) have suggested minor modifications to the details which results in a higher ejection velocity than that given by Whipple, but all result in speeds between a few tens and a few hundred meters per second. It is also possible that meteoroids can be ejected off the surface of asteroids, collisions being the most obvious driving mechanism. A review of this topic was given, for example, by Štohl and Porubčan (1993). Within the context of this discussion, the mechanism for ejecting meteoroids from their parent does not matter, we are concerned more with the dynamics of the consequential motion, based on the changes generated in the orbital energy and angular momentum.

2. The Dynamical Evolution of Meteoroid Streams-Historical Overview

The most dominant force acting on a meteoroid is solar gravity, expressed in the usual way as

$$F_{\text{grav}} = \frac{-GM_\odot m}{r^2},$$

where m is the mass of the meteoroid, r the heliocentric distance and $M_\odot$ is the mass of the Sun. To a reasonable approximation, meteoroids move on Keplerian ellipses about the Sun, so that standard theory gives us

$$E = \frac{-GM_\odot}{2a}, \tag{1}$$

and that

$$P^2 = a^3, \tag{2}$$

where a is the semi-major axis of the orbit in astronomical units and P the orbital period in years. Hence we can obtain

$$\frac{\Delta E}{E} = \frac{-\Delta a}{a} = \frac{-2\Delta P}{3P}. \tag{3}$$

Similarly, consideration of the angular momentum per unit mass, h, gives h,

$$h^2 = GM_\odot p,$$

where p is the semiparameter of the orbit, that is $p = a(1 - e^2)$. This yields

$$\frac{\Delta h}{h} = \frac{\Delta p}{2p} = \frac{\Delta a}{2a} - \frac{e\Delta e}{(1 - e^2)}. \tag{4}$$

Meteoroids are in general small so that solar radiation has some effect on their motion. Firstly radiation pressure weakens gravity. The simplest way of dealing with this is to regard the gravitational constant G as being replaced by $G(1 - \beta)$, where β is the ration of the radiation pressure to the gravitational force. For a spherical meteoroid of radius b and bulk density σ,

$$\beta = 5.75 \times 10^{-5}/(b\sigma),$$

with b measured in centimeters and σ also in cgs units.

Kresak (1974) showed that the effect is to cause the meteoroid to apparently move on an elliptical orbit with semi-major axis a_1 and eccentricity e_1 that are related to the Keplerian values a and e by

$$a_1 = ar(1 - \beta)(r - 2a\beta)^{-1}, \tag{5}$$

$$e_1 = 1 - (1 - e^2)(r - 2a\beta)(1 - \beta)^{-2}r^{-1}. \tag{6}$$

Radiation also produces a drag, known as the Poynting–Robertson effect (Poynting, 1903; Robertson, 1937), through the radiation being absorbed in the solar rest frame but re-emitted in the moving frame of the meteoroid. Mathematical expressions for the rate of change of the orbital parameters have been derived by Wyatt and Whipple (1950), Williams (1983), in particular,

$$a\frac{da}{dt} = -\gamma(2 + 3e^2)(1 - e^2)^{3/2} \tag{7}$$

and

$$\frac{de}{dt} = 2.5\gamma(1 - e^2)^{-1/2}a^{-2}. \tag{8}$$

Here,

$$\gamma = \frac{GM_\odot\beta}{c},$$

where c denotes the speed of light. For orbits in the inner solar system, this gives a timescale for significant change of the order of a^2/γ, or of the order of $b\sigma\,10^7$ years. Major changes due to the Poynting–Robertson effect thus probably requires longer than the age of most streams. However, since the only relevant criterion for a meteoroid is whether or not it hits the Earth, then changes may not have to be that significant so that an important change may take a factor of order 1000 less than implied above.

An other very significant effect is, however, that due to the gravitational perturbations by the planets and, unfortunately, this does not allow closed algebraic solutions. The force on any meteoroid of known position, can be calculated at any instant provided we know all the planetary positions at the same instant. Hence, the instantaneous change in the motion due to this force field can also be computed. Sufficient repetition of this calculation allows the meteoroid position to be obtained at any future time. Brouwer (1947) produced a mathematical algorithm based on this which was used, for example, by Whipple and Hamid (1950) to follow the evolution of the mean Taurid stream over an interval of 4700 years, demonstrating a similarity with the evolution of comet Encke's orbit. At this time, primarily due to a lack of available computing power, secular perturbation methods were popular and used, for example, by Plavec (1950), Babadzhanov and Obrubov (1980, 1983).

With the improvement in both the speed and memory of computers, direct numerical integration methods gained in popularity. In essence, the problem is well suited for a computer simulation involving a numerical integration of the equations of motion, for, if some model for the initial ejection velocity is assumed, then the initial position and velocity is known, the perturbations to be included in the equations of motion are known and so the final state can be predicted and compared with the observations. The first to use such a method was probably Hamid and Youssef (1963), where the motion of six Quadrantid meteoroids was investigated. Nearly two decade later, Hughes et al. (1981) had increased the number of test particles to over 200, while Fox et al. (1983) increased this to 500 000. After this, the use of direct methods became widespread (see e.g., Hunt et al., 1985; Jones and

McIntosh, 1986; Gustafson 1989b; Asher et al., 1993; Williams and Wu, 1994; Wu and Williams, 1995, 1996; Brown and Jones, 1996; Steel and Asher, 1996; Arter and Williams, 1997; Jenniskens and van Leeuwen, 1997)

As mentioned already, by using such numerical techniques, the evolution of a given stream over a given time interval can easily be modelled and this has already been done for most major streams. There are, however, a number of problems with such numerical simulations. The first, and possibly the most important is that the effect of increasing the dispersion in the initial velocity distribution has the same effect as increasing the time over which the perturbations have been acting. The second problem is that a numerical simulation can never actually include the correct number of test particles, (see conclusions for a discussion of stream mass which implies that in a stream there are at least 10^{16} meteoroids). Third, while many meteoroid destruction mechanisms are known, quantifying them within a realistic evolutionary model is much harder, so that deciding how many meteoroids to include from each past apparition of the parent comet is almost a matter of guesswork. Since it is possible to obtain an insight into the behaviour of streams without resorting to a large computational effort, we devote the remainder of this article to discussing this.

3. Meteor Showers, General Analytical Principles

There are two facts regarding meteor showers that simplifies any discussion of the event. First, the longitude of the ascending node, Ω, is known exactly from the time of observation of the shower. Second one node of the meteoroid stream that causes the shower must be exactly at the Earth's heliocentric distance which, in this discussion we will take to be $1\,AU$. Since the orbit of the parent comet does not satisfy this, some changes from the cometary orbit must have taken place. We will consider these in turn.

3.1. THE LONGITUDE OF THE ASCENDING NODE, Ω

Showers have not always appeared when expected. In 1998 a strong component, rich in bright meteors, appeared about 16 h before the expected maximum of the main shower. An explanation for this has been given by Asher et al. (1999). Other examples of a display taking place at an unexpected time is the new peak in the Perseids that appeared in the early nineties slightly separated in time from the traditional main peak or the variability on a short time-scale in the Quadrantids. An other example was the unusual level of ionization detected in the ionosphere somewhat after the main Leonid activity in 1998 (Ma et al., 2001), and thought to be due to the impact of a swarm of very small particles. Since, as already stated, the appearance time is really a measure of Ω, all of these represent changes in Ω.

The longitude of the ascending node denotes the position of the line of nodes, that is the intersection line of the orbital plane and the ecliptic, with respect to the

108   I. P. WILLIAMS

first point of Aries. Any change in this angle must thus represent a change in the position of the line of nodes, which means the orbital plane must have changed.

Radiation pressure and all drag forces (including the Poynting–Robertson effect) act in the plane of the orbit and so can not change Ω. Gravitational perturbations can change the orbital plane but can not differentiate between masses, a large comet and a small meteoroid are equally affected. Hence, the most obvious explanation is that we are observing the effects of the initial meteoroid ejection process – meteoroids had a component of the velocity perpendicular to the orbital plane. Indeed, common sense suggests that this would be the case in most situations. Added weight to this supposition is the fact that the phenomenon is more pronounced in very young streams. Let us then consider the orthogonal component of this ejection process.

$$\mathbf{h} = (h_x, h_y, h_z) = \mathbf{r} \times \mathbf{V} \tag{9}$$

or

$$\mathbf{h} = (y\dot{z} - z\dot{y}, z\dot{x} - x\dot{z}, x\dot{y} - y\dot{x}). \tag{10}$$

But, from normal orbital theory,

$$\pm h_x = h \sin i \sin \Omega, \qquad \mp h_y = h \sin i \cos \Omega,$$

so that

$$\tan \Omega = -\frac{h_x}{h_y}$$

and

$$\sec^2 \Omega \Delta \Omega = -\left(\frac{\Delta h_x}{h_y} - \frac{h_x \Delta h_y}{h_y^2} \right) = \tan \Omega \left(\frac{\Delta h_x}{h_x} - \frac{\Delta h_y}{h_y} \right)$$

so

$$\Delta \Omega = \sin \Omega \cos \Omega \left(\frac{\Delta h_x}{h_x} - \frac{\Delta h_y}{h_y} \right). \tag{11}$$

Assuming that a velocity component perpendicular to the orbital plane is responsible for the change in the angular momentum change,

$$\Delta h_x = y\Delta\dot{z} - z\Delta\dot{y}, \tag{12}$$

$$\Delta h_y = z\Delta\dot{x} - x\Delta\dot{z}. \tag{13}$$

Substituting Δh_x and Δh_y into Equation (10) gives, after some algebraic manipulations,

$$\Delta \Omega = -\sin \Omega \cos \Omega \frac{z}{h_x h_y} (h_x \Delta\dot{x} + h_y \Delta\dot{y} + h_z \Delta\dot{z}) \tag{14}$$

or

$$\Delta\Omega = \frac{z}{h^2 \sin^2 i}(\mathbf{h} \cdot \Delta\mathbf{V}). \tag{15}$$

We know $z = r \sin i \sin(\omega + f)$, hence

$$\Delta\Omega = \frac{r_0 \sin(\omega + f_0)}{h \sin i} v \sin\phi, \tag{16}$$

where $\Delta\Omega$ is the longitude change of the ascending node caused by the ejection velocity, r_0 and f_0 are the heliocentric distance and the true anomaly of the ejection point, i is the inclination of the orbit, and ϕ is angle between the direction of ejection and the orbital plane so that $v \sin\phi$ is the component of the ejection velocity perpendicular to the orbital plane.

In practice, Equation (14) will be more useful to determine the ejection velocity, than to predict the change in the longitude of the node since $\Delta\Omega$ will be determinable from direct observations.

3.2. THE NODAL DISTANCE r_N

The nodal distances are derived from the standard equation for an ellipse with the true anomaly being taken as $-\omega$ or $\pi - \omega$, that is $(1 - e\cos\omega)r_N = p$ and $(1 + e\cos\omega)r_N = p$.

Hence, we can obtain

$$(1 - e\cos\omega)\frac{\Delta r_N}{r_N} = \frac{1 - e^2}{2e}\cos\omega\frac{\Delta a}{a} - \frac{(e^2\cos\omega + \cos\omega - 2e)}{2e}\frac{\Delta p}{p} - \\ -e\sin\omega\Delta\omega \tag{17}$$

for the first node with a similar equation for the other node.

From the fact that the point where the change occurred must be on both the old and the new orbit, we can obtain

$$e\cos(f_0 + \Delta\omega) - e\cos f_0 = \frac{(2e + e^2\cos f_0 + \cos f_0)}{2e}\frac{\Delta p}{p} - \\ -\frac{1 - e^2}{2e}\cos f_0\frac{\Delta a}{a} \tag{18}$$

from which $\Delta\omega$ can be determined for a given situation. However, Equations (3) and (4) show that the relevant quantities above can be related directly to changes in energy and angular momentum. The formulation above has been in terms of an initial velocity being given to a meteoroid, but this need not be so, drag forces and the effects of radiation pressure can also be expressed in terms of changes in angular momentum and energy and so can be incorporated into the above formalism, as can in principle perturbation effects. Note, however, that it is assumed that these

110 I. P. WILLIAMS

changes are small, so that the equations could become invalid after a long period of evolution.

4. Meteoroid Loss from Streams

The most obvious loss mechanism from a meteoroid stream is the production of a meteor shower. Every dust grain that is seen as a meteor has burnt up in the Earth's atmosphere and so has been lost from the stream. However, even the simplest of calculations will show that this loss is insignificant compared to the total content of a stream. A number of authors (e.g. Lovell, 1954; Hughes and McBride, 1989; Jenniskens, 1994) have estimate the total mass of various meteoroid streams and all arrive at a typical value for major streams of the order of 10^{16} g. Since the total mass input from a shower is no more than 100 kg, it is self evident that in the short to medium term this hardly makes a dent in the total stream. More important are those meteoroids that had a near miss since they will be scattered by the gravitational field of the Earth, though when moving at typical speeds of 30 kms^{-1} meteoroids have to pass very close to the Earth (escape velocity of order 10 kms^{-1}) to be perturbed out of the stream. Other mechanisms that have been proposed are inter-meteoroid collisions, in particular high velocity collisions as discussed by Williams et al. (1993). Again unlikely to be important to the stream as a whole. Fragmentation following collisions with solar wind electrons, which leads to an increased efficiency of radiation forces also leads to meteoroid loss. A mechanism that has not received much attention is the sublimation of residual ices which again leads to fragmentation. A much less dramatic effect is the combined perturbation of the planets that slowly change the orbital parameters so that coherence is gradually lost and the stream appears to get weaker and weaker and of longer and longer duration. From the point of view of a stream none of these effects may appear dramatic, but they all do the same thing, they feed the interplanetary dust complex with small grains. All streams do this and so the cumulative effect is significant.

5. Conclusions

In its broadest sense, the evolution of meteoroid streams and the generation of meteor showers has been understood for some considerable time. However, it is only in recent years that the computational capabilities have been available to allow realistic models of meteoroid streams to be developed and much success has been obtained in doing this. References to some of this work has been given here. More details of this including numerical methods used can also be found in recent reviews (e.g. Williams, 1995). The aim of this review was to discuss more fully the underlying celestial mechanics and physics so that an understanding of the phenomenon can be gained as well as the ability to reproduce observed events.

References

Adams, J. C.: 1867, 'On the orbit of the November meteors', *Mon. Not. R. Astr. Soc.* **27**, 247–252.

Arter, T. R. and Williams, I. P.: 1997, 'Periodic behaviour of the April Lyrids', *Mon. Not. R. Astr. Soc.* **286**, 163–172.

Asher, D., Clube, S. V. M. and Steel, D. I.: 1993, 'The Taurid complex asteroids', In: J. Stohl and I. P. Williams (eds), *Meteoroids and their Parent Bodies*, Slovak Academy of Sciences, Bratislava, pp. 93–96.

Asher, D., Bailey, M. E and Emel'yanenko, V. V.: 1999, 'Resonant meteoroids from comet Temple–Tuttle in 1333: the cause of the unexpected Leonid outburst in 1998', *Mon. Not. R. Astr. Soc.* **304**, L53–57.

Babadzhanov, P. B. and Obrubov, Y. Y.: 1980, 'Evolution of orbits and intersection conditions with the Earth of Geminid and Quadrantid meteor Streams', In: I. Halliday and B. A. McIntosh (eds), *Solid Particles in the Solar System*, D. Reidel, Dordrecht, pp. 157–162.

Babadzhanov, P. B. and Obrubov, Y. Y.: 1983, 'Some features of evolution of meteor streams', In: R. M. West (ed.), *Highlights in Astronomy*, D. Reidel, Dordrecht, pp. 411–419.

Benzenberg, J. F. and Brandes, H. W.: 1800, 'Versuch die entfernung, die geschwindigkeit und die bahn der sternschnuppen zu bestimmen', *Annal. Phys.* **6**, 224–232.

Betlem, H., Kuile, C. R., Ligne, M., van't Leven, J., Jobse, K., Miskotte, K. and Jenniskens, P.: 1997, 'Precission meteor orbits obtained by the Dutch Meteor Society – photographic meteor survey (1981–1993)', *Astron. Astrophys. Suppl. Ser.* **128**, 179–185

Brouwer, D.: 1947, *Astron.Jl.* **52**, 190.

Brown, P. and Jones, J.: 1996, 'Dynamics of the Leonid meteoroid stream: a numerical approach', In: B. A. S. Gustafson and M. S. Hanner (eds), *Physics, Chemistry and Dynamics of Interplanetary Dust*, ASP Conf. Ser pp. 113–116.

Brown, P. and Rendtel, J.: 1996, 'The Perseid meteoroid stream: characterization of recent activity from visual observations', *Icarus* **124**, 414–428.

Cook, A. F.: 1973, 'A working list of meteor streams, In: C. L. Hemenway, P. M. Millman and A. F. Cook (eds), *Evolutionary and Physical Properties of Meteoroids*, NASA SP-319, Washington DC, pp. 183–191.

Drummond, J.D. :1981, 'A test of comet and meteor shower associations, *Icarus* **45**, 545–553.

Fox, K., Williams, I. P. and Hughes, D. W.: 1983, 'The rate profile of the Geminid meteor shower', *Mon. Not. R. Astr. Soc.* **205**, 1155–1169.

Gustafson, B. Å. S.: 1989a, 'Comet ejection and dynamics of nonspherical dust particles and meteoroids', *Astrophys. Jl.* **337**, 945–949.

Gustafson, B. Å. S.: 1989b, 'Geminid meteoroids traced tocometary activity on Phaethon', *Astron. Astrophys.* **225**, 533–540.

Hamid, S. E. and Youssef, M. N.: 1963, 'A short note on the origin and age of the quadrantids', *Smith. Cont. Astrophys.* **7**, 309–311.

Hasegawa I.: 1993, 'Historical records of meteor showers, In: J. Stohl and I. P. Williams (eds), *Meteoroids and their Parent Bodies*, Slovak Academy of Sciences, Bratislava, pp. 209–223.

Harris, N. W. and Hughes, D. W.: 1995, 'Perseid meteors – the relationship between mass and orbital semi-major axis, *Mon. Not. R. Astr. Soc.* **273**, 992–998.

Herrick, E. C.: 1837, 'On the shooting stars of August 9th and 10th, 1837, and on the probability of the annual occurrence of a meteoric shower in August', *Amer. Jl. Sci.* **33**, 176–180.

Herrick, E. C.: 1838, 'Further proof of an annual meteoric shower in August, with remarks on shooting stars in general', *Amer. Jl. Sci.* **33**, 354–364.

Hughes, D. W. and McBride, N.: 1989, 'The mass of meteoroid streams', *Mon. Not. R. Astr. Soc.* **240**, 73–79.

Hughes, D. W., Williams, I. P. and Fox, K.: 1981, 'The mass segregarion and nodal retrogression of the Quadrantid meteor stream', *Mon. Not. R. Astr. Soc.* **195**, 625–637.

Hunt, J., Williams, I. P. and Fox, K.: 1985, 'Planetery perturbations on the Geminid meteor stream', *Mon. Not. R. Astr. Soc.* **217**, 533–538.

Jenniskens, P. :1994, 'Meteor stream activity I. The annual streams', *Astron. Astrophys.* **287**, 990–1013.

Jenniskens, P. and van Leeuwen, G. D.: 1997, 'The α-Monocerotid meteor outburst: the cross section of a comet dust trail', *Planet.Space Sci.* **45**, 1649–1652.

Jones, J. and McIntosh, B. A.: 1986, 'On the structure of the Halley comet meteor stream', In: *Exploration of Comet Halley*, ESA-SP 250, Paris, pp. 233–243.

Jopek, T. J.: 1993, 'Remarks on the meteor orbital similarity D-criterion', *Icarus* **106**, 603–607.

Kresák, L.: 1974, *Bull. Astron. Inst. Czechos.* **13**, 176.

Le Verrier, U. J. J.: 1867, 'Sur les etoiles filantes de 13 Novembre et du 10 Aout', *Comp. Rend.* **64**, 94–99.

Lindblad, B. A. and Porubcan, V.: 1992, 'Activity of the Lyrid meteoroid stream', In: A. W. Harris and E. Bowell (eds), *Asteroids Comets Meteors 91*, Lunar and Planetary Institute, Tucson, pp. 367–370.

Lindblad, B. A. and Steel, D. I.: 1994, 'Meteoroid orbits available from the IAU meteor data center', In: A. Milani, M. DiMartino and A. Cellino (eds), *Asteroids, Comets, Meteors 1993*, Kluwer, Dordrecht, pp. 497–501.

Lovell, A. C. B.: 1954, *Meteor Astronomy*, O.U.P., New York.

Ma, Y., He, Y. and Williams I. P.: 2001, 'A micrometeor component of the 1998 Leonid shower', *Mon. Not. R. Astr. Soc.* **325**, 457–462.

Olmstead, D.: 1834, 'Observations on the meteors of 13 Nov.1833', *Amer. Jl. Sci.* **25**, 354–411.

Plavec, M.: 1950, *Nature* **165**, 362.

Poynting, J. H.: 1903, 'Radiation in the solar system: its effect on temperature and its pressure on small bodies', *Proc. R. Soc. London* **72**, 265–267.

Robertson, H.P.: 1937, 'Dynamical effects of radiation in the solar system', *Mon. Not. R. Astr. Soc.* **97**, 423–438.

Schiaparelli, G. V.: 1867, 'Sur la relation qui existe entre les cometes et les etoiles filantes', *Astron. Nach.* **68**, 331–332.

Steel, D. S. and Asher, D. J.: 1996, In: B. Å. S Gustafson and M. S. Hanner (eds), *Physics, Chemistry and Dynamics of Interplanetary Dust*, Pub. Astron. Soc Pacific Conference Series, 125.

Štohl, J. and Porubčan V.: 1993, 'Meteor streams of asteroidal origin', In: J. Štohl and I. P. Williams (eds), *Meteoroids and their Parent Bodies*, Slovak. Acad. Sci., Bratislava, 41.

Southworth, R. B., and Hawkins, G. S.: 1963, 'Statistics of meteor streams', *Smith. Cont. Astrophys.* **7**, 261–285.

Twining, A. C.: 1834, 'Investigations respecting the meteors of Nov. 13th, 1833', *Amer. Jl. Sci.* **26**, 320–352.

Whipple, F. L.: 1950, 'A comet model 1: the acceleration of comet Encke'. *Astrophys. Jl.* **111**, 375.

Whipple, F. L.: 1951, 'A comet model II. Physical relations for comets and meteors', *Astrophys. Jl.* **113**, 464–474.

Whipple, F. L. and Hamid, S. E.: 1950, *Sky Teles.* **9**, 248.

Williams, I. P.: 1983, 'Physical processes affecting the motion of small bodies in the Solar System and their application to the evolution of meteor streams', In: V. V. Markellos and Y. Kozai (eds), *Dynamical Trapping and Evolution in the Solar System*, D. Reidel, pp. 83–87.

Williams, I. P.: 1995, 'Meteoroid Stream dynamics', In: A. E. Roy and B. A. Steves (eds), *From Newton to Chaos*, Plenium Press, New York, pp. 199 208.

Williams, I. P. and Wu, Z.: 1994, 'The current Perseid meteor shower', *Mon. Not. R. Astr. Soc.* **269**, 524–528

Williams, I. P., Hughes, D. W., McBride, N and Wu, Z.: 1993, 'Collision between the nucleus of comet Halley and dust from its own stream', *Mon. Not. R. Astr. Soc.* **260**, 43–48.

Wu, Z. and Williams, I. P.: 1995, 'P/Giacobini-Zinner and the Draconid meteor shower', *Plan. Sp. Sci.* **43**, 723–731.

Wu, Z. and Williams, I. P.: 1996, 'Leonid meteor storms', *Mon. Not. R. Astr. Soc.* **264**, 980–990

Wyatt, S. P. and Whipple, F. L.: 1950, 'The Poyntin–Robertson effect on meteor orbits', *Astrophys. Jl.* **111**, 134–141.

NEW ESTIMATION OF USUALLY NEGLECTED FORCES ACTING ON GALILEAN SYSTEM

V. LAINEY, A. VIENNE and L. DURIEZ

IMCCE/Université de Lille, 1 impasse de l'Observatoire 59000 Lille, France,
e-mail: lainey,vienne,duriez@bdl.fr

Abstract. We studied small perturbations acting on Galilean satellites. Most of them are still not computed in the analytical theories and could probably improve the ephemeris of these satellites which are outside the precision of the observations. We used a numerical method to test the effect of such perturbations. Here are reporting the main results we obtained.

Key words: Galilean satellites, analytical theories, small perturbations, satellites' oblateness, ephemeris

1. Introduction

The study of small perturbations acting on the Galilean system (composed of the satellites Io, Europa, Ganymede and Callisto) is interesting for two reasons. First of all, with the PHEMU campaign, the observations of these satellites reach few tens of kilometers of precision. Hence we need ephemeris with a very high accuracy. The second reason concerns the opportunity that we have to improve in a very strengthening way the modeling of this system (accurate J_2 values for the satellites, etc.), thanks to Galileo spacecraft.

In this work we tried to find out the still ignored perturbations, which would be important to include for improving the computation of ephemeris.

We used a numerical method to test the effect of such perturbations. Using an integrator based on Everhart method (Everhart, 1985), we integrated the Galilean system (basically the four satellites and J_2, J_4, J_6 Jovian coefficients) using each time a new perturbation. The influence of each perturbation was tested by looking at differences on positions between two integrations (with and without the perturbation) over a time span of about one century. It appears that most of the perturbations tested look more influent than what we expected, maybe because of the strength of masses, equatorial radii and J_2 coefficients of the Galilean satellites.

We present in the two next sections mathematical results and graphs concerning the two most important perturbations presented here. Then, are expressed in a summary table the results of the variations on positions, of each perturbation we tested, in decreasing order of magnitude.

Celestial Mechanics and Dynamical Astronomy **81:** 115–122, 2001.
© 2001 *Kluwer Academic Publishers.*

2. Inertial Forces Usually Neglected

2.1. EQUATIONS

Most theories of the Galilean system use a jovicentric frame. Using such a frame, we have the classical equality

$$\ddot{\mathbf{r}}_i = \frac{\mathbf{F}_i}{m_i} - \frac{\mathbf{F}_J}{m_J}, \tag{1}$$

where $\mathbf{F}_i$ represents the forces acting on a satellite P_i and $\mathbf{F}_J$ the forces acting on Jupiter. Hence, the equation generally used for $\mathcal{N}$ satellites in the gravitational field of an oblate Jupiter is

$$\ddot{\mathbf{r}}_i = -\frac{G(m_J + m_i)\mathbf{r}_i}{r_{Ji}^3} + \sum_{j=1, j\neq i}^{\mathcal{N}} Gm_j \left(\frac{\mathbf{r}_j - \mathbf{r}_i}{r_{ij}^3} - \frac{\mathbf{r}_j}{r_{Jj}^3} \right) +$$

$$+ Gm_J \nabla_i U_{\bar{i}j}, \tag{2}$$

where

$$U_{\bar{i}j} = \sum_{n=2}^{\infty} \frac{(E_r)^n}{r_i^{n+1}} \{ -J_n P_n(\sin \phi_i) +$$

$$+ \sum_{p=1}^{n} P_n^{(p)}(\sin \phi_i)[c_{np} \cos p\beta_i + s_{np} \sin p\beta_i] \}, \tag{3}$$

and (ϕ_i, β_i) are the equatorial coordinates of the satellite P_i in a frame connected to Jupiter.

In fact the Equation (2) is incomplete, because some inertial forces are missing. As a matter of fact, product of satellites' masses by oblateness coefficients of Jupiter have to appear, as long as the Galilean satellites act on Jupiter supposed to be oblate.

The complete equation is

$$\ddot{\mathbf{r}}_i = -\frac{G(m_J + m_i)\mathbf{r}_i}{r_{Ji}^3} + \sum_{j=1, j\neq i}^{\mathcal{N}} Gm_j \left(\frac{\mathbf{r}_j - \mathbf{r}_i}{r_{ij}^3} - \frac{\mathbf{r}_j}{r_{Jj}^3} \right) +$$

$$+ G(m_J + m_i)\nabla_i U_{\bar{i}j} + \sum_{j=1, j\neq i}^{\mathcal{N}} Gm_j \nabla_j U_{\bar{j}j}, \tag{4}$$

where $\sum_{k=1}^{\mathcal{N}} Gm_k \nabla_k U_{\bar{k}j}$ are the terms we were looking for.

2.2. NUMERICAL RESULTS

We can look at differences between the two integrations using respectively Equations (2) and (4). On Figure 1 we can see variations on positions (km) and on mean longitudes (factored by the semi-major axis in km) over a time span of one century.

The variations obtained can be essentially fit by straight lines. In fact, it is the mean longitude which is the most perturbed. Indeed, the perturbation changes the mean mean motions. These differences imply straight variations on mean longitudes, so on positions. The other elliptical elements are changing less than few tens of kilometers.

3. Satellites' Oblateness

3.1. EQUATIONS

We can write the action of two oblate bodies into four components (for details see Krivov, 1993)

$$\mathbf{F}_{kl} = \mathbf{F}_{\bar{k}\bar{l}} + \mathbf{F}_{\hat{k}\bar{l}} + \mathbf{F}_{\bar{k}\hat{l}} + \mathbf{F}_{\hat{k}\hat{l}}, \tag{5}$$

where a line over an index denotes the spherical part and a hat the oblate part, of the corresponding body (P_k or P_l).

We eliminate the force $\mathbf{F}_{\hat{k}\hat{l}}$ which is negligible. Hence after computing the three first components, arrive the terms

$$\mathbf{A} = -(m_J + m_i)\frac{\mathbf{F}_{\bar{J}\hat{\imath}}}{m_i m_J} = -G(m_J + m_i)\nabla_J U_{\bar{J}\hat{\imath}}, \tag{6}$$

$$\mathbf{B} = \sum_{j\neq i, j\neq 0}^{\mathcal{N}} G\left(m_j \nabla_i U_{\bar{\imath}\hat{\jmath}} - m_j \nabla_j U_{\hat{\jmath}\bar{\imath}}\right), \tag{7}$$

$$\mathbf{C} = -\sum_{j\neq i, j\neq 0}^{\mathcal{N}} G m_j \nabla_J U_{\bar{J}\hat{\jmath}}, \tag{8}$$

which must be added into Equation (4). The term $\mathbf{A}$ is the most important one because it gets the mass of Jupiter in factor. It correspond to the action of Jupiter (Keplerian part) over the oblate part of the satellite P_i. Finally we replace Equation (4) by the new equation

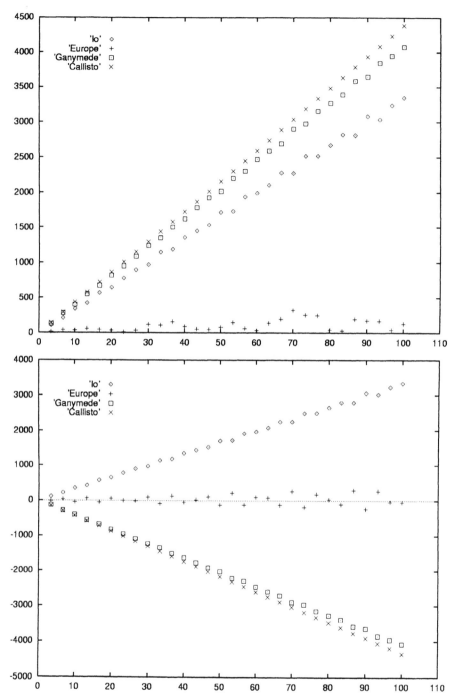

Figure 1. Variations in km on positions and mean longitudes induced by neglect inertial forces.

$$\ddot{\mathbf{r}}_i = -G(m_J + m_i)\left[\frac{\mathbf{r}_i}{r_{Ji}^3} - \nabla_i U_{\bar{i}\hat{j}} + \nabla_J U_{\bar{j}\hat{i}}\right] +$$

$$+ \sum_{j=1, j \neq i}^{\mathcal{N}} Gm_j \left[\frac{\mathbf{r}_j - \mathbf{r}_i}{r_{ij}^3} + \nabla_j U_{\bar{j}\hat{i}} - \nabla_J U_{\bar{J}\hat{j}} - \right.$$

$$\left. - \frac{\mathbf{r}_j}{r_{Jj}^3} - \nabla_j U_{\bar{j}\hat{i}} + \nabla_i U_{\bar{i}\hat{j}}\right]. \tag{9}$$

3.2. NUMERICAL RESULTS

We can look at differences on positions (using respectively equations (4) and (9)), in kilometers over a time span of one century (Figure 2 (a)). The oblateness coefficients of the satellites were taken from (Anderson et al., 1996) and (Schubert et al., 1994). It may be also interesting to compare an integration with all the terms, and another one with only the term **A** (Figure 2 (b)).

We observe once again the straight lines of variations on positions (which should be also observed on mean longitudes). In other respects, the terms **B** and **C** appear to be negligible. So only the term **A** has to be use definitely.

4. Summary Table

Now we present, in Table I, the variations induced by each perturbation we tested, in decreasing order of magnitude.

During the integration of the inner satellites Amalthea and Thebe, their masses have been get out of the Jupiter's one. Indeed we know that the inner satellites are usually taken into account by adding their mass to the mass of Jupiter. That means the results presented here are, in fact, an error of modeling by putting the mass of the inner satellites into the mass of Jupiter (and not exactly the total influence of the satellites themselves).

We can compare these results with the influence of the perturbations retained in current theories (see Table II), again with our numerical method. Here, the differences on positions are expressed only over 1 year (instead of one century).

Remark. The results of Table I add each other. That way, the differences between integrating no small perturbations at all, and the three most important ones can give more than 10 000 kilometers over one century.

5. Conclusion

We have classified the small perturbations in order of magnitude. This allows us to know what perturbations must be included in the computation of ephemeris.

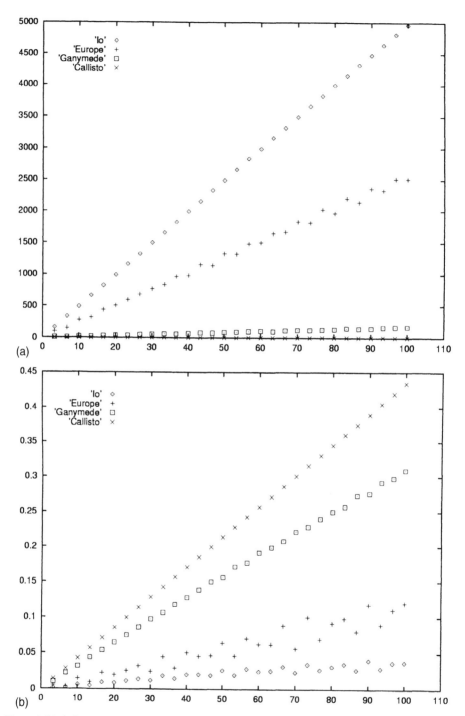

Figure 2. Variations in km on positions induced by the satellites' oblateness (a), and the contribution of the terms **B** and **C** seen when subtracted from Equation (9) (b).

NEGLECTED FORCES ACTING ON GALILEAN SYSTEM

TABLE I

Maximum of the variations on positions induced by secondary perturbations over a time span of one century

Perturbation	Description	Amplitude
Satellites' oblateness (see Figure 2)	Introduction of J_2 coefficients of the satellites	5,000 km (Io)
Inertial forces (see Figure 1)	Inertial forces usually neglected induced by the Jovian oblateness	4,500 km (Ca)
Amalthea	The biggest satellites in the inner satellites family	680 km (Io)
Saturn	Influence of Saturn over the satellites (direct and UN direct part)	226 km (Ca)
J_6 coefficient	Jupiter's oblateness	150 km (Io)
Precession *	Jovian equatorial plane moving	80 km (Ca)
Terms c_{22}, s_{22}	Asymetry in respect to the rotation axis of Jupiter	75 km (Io)
Thebe	The nearest inner satellite to the Galilean system	70 km (Io)
Relativistic effects	Influence of the mass of Jupiter	2 km (Io)
Term J_3	Asymmetry between the two poles	1.5 km (Io)

The star indicates the perturbations which are partially or completely included in analytical theories.

TABLE II

Variations on positions induced by usual perturbations over 1 year

J_2 coefficient	Jupiter's oblateness	600 000 km (Io)
Sun	Solar influence	120 000 km (Eu)
Mutual perturbations (UN direct part)	Action of the satellites upon Jupiter	100 000 km (Eu)
Mutual perturbations (direct part)	Satellites interactions	72 000 km (Ga)
J_4 coefficient	Jupiter's oblateness	920 km (Io)

We have seen also that most of the perturbations affect essentially the mean longitudes. Hence, when comparing to observations, they can be roughly absorbed by changing constants of the theory (such as mean mean motions, mass of Jupiter, etc.). That means the values of the constants of the theory cannot be consistent.

Moreover, simplifications for future analytical work have been done, such as for the satellites' oblateness where only one term is important.

To conclude, our study has revealed a little incoherence in the Sampson–Lieske theory (Lieske, 1977), which is the analytical theory still used today. As a matter of fact, this theory takes into account Jupiter's precession but not the satellites' oblateness or every inertial forces which are in fact really more important.

References

Anderson, J. D., Sjogren, W. L. and Schubert, G.: 1996, 'Galileo gravity results and the internal structure of Io', *Science* **272**, 709–712.

Everhart, E.: 1985, 'An efficient integrator that uses Gauss–Radau spacings', In: *Proceedings of the 83rd colloquium of UAI*, pp. 185–202.

Krivov, A. V.: 1993, 'Indirect influence of the external bodies on the motion of artificial earth satellites', In: *Dynamics and Astrometry of Natural and Artificial Celestial Bodies*, Poznań, Poland, pp. 353–358.

Lieske, J. H.: 1977, 'Theory of motion of Jupiter's Galilean satellites', *Astron. Astrophys.* **56**, 333–352.

Schubert, G., Limonadi, D., Anderson, J. D., Campbell, J. K. and Giampieri, G.: 1994, 'Gravitational coefficients and internal structures of the icy Galilean satellites: an assessment of the Galileo orbiter mission', *Icarus* **111**, 433–440.

OCCURRENCE OF PLANETARY RINGS WITH SHEPHERDS

LUIS BENET*

Max-Planck-Institut für Kernphysik, Postfach 103980, D-69029 Heidelberg, Germany,
e-mail: luis.benet@mpi-hd.mpg.de

Abstract. We consider the system of planetary rings with shepherds as a restricted three or four-body problem, neglecting interactions between ring particles. We show that the generic occurrence of rings for the case of rotating short-range potentials can be extended to the case of gravitational potentials. The consecutive collision periodic orbits created by saddle-center bifurcations are of central importance.

Key words: planetary rings, shepherd moons, periodic orbits

Abbreviation: RTBP – restricted three-body problem

1. Introduction

In 1977, the Uranian ring system was first observed during a stellar occultation by the planet (Elliot and Nicholson, 1984). The rings of Uranus turned to be very different from those of Saturn, being opaque, sharp-edged, eccentric and narrow. Goldreich and Tremaine (1979) proposed that such narrow rings are confined by the existence of pairs of small shepherd satellites orbiting around the rings. They conjectured that these satellites, through gravitational perturbations, prevent the ring from spreading, as caused by the internal viscous stress that results from Keplerian shear. The Voyager missions confirmed the existence of two shepherd moons around the F-ring of Saturn (Prometheus and Pandora) and around the ϵ-ring of Uranus (Cordelia and Ophelia), offering details of the structure of such rings. Revisions to the original shepherding argument were soon triggered to account several unexplained observations (Borderies et al., 1984). For instance, the ϵ-ring is also shepherded by resonances and is strongly perturbed; it is not clear how the rich structure found in Saturn's F-ring (braids, clumps and kinks) is related dynamically to the shepherd satellites; the eccentric character of all shepherd rings is not fully understood.

Recently, we showed that well-defined narrow rings (qualitatively similar to the shepherd planetary rings) do occur generically in independent-particle systems, where the ring particles interact with external rotating (attractive or repulsive) short-range potentials (Benet and Seligman, 2000). Potentials defined like this represent a generalization of the shepherd interactions. However, the physically most interesting case, the gravitational potential was only mentioned briefly. In this paper, we shall consider the effect of $1/r$ interactions, first from the planet

---

* On leave from Centro de Ciencias Físicas, UNAM, México.

and then from the shepherds considered separately. In particular, we show that certain periodic orbits appear through saddle-center bifurcations, which implies the existence of regions of trapped motion. This establishes the existence of rings in the planetary case within a purely Hamiltonian formulation.

2. Generic Occurrence of Rings in Rotating Systems

We consider first the planar motion of point particles under the influence of an external compact short-range potential $V_1(X, Y, t)$. The potential rotates in a circular orbit of radius R_1, with constant angular frequency ω_1, and its finite range is characterized by $d_1 < R_1$: particles at a distance larger than d_1 from the center of the potential move on rectilinear trajectories (free motion). We shall concentrate on the trapped motion of non-interacting particles initially outside the range of the potential, which is the situation of interest in the context of rings. The case of particles confined within the range of the rotating potential mimics the situation of the Trojan asteroids.

In the sidereal (fixed) frame, a particle initially outside the range of the potential will move on a rectilinear trajectory until it encounters the potential. Without such an encounter or *collision*, the particle will abandon the interaction region, being thus scattered to infinity. If a collision takes place, the motion of the particle is due to the rotation of the potential and its actual shape. Except for a set of initial conditions of measure zero, the particle will be eventually ejected to the region of free motion (Liouville theorem), and then move again on a rectilinear trajectory which can be treated as a new initial condition. Therefore, consecutive encounters with the short-range potential lead to trapped motion. In general, the only integral of motion is the Jacobi integral. In a frame which rotates with the potential (synodic frame), the Jacobi integral is given by $J = 1/2(p_x^2 + p_y^2) + V_1(x, y) - \omega_1(xp_y - yp_x)$, where the potential $V_1(x, y)$ vanishes if the distance to its center is larger than d_1.

The Hamiltonian flow described above defines a scattering process, and trapped trajectories can only result from consecutive encounters with the rotating potential. The underlying dynamics of a scattering system are determined by the manifolds of the unstable periodic orbits that reach the asymptotic region (Jung and Scholz, 1987). The homoclinic and heteroclinic transversal intersections of these manifolds lead at most to a Cantor set of trapped orbits. This is a set of measure zero in the space of initial conditions, and therefore does not contribute to the occurrence of rings. On the other hand, the existence of stable periodic orbits defines in phase space dense sets of initial conditions that do not escape to infinity. These structures of trapped motion are robust, in the sense that they persist under generic small perturbations, in particular under small changes of the Jacobi integral. Consider the evolution of an ensemble of non-interacting particles whose initial conditions are distributed everywhere in phase space. Most particles will be scattered to infinity, except for those whose initial conditions belong to the stable structures and remain trapped. These particles form rings in the sidereal frame, which rotate together with the rotating potential. For the shepherd systems defined above, the periodic orbits are created by saddle-center bifurcations with respect to variations of the Jacobi integral. This establishes generically the existence of stable periodic orbits. The bifurcation scenario is related to the Coriolis term which appears in the expression of the Jacobi integral. Notice that we have made no explicit assumption whether the potential is attractive or repulsive.

The rings constructed above are clearly eccentric. They are narrow as a consequence of the small portion of phase space occupied by the regions of trapped motion. This is due to the typical fast development of the underlying Smale horseshoe, where cascades of period-doubling bifurcations transform the elliptic fixed-point into an inverse hyperbolic, thus wiping out rapidly the regions of trapped motion. The sharp-edged character of the rings is related to the sudden change between trapped orbits and scattering trajectories; in fact, the sticky regions found at the border region are even smaller than the regions of trapped motion. Finally, the rings are hierarchically organized in distinct ring components; each of them is associated to an independent stable periodic orbit, that is, to the occurrence of distinct saddle-center bifurcations. These different ring components are, in a qualitative sense, similar to the observed braids.

We now consider a second short-range potential $V_2(X, Y, t)$, which rotates on a smaller circular orbit of radius $R_2 < R_1$, with frequency ω_2 and range d_2. We also assume no overlapping of the potentials, $R_2 + d_2 < R_1 - d_1$. Notice that in this case, the Jacobi integral is no longer an invariant of the motion, since the frequencies of rotation are typically incommensurable; we deal thus with a phase space of more dimensions. Ring particles generated by the outer potential, whose trajectories intersect the orbit of the second shepherd, will eventually encounter the inner potential. Due to this collision, they lose the correlations that kept them within the region of trapped motion, and are eventually scattered to infinity. This implies that the corresponding ring disappears after a few periods of the second shepherd. In turn, all rings that do not cross the orbit of the inner potential will persist. Clearly, the important parameter that distinguishes these situations is defined by the tangential encounters with the inner shepherd. Therefore, the introduction of the second potential defines dynamical mechanisms of selection of the rings that destroy several, none or all ring components.

3. The Case of Gravitational Rotating Potentials

We shall now extend the ideas of the previous section to the case of gravitational interactions between non-interacting ring particles with the central planet and with the shepherd moons. Let us consider first the motion of a particle under the gravitational attraction from a massive planet at the origin and the influence of a rotating short-range potential as before. The rectilinear free motion is replaced now by motion along solutions of the Kepler problem. Periodic orbits that interact with the rotating potential are created once again by saddle-center bifurcations. This ensures the existence of regions of trapped motion in phase space that sustain the rings. Due to the attractive character of the central potential, new regions of trapped motion and therefore new rings occur that are not confined within the orbit of the rotating potential. Notice that circular and elliptic periodic orbits which do not interact with the rotating potential are also created; these are not of interest in the present context.

Now let the range of the rotating potential d_1 tend to zero, but keeping a rotating singularity. We observe that this situation corresponds to the restricted three-body problem (RTBP) for $\mu = 0$, the Kepler problem with a rotating collision singularity, as defined by Hénon (1968). Clearly, in the context of rings the periodic orbits of interest, that is, that interact with the shepherd, are precisely the consecutive collision periodic orbits of this problem. These orbits are the generating orbits for the second species solutions of the RTBP for small non-zero μ. In contrast to the case $\mu = 0$, where the generating orbits manifestly exhibit collisions with the rotating singularity, for non-zero μ the periodic orbits are essentially arcs of hyperbola or ellipses locally deformed by a close approach (without physical collision) with the shepherd moon. Although for $\mu = 0$ the meaning of stability of these orbits is ill defined, Figure 1 shows the appearance of minima and maxima in the chart of consecutive collision orbits, which is characteristic of saddle-center bifurcations. Numerical investigations for small non-zero μ have shown that these are the fundamental periodic orbits in the construction of the chaotic saddle, which determine the properties of the scattering process (Benet et al., 1999). Moreover, the creation or annihilation of these periodic orbits are saddle-center bifurcations.

This scenario certainly carries over to the restricted four-body problem, where the influence of a second shepherd satellite is taken into account. In this case, the second shepherd may also generate regions of trapped motion and thus contribute to the occurrence of rings. The mechanism of selection of rings is defined by each shepherd acting on the other generated stable periodic orbits. In comparison to the case of short-range potentials, it is enhanced by the long-range character of the gravitational potential. The robust character of the stable regions implies that they exist under small perturbations. Again, the selection mechanism depends in a subtle manner on the orbital parameters of the shepherds, and may allow only the existence of a certain number of regions of trapped motion, or none. An ensemble of initial conditions defined in the existent regions of trapped motion will form stable *planetary* rings. These rings are narrow, eccentric, sharp-edged and hierarchically organized.

4. Summary and Discussion

In this paper we have extended the arguments for the generic occurrence of rings in rotating systems with short-range potentials, to the case which includes

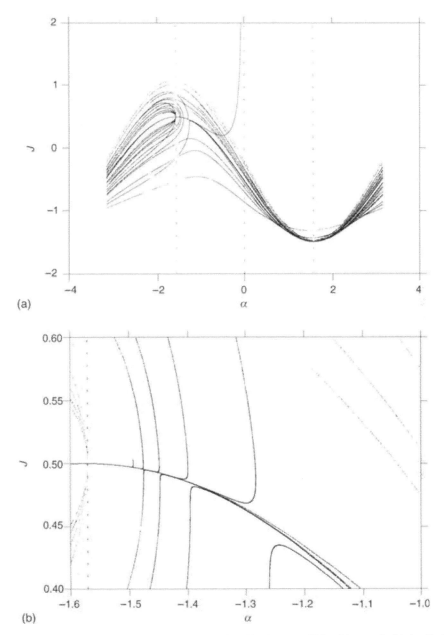

Figure 1. (a) Chart of consecutive collision periodic orbits for the RTBP with $\mu = 0$; (b) detail of a region. The outgoing angle after collision α is plotted against the Jacobi integral J. The maxima and minima displayed are related to the appearance through saddle-center bifurcations of second species periodic orbits for small non-zero μ.

gravitational interactions from a massive central planet and two shepherd satellites. We showed first that the families of periodic orbits that sustain the rings in the former case correspond to the consecutive collision orbits of the restricted three and four-body problems of zero mass parameters. These periodic orbits appear through saddle-center bifurcations, which establishes the existence of stable periodic orbits and the associated region of trapped motion. Rings are formed by an ensemble of non-interacting particles whose initial conditions belong to the regions of trapped motion. These planetary rings are stable, eccentric, narrow and display a hierarchic organization. We believe that the latter accounts for the individual ring components known as braids. Since this scenario is robust and the problem is explicitly time-dependent, we expect that small eccentricity of the orbits of the shepherds can be incorporated along the same lines; work on this direction is in progress.

An important assumption made here is to neglect interactions among the ring particles. This allowed us to consider only the motion of a single ring particle; the motion of an ensemble of such particles corresponds to different initial conditions in the one-particle problem. This assumption is certainly unrealistic, in the sense that ring particles do collide a few times each period of motion introducing important dissipation processes in the dynamics of the ring (Brahic, 1977). Although this is a quite restrictive assumption in terms of our treatment, our results emphasize the dynamical origin of the different families of ring-particle trajectories, which can be used in more realistic computations.

Acknowledgement

I am grateful to T. H. Seligman for the many stimulating discussions and comments.

References

Benet, L., Seligman, T. H. and Trautmann, D.: 1999, 'Chaotic scattering in the restricted three–body problem II: small mass parameters', *Celest. Mech. & Dyn. Astr.* **71**, 167–189.

Benet, L. and Seligman, T. H.: 2000, 'Genetic occurrence of rings in rotating systems', *Phys. Letts.* A **273**, 331–339.

Borderies, N., Goldreich, P. and Tremaine, S.: 1984, In: R. Greenberg and A. Brahic (eds), *Planetary Rings*, University of Arizona Press, Tucson.

Brahic, A.: 1977, 'System of colliding bodies in a gravitational field I: numerical simulation of the standard model', *Astron. Astrophys.* **54**, 895–907.

Elliot, J. L. and Nicholson, P. D.: 1984, In: R. Greenberg and A. Brahic (eds), *Planetary Rings*, University of Arizona Press, Tucson.

Goldreich, P. and Tremaine, S.: 1979, 'Towards a theory for the uranian rings', *Nature* **277**, 97–99.

Hénon, M.: 1968, 'Sur les Orbites Interplanétaires qui Rencontrent Deux Fois la Terre', *Bull. Astron.* (Serie 3) **3**, 337–402.

Jung, C. and Scholz, H. J.: 1987, 'Cantor set structures in the singularities of classical potential scattering', *J. Phys. A: Math. Gen.* **20**, 3607–3617.

ON THE RELATIONSHIP BETWEEN FAST LYAPUNOV INDICATOR AND PERIODIC ORBITS FOR SYMPLECTIC MAPPINGS

ELENA LEGA and CLAUDE FROESCHLÉ

Observatoire de Nice, Bv. de l'Observatoire, B.P. 4229, 06304 Nice cedex 4, France

Abstract. The computation on a relatively short time of a quantity, related to the largest Lyapunov Characteristic Exponent, called Fast Lyapunov Indicator allows to discriminate between ordered and weak chaotic motion and also, under certain conditions, between resonant and non resonant regular orbits. The aim of this paper is to study numerically the relationship between the Fast Lyapunov Indicator values and the order of periodic orbits. Using the two-dimensional standard map as a model problem we have found that the Fast Lyapunov Indicator increases as the logarithm of the order of periodic orbits up to a given order. For higher order the Fast Lyapunov Indicator grows linearly with the order of the periodic orbits. We provide a simple model to explain the relationship that we have found between the values of the Fast Lyapunov Indicator, the order of the periodic orbits and also the minimum number of iterations needed to obtain the Fast Lyapunov Indicator values.

Key words: chaotic motion, Fast Lyapunov Indicator, periodic orbits

1. Introduction

In the last 10 years many efforts have been made in order to have fast indicators for distinguishing between chaotic and regular motion, and also, in the case of regular motion, between resonant and non resonant orbits. Although the computation of the Lyapunov Characteristics Indicators (for a detailed review see Benettin et al., 1980; Froeschlé, 1984) represents, from theoretical and from numerical point of view, the best suited method for distinguishing between ordered and chaotic motion, it doesn't allow to separate resonant from non resonant orbits.

This last task is very important, for example, when studying the global stability of a system. It is well known, from the celebrated Nekhoroshev's theorem (1977), that in a Hamiltonian system, satisfying some non-degeneracy condition, if the perturbation is small, orbits with initial conditions in a chaotic region do not diffuse in the action space. More precisely, given a perturbation parameter ϵ, only diffusion with a velocity exponentially small with respect to $-1/\epsilon$ can be expected.

When increasing the perturbation parameter the dynamic is no more controlled by the Nekhoroshev theorem. The system is described by the well-known Chirikov (1960) overlapping criterium, which allows the resonant chaotic orbits to pass from one resonance to the other one, thus possibly giving large scale diffusion.

By a graphical representation of the distribution of tori, chaotic orbits and resonant regular orbits, that is of the Arnold web, it is possible to determine if the dynamics is controlled by Nekhoroshev theorem or by the Chirikov criterium.

Celestial Mechanics and Dynamical Astronomy **81:** 129–147, 2001.
© 2001 *Kluwer Academic Publishers.*

Already Laskar (1993) and Lega and Froeschlé (1997) have managed, using respectively the frequency and the sup map analysis, to obtain a graphical representation of the Arnold web.

By computing, on a relatively short time, a quantity related to the largest Lyapunov Characteristic Exponent, Froeschlé and Lega (2000) have shown that this quantity, called fast lyapunov indicator (FLI), distinguishes very well not only between chaotic and regular orbits but also between resonant regular orbits and non resonant ones.

Using the FLI, Froeschlé et al. (2000) have recovered easily and with a lot of details the Arnold web in a Hamiltonian system and also showed the transition from Nekhoroshev's regime to Chirikov's one. In the case of symplectic maps it is less clear if all the hypothesis of the Nekhoroshev theorem are satisfied. However, the numerical experiments (Froeschlé and Lega, 2000; Guzzo et al., 2000) have shown structure and evolution very similar to those arising in Hamiltonian systems justifying the appellation of Nekhoroshev's like regime for the case of a map and the study of the transition from Nekhoroshev's like regime to Chirikov's one.

To summarize the present state of the art we can then say that the FLI is an indicator very sensitive to the dynamics of the orbits and because it is easy to be implemented and rapid in computation it turns out to be well adapted for studies concerning the global stability of a system.

Moreover, since the FLI have been shown to be very sensitive for the detection of resonant regular orbits we think that it is interesting to study numerically the relation existing between the FLI values and the order of periodic orbits. This is the aim of the present paper.

Taking the standard map as a model problem we show that there exists a relation $FLI(q) \simeq \log(q)$ between the FLI and the order q of the periodic orbits up to a given order q. Studying the Fibonacci sequence of the rational approximants to the golden torus we show that the FLI of periodic orbits becomes constant after a transient linear increase for an interval of time which grows with the order q.

We have also obtained that both the FLI values and the number of iterations which are necessary to obtain them are proportional to $\log(q)$ up to a given order. For higher order periodic Fibonacci orbits, both, the FLI values and the number of iterations to obtain them, increase linearly with the order.

The difference between the FLI values obtained for resonant regular orbits and the FLI values of circulation tori is explained through the properties of the differential rotation (Froeschlé and Lega, 2000). In the case of periodic orbits there is no differential rotation and we have introduced a very simple model of linear elliptic rotation for explaining the relationship that we have found between the values of the FLI, the order of the periodic orbits and also the minimum number of iterations needed to obtain the FLI values.

In the light of the results of the simple model we have found that the FLI values obtained on the Fibonacci sequence are in very good agreement with the measure of the size of the Fibonacci islands (Lega and Froeschlé, 1996; Locatelli et al., 2000),

RELATIONSHIP BETWEEN FAST LYAPUNOV INDICATOR AND PERIODIC ORBITS 131

and hence with the description of the structure around an invariant KAM torus, the golden torus in this case, theoretically predicted by Morbidelli and Giorgilli (1995). We remark that the study of the neighborhood of the golden torus has been much easier with the FLI than with the frequency-map analysis as in Lega and Froeschlé (1996).

The paper is organized as follows. In Section 2 we recall the definition of the FLI for a mapping; in Section 3 we compute this indicator on some orbits of the two-dimensional standard map. In Section 4 we obtain a relation between the FLI and the order of the periodic orbits. We compute the FLI on the sequence of Fibonacci periodic orbits in Section 5. In the same Section we introduce a simple model based on linear elliptic rotation which help us for discussing the results on the Fibonacci sequence. Conclusion and future developments are provided in 6.

2. The FLI Revisited

When computing the Lyapunov Characteristics Indicators the attention is focused on the length of time necessary to get a reliable value of their limit, that is of the Lyapunov Characteristic Exponents, but very little importance has been given to the first part of the computation. In fact, this part was considered as a kind of transitory regime depending, among other factors, on the choice of an initial vector of the tangent manifold.

Already in 1997 Froeschlé et al., have remarked that the intermediate values of a quantity related to the largest Lyapunov Characteristic Indicator taken at equal times for chaotic, even slow chaotic, and ordered motion, allows to distinguish between them. This indicator, called FLI, turned out to distinguish also among ordered motions of different origins, like resonant and non resonant motion (Froe schlé and Lega, 2000), despite the fact that in both cases the Lyapunov Character-istic Indicators converge to zero when t goes to infinity.

Given a mapping M from $\Re^n$ to $\Re^n$, an initial condition $\vec{x}(0) \in \Re^n$, and an initial vector $\vec{v}(0) \in \Re^n$ of norm 1, the FLI function $\bar{F}(\vec{x}(0), \vec{v}(0), t)$, t belonging to Z^+, is defined as:

$$\bar{F}(\vec{x}(0), \vec{v}(0), t) = \frac{1}{2N} \sum_{k=t-N}^{k=t+N-1} \log \|\vec{v}(k)\|, \tag{1}$$

where the evolution of the vector $\vec{v}(t)$ is given by the set of coupled equations

$$\begin{cases} \vec{x}(t+1) = M\vec{x}(t), \\ \vec{v}(t+1) = \dfrac{\partial M}{\partial \vec{x}}(\vec{x}(t))\vec{v}(t). \end{cases} \tag{2}$$

The running average on a set of N data was introduced in Froeschlé and Lega (2000) in order to smooth the oscillations of the norm of $\vec{v}(t)$ due to the distortion of the orbits.

For the same reason, in view to the study of very small resonant zones, like for instance the set of Fibonacci islands which approximate the golden torus, we slightly modify the definition of the FLI function F taking the maximum value that the norm of the tangent vector reaches up to time t

$$F(\vec{x}(0), \vec{v}(0), t) = \sup_{0 \leqslant k \leqslant t} \log ||\vec{v}(k)||. \tag{3}$$

The advantage of the last definition when comparing different orbits is to be independent from the parameter N, that is, from the portion of the orbit visited in the interval $[t - N, t + N - 1]$ (Eq. (1)). Although the same N can be used when comparing resonances of small order, like in Froeschlé and Lega (2000), the parameter N should be adapted to the resonance order in view of fine studies like the analysis of the set of Fibonacci islands. The new definition (Eq. (3)) avoids us tedious calibrations of N as a function of the resonance order.

3. Application to the Standard Map

We consider as a model problem the two-dimensional standard map (Froeschlé, 1970; Lichtenberg & Lieberman, 1983):

$$M = \begin{cases} x(t+1) = x(t) + \epsilon \sin(x(t) + y(t)) & (\text{mod} 2\pi), \\ y(t+1) = x(t) + y(t) & (\text{mod} 2\pi). \end{cases} \tag{4}$$

Figure 1 displays orbits of the standard map of Equation (4) for $\epsilon = 0.9$.

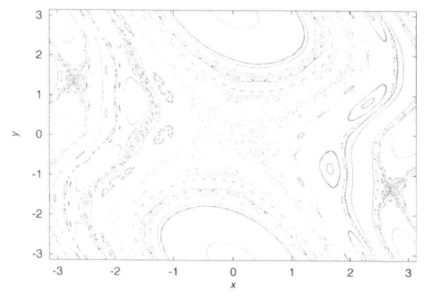

Figure 1. A set of orbits of the standard map for $\epsilon = 0.9$.

For this value of the perturbing parameter the majority of tori have disappeared and the phase space shows a rich structure of chains of islands and chaotic orbits. Let us recall that the value of the perturbation parameter for which the last KAM torus is broken down is $\epsilon = 0.971635$ (Greene, 1979; Olivero and Simó, 1987; Laskar et al., 1992; Mackay, 1993; Laskar, 1995).

For $\epsilon = 0.9$ we have computed the FLI (as in Eq. (3)) for four orbits of the standard map. For the chaotic orbit (Figure 2) with initial conditions $x(0) = 0.00001$, $y(0) = 0$, the FLI shows an exponential increase with time. In the case of a regular non resonant orbit with $x(0) = 2.13$, $y(0) = 0$ and of a regular resonant orbit with $x(0) = 1.2$, $y(0) = 1$ (the small chain of four islands plotted in Figure 1) the FLI grows linearly with time but with a different speed (Figure 2). In fact, because of the differential rotation, the norm of the vector $\vec{v}$ asymptotically grows as $||\vec{v}(t)|| \simeq \alpha t$, the coefficient α depending on the nature of the invariant curve (torus or libration island). In particular, inside a libration island the variation of the differential rotation is lower than inside a region of tori (for an heuristic analytic demonstration using an Hamiltonian system see Guzzo et al., 2001 and for numerical experiments see Froeschlé and Lega, 2000). Therefore, since we plot the logarithm of the norm of $\vec{v}(t)$ against the logarithm of the time t we get parallel lines for the resonant and non resonant cases and these two regular dynamics are clearly distinguished.

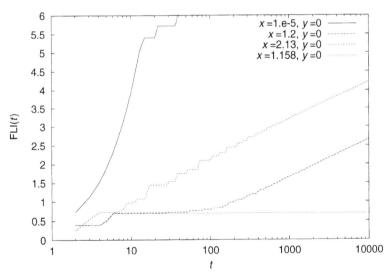

Figure 2. Variation of the FLI with time for four orbits of the standard map with $\epsilon = 0.9$. The upper curve is for a chaotic orbit with initial conditions $x(0) = 0.00001$, $y(0) = 0$, the second one is for a regular non resonant orbit with $x(0) = 2.13$, $y(0) = 0$ and the third one is for a regular resonant orbit with $x(0) = 1.2$, $y(0) = 1$. This orbit is plotted in Figure 1 and corresponds to a chain of four small islands. The curve with constant FLI is obtained for the corresponding periodic orbit of order $q = 4$ of initial conditions $x(0) = 1.158$, $y(0) = 0$.

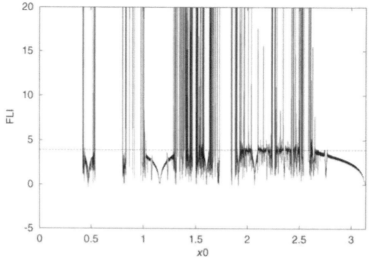

Figure 3. Variation of the FLI computed at $t = 10\,000$ iterations of the mapping, as a function of the initial action $x(0)$ for a set of $10\,000$ orbits regularly spaced in the interval $[0, \pi]$. The perturbation parameter is $\epsilon = 0.9$. The initial vector is $\vec{v}(0) = (1, 0)$ for the whole set of orbits. The threshold for the detection of islands FLI $= 3.9$ is also plotted.

The curve with constant FLI is obtained for the periodic orbit of order $q = 4$ of initial conditions $x(0) = 1.158$, $y(0) = 0$. The aim of this paper is to understand such peculiar behavior of the FLI for periodic orbits.

We have started computing the FLI for a set of 10 000 initial conditions regularly spaced on the x-axis in the interval $[0, \pi]$. For each orbit $y(0) = 0$. Figure 3 shows what we call the FLI-map, that is the value of F after $t = 10\,000$ iterations against $x(0)$, $\vec{v}(0)$ being always taken in the direction of the x-axis. As shown by Foreschlé and Lega (2000) the values of F depend on the direction of $v(0)$ and the authors suggest to take the same $v(0)$ for the whole set of orbits in order to compare their dynamical character.

In the same article it is explained, by considerations on the differential rotation, that the tori have all the same FLI value which is of the order of the logarithm of the number t of iterations since the norm of $v(0)$ grows linearly with t. Actually, a small number of orbits of Figure 3 appear to have $F \simeq 4$ in agreement with the fact that the majority of tori have disappeared for this value of the perturbation parameter.

Islands, as already observed in Figure 2, are identified by values of the FLI lower than the reference value of tori. In fact, the differential rotation inside a chain of islands turns out to be lower than between tori and its derivative becomes close to zero (Froeschlé and Lega, 2000) towards the center of the islands. The minima of the FLI observed in Figure 3 are a quite good approximation for the center of the crossed islands.

Chaotic orbits have FLI greater than the reference value of tori. The computation has been stopped when $F > 20$. Such orbits have a large positive Lyapunov Characteristic Indicator and for the purpose of this paper we do not need other information about them.

4. Relationship Between FLI and Periodic Orbits

As we said above we can associate regular resonant motion to all orbits of Figure 2 having FLI lower than the reference value of tori. Taking as a threshold the value FLI = 3.9, we obtain that 3772 of the previous analyzed set of 10 000 orbits are resonant regular orbits. For the corresponding set of initial conditions we have computed, using a fast method introduced in Lega and Froeschlé (1996), the frequency or rotation number.

Figure 4 shows the values of the frequency f computed using 10 000 iterations of the mapping for the set of 3772 orbits. As shown by Laskar et al. (1992) in the plot of the frequency as a function of the initial conditions, that is in the plot of the frequency-map, regular resonant orbits are identified by 'plateaux' of constant frequency. Such 'plateaux' appear clearly in Figure 4, for example the large one for orbits with initial conditions in the interval $0.95 < x(0) < 1.3$, $y(0) = 0$, corresponds to the chain of four islands of Figure 1 surrounding the periodic orbit $p/q = 1/4$ of initial conditions $x(0) = 1.158$, $y(0) = 0$. The sensitivity of the FLI allows also to find very small chains of islands, the small 'plateaux' of Figure 4.

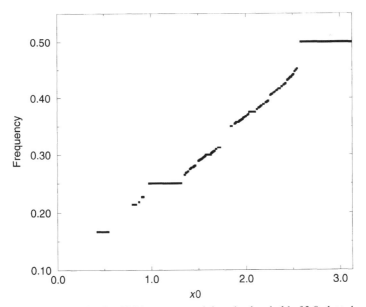

Figure 4. Frequency of orbits having FLI lower or equal than the threshold of 3.9 plotted as a function of the initial condition x_0.

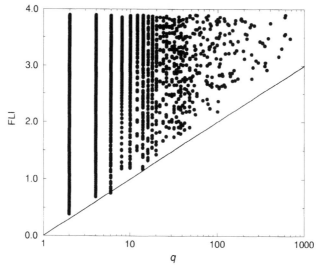

Figure 5. Values of the FLI corresponding to resonant orbits, that is to orbits of Figure 2 having FLI lower or equal to 3.9, plotted against the order of the resonance. This latter has been obtained by a development in continued fraction of the rotation number of the orbit. The straight line is FLI = log(q).

Actually, at this resolution it is quite difficult, at least visually, to distinguish on the frequency-map small 'plateaux' from a monotonic increase of the frequency associated to invariant tori. We have developed in continued fraction the set of 3772 frequencies obtaining for each of them a rational representation p/q, $p, q \in Z^+$ thus confirming the detection of regular resonant motion by the FLI method.

Figure 5 shows the relation between the FLI values and the order q of the resonance obtained by the development in continued fraction of the orbits's frequencies. The vertical lines correspond to the set of values of the FLI of a given chain of islands. For example the line at $q = 2$ is for the set of initial conditions of the chain of two islands of Figure 1 whose center is the periodic orbit of coordinates [π, $-\pi$], [π, 0]. The minimum value of the FLI for each chain of islands corresponds to an orbit close to the periodic one. For the case considered, $p/q = 1/2$, the FLI is 0.39 for an orbit with initial conditions: $x(0) = 3.1396$ $y(0) = 0$ very close to the periodic orbit: $x(0) = \pi$ $y(0) = 0$.

Considering the minima of the FLI against the order q of the periodic orbits we obtain numerically that the such values behaves like log(q) (the straight line on Figure 5).

5. A Fine Study of the Fibonacci Sequence

In order to investigate more precisely the relationship obtained above between the FLI and the order of periodic orbits, we have computed the FLI as a function of time for a particular set of periodic orbits, the Fibonacci sequence.

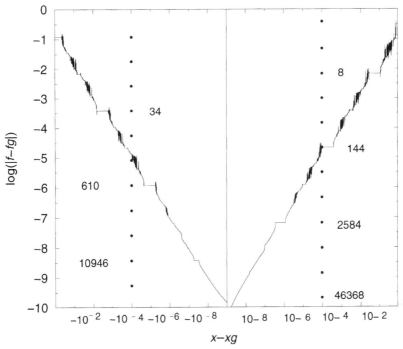

Figure 6. Logarithm of the absolute value of the difference between the frequency f of an orbit and the frequency fg of the golden torus as a function of the distance to the golden torus for the standard map with $\epsilon = 0.9715$. The filled circles in the figure correspond to the values $\log|f - fg|$ of the Fibonacci sequence. The set of Fibonacci terms that we analyze in the present paper is labeled in the figure by the corresponding values of q_k.

Let us recall that the Fibonacci sequence is the set of the successive terms p_k/q_k obtained when developing the golden number through the continued fraction process and therefore is the sequence which best approximates[1] the golden number.

As explained in Lega and Froeschlé (1996), because of the smallness of the denominators the amplitude of the perturbation associated to the Fibonacci terms is large. This make the sequence a good candidate in view to the study of regular resonant motion, and in particular of the corresponding periodic orbits, up to a large order q. Such terms are also further from the golden number than any other set p_k/q_k from its corresponding irrational number. This is the origin of the strength of the golden torus, that is, of the last KAM torus of the standard map of Equation (4) which, as we said in Section 3, is broken down for $\epsilon = 0.971635$.

Figure 6 shows the variation of the frequency as a function of the distance to the golden torus for $\epsilon = 0.9715$, that is very close to the perturbation parameter of the breakdown of the golden torus. More precisely, we plot the logarithm of

---

[1] We say that p/q is a best approximation of x if $|qx - p| < |q'x - p'|$ for all p', q' such that $0 < q' \leq q$ and $p'/q' \neq p/q$.

the absolute value of the difference between the frequency f of an orbit and the frequency fg of the golden torus.

The frequency is computed, using a fast method introduced in Lega and Froeschlé (1996), for a set of 20 000 orbits with $y(0) = 0$ and $x(0)$ regularly spaced on a logarithmic scale from $|x(0) - xg| = 10^{-9}$ to $|x(0) - xg| = 1$, xg being the action of the golden torus (when the angle yg is 0). The number of iterations used is suited to the precision required for the computation of the frequency and goes from $5\,10^6$ iterations for the orbits with $10^{-9} < |x(0) - xg| < 10^{-8}$ to 10^4 iterations for the orbits with $10^{-4} < |x(0) - xg| < 1$.

The set of Fibonacci terms that we analyze in the present paper is labeled in the figure by the corresponding values of q_k. Far from the golden torus we can see only libration islands (the 'plateaux' in the figure) and chaotic zones (the noisy variations of the frequency). Starting from the resonance term 1597/2584 a surprisingly clear change of regime occurs: only tori remain (the monotonic variation of the frequency) sometimes interrupted by small libration islands and small vertical jumps (the crossing of thin separatrices). This change of regime perfectly agrees

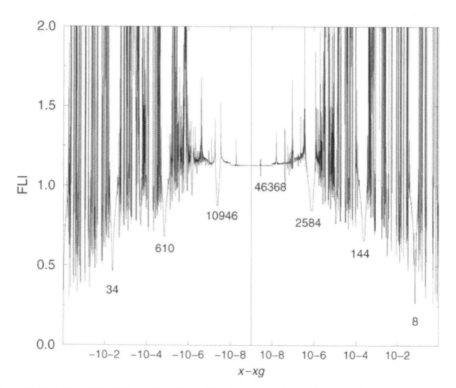

Figure 7. Variation of the FLI as a function of the distance to the golden torus for the standard map with $\epsilon = 0.9715$. The set of Fibonacci terms that we analyze in the present paper is labeled in the figure by the corresponding values of q_k.

with the representation given by a theorem of Morbidelli and Giorgilli (1995) about the dynamics in a small neighborhood of a noble torus.

The same features appear in Figure 7 where the FLI is plotted as a function of the distance to the golden torus for $\epsilon = 0.9715$, and for the same number of iterations used for the frequency analysis of Figure 6. The structures detected seem to be even more rich (because this indicator is very sensitive even to very small chaotic zones) and the change of regime appears very clearly in agreement with the results on the frequency map.

Figure 8 shows the variation with time of the FLI for a set of periodic orbits belonging to the Fibonacci sequence for the standard map with $\epsilon = 0.9715$. As obtained in Figure 2 the FLI of periodic orbits, after a transitory linear increase (which is very short for the periodic orbit of order four of Figure 2) becomes constant, that is the norm of $\vec{v}(t)$ reaches some maximum value.

Let us consider the trivial case of the two-dimensional standard map with $\epsilon = 0$. In this case the norm of the vector $\vec{v}$ is equal to

$$||(\vec{v}(t)|| = \sqrt{t^2 v_x(0)^2 + v_y(0)^2}, \tag{5}$$

independently on the initial conditions, that is for a given initial vector of components $\vec{v}(0) = (v_x(0), v_y(0))$ the FLI is the same for both a periodic and a quasi-periodic orbit. The difference with the perturbed case is that in this last case a

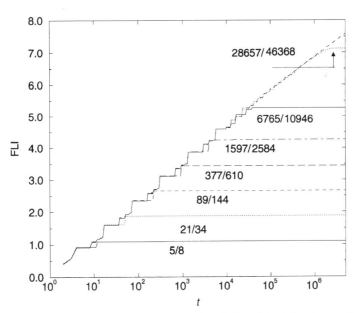

Figure 8. Variation with time of the FLI for a set of periodic orbits belonging to the Fibonacci sequence for the standard map with $\epsilon = 0.9715$. The rational ratios p/q of each term are written in the figure near the corresponding FLI curve. The FLI curve which grows linearly with time all over the interval $0 < t < 5\,10^6$ iterations is obtained for the golden torus.

periodic orbit is surrounded by quasi-elliptic regular resonant orbits. We think that the reason why the FLI becomes constant for the periodic orbits in the perturbed case must be found exactly on the fact that the linearized orbits in the phase space around the periodic orbits are elliptic. Therefore, in order to study the behavior of the FLI for periodic orbits we will introduce below a simple are-preserving mapping of linear elliptic rotation.

In Figure 8 the FLI curve which grows linearly with time all over the interval $0 < t < 5\,10^6$ iterations is obtained for the golden torus.

It is interesting to observe that, during the transitory linear increase, the FLI values obtained for the whole set of Fibonacci terms coincide with those of the golden torus. In fact, the ellipses surrounding the periodic orbits are more and more elongated for increasing order q, they become almost tangent to the near invariant tori, while the chaotic zone surrounding the islands shrinks to zero. Therefore, at least for high order resonances ($q \geqslant 610$) the FLI for a transient period of time is the same for the periodic orbits and for the near tori because of the continuity of the tangent manifold of Equation (2) (we recall that the FLI is the same for all tori, so the curve of Figure 8 obtained for the golden torus is valid for all invariant tori). Surprisingly, this argument based on continuity seems to hold also for lower order periodic orbits, although the chaotic zone surrounding the libration islands is large (Figures 6 and 7) and probably a very small number of tori remains.

We can observe in Figure 8 that both the duration of the transitory linear increase of the FLI and its maximum and constant value seem to be related to the order q of the periodic orbits.

In what follows we introduce a simple model in order to explain the observed features: the linear increase of the FLI, the constancy after a transient interval of time and the dependence on the order q.

5.1. A SIMPLE MODEL

A small libration island close to the periodic orbit can be interpreted, through linearization, by an ellipse of ratio $\gamma = a/b$ between the semi-major axis a and the semi-minor axis b. Since there is no differential rotation in the case of periodic orbits, we consider a model without differential rotation. Let us define the following area-preserving mapping:

$$
N = \begin{cases} x(t+1) = x(t)\cos(\alpha) - \gamma\, y(t)\sin(\alpha), \\ y(t+1) = \frac{1}{\gamma}x(t)\sin(\alpha) + y(t)\cos(\alpha). \end{cases} \tag{6}
$$

where α is the mean rotation angle of the radius vector of the ellipse.

In fact, since this mapping is area preserving, for a given number of iterations the density of the points will be higher for the regions further from the center and the angle α turns out to be a mean rotation angle as it is for the islands of the standard mapping.

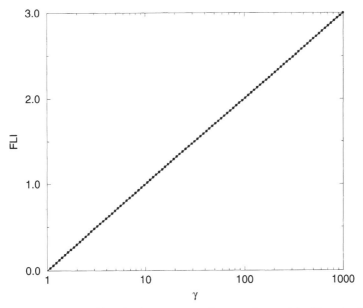

Figure 9. Variation of the FLI with γ for the area-preserving mapping model of Equation (6). Results are in perfect agreement with the expected behavior FLI $= \log(\gamma)$.

A vector of the tangent manifold corresponding to an elliptic orbit of Equation (6) is contracted and dilated along the ellipse, but its norm, because there is no differential rotation, cannot grow more than a maximum value which should be exactly γ. Figure 9 shows the variation of the FLI with the parameter γ for $t = 10\,000$ and γ belonging to the interval $1 < \gamma < 1000$. Results are in perfect agreement with the expected behavior FLI $= \log(\gamma)$.

Let us remark that results of Figure 9 are invariant for increasing values of the time t, that is the FLI values plotted correspond to the stable values of Figure 8.

A question remains about the transitory regime in which the FLI grows linearly with time.

Actually, in order to reach its maximum value, the vector $\vec{v}(t)$ has to 'visit' the ellipse from the the semi-minor to the semi-major axis, that is has to rotate of at least an angle of 90°. In Figure 10 we have plotted the evolution of the FLI with time for seven orbits of the mapping of Equation (6) having γ values such that $\log(\gamma) \equiv$ FLI of the seven Fibonacci periodic orbits analyzed in Figure 8. For the rotation angles we have considered $\alpha = 90/t$FLI degrees, where tFLI is the time necessary for the FLI to become constant for the seven orbits of Figure 8. Results of Figure 10 reproduce very well the evolution of the FLI with time of the seven Fibonacci periodic orbits, confirming the validity of the simple model for explaining the behavior of the FLI for periodic orbits.

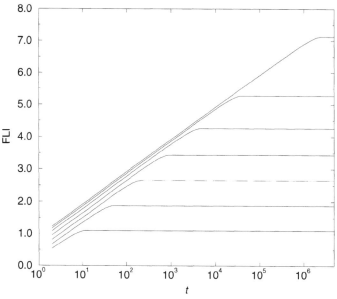

Figure 10. Variation of the FLI with time for the area-preserving mapping model of Equation (6). The values of γ correspond to the final FLI values of the Fibonacci periodic orbits of Figure 8. The rotation angles are the inverse of the time necessary for the FLI to reach its maximum value for the Fibonacci terms analyzed in Figure 8. Results of Figure 8 are well reproduced by the model.

5.2. THE RESULTS OF THE SIMPLE MODEL AS A GUIDELINE FOR THE STUDY OF THE FLI ON THE FIBONACCI SET

In the light of the results obtained with the simple model, we can test if the maxima of the FLI for periodic orbits correspond to the geometric parameter $\gamma = a/b$ and if the time of transition to reach the maximum FLI value agrees with the kinematic parameter α corresponding to the mean rotation angle.

Figure 11 shows the relation between the FLI and the order q of the periodic orbits. The behavior of the FLI as $\log(q)$ is quite well respected up to the order $q = 610$, that is by the first four periodic orbits. Then a sort of change of regime occurs.

The semi-major axis a decreases with the order like $1/q$. Concerning the semi-minor axis b, it has been obtained by numerical experiments (Lega and Froeschlé, 1996) on the two-dimensional standard map that its size decreases like the distance from the golden torus up to $q = 610$. Since the distance of a Fibonacci island from the golden torus behaves like $1/q^2$ we expect $a/b \simeq q$, that is FLI $\simeq \log(q)$, in agreement with the results of Figure 11.

Then, for higher order resonances, $q \geqslant 2584$, in a very small neighborhood of the golden torus, the size b of the Fibonacci islands has been shown (Lega and Froeschlé, 1996; Locatelli et al., 2000) to decrease exponentially as theoretically predicted in a theorem by Morbidelli and Giorgilli (1995). Taking the values of

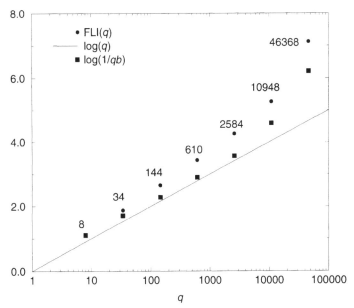

Figure 11. FLI values plotted against the order q for the set of Fibonacci periodic orbits of the standard map with $\epsilon = 0.9715$ analyzed in Figure 8. The straight line is the line FLI $= \log q$. The points labeled by a square correspond to the $\log(a/b)$, that is the expected values for the FLI, with $a = 1/q$ and b measured with the frequency-map analysis of the Fibonacci sequence (from Lega and Froeschlé, 1996).

b measured by Froeschlé and Lega (1996) using the frequency map analysis we have plotted in Figure 11 the values $\log \gamma(q) \equiv \log(a(q)/b) \equiv \log(1/qb)$. It turns out that the relation FLI$(q) = \log \gamma(q)$ expected from the simple model of Equation (6) is quite well respected from the results of Figure 11. Although the behavior is the expected one, we can observe that the difference between the FLI values and the numerical values $\log(a(q)/b)$ with b taken from Lega and Froeschlé (1996) increases in the exponential regime ($q \geqslant 2584$). This fact suggests to us to make finer studies, out of the purpose of this paper, of the exponential decrease of the volume of high order resonances.

However, the change of regime on b from a power low to an exponentially small b occurs also for the FLI, but the results are very much simpler to obtain than the measure of 'plateaux' by the frequency analysis. Although a finer study is out of the purpose of this paper we think that the FLI method is well suited for a better comprehension of the dynamical properties of the standard map.

Figure 12 shows the time necessary for the FLI to become constant against the order q of the periodic Fibonacci orbits. A linear relation between the FLI and the order q holds up to the order $q = 610$ of the Fibonacci term $p/q = 377/610$. Then, again a sort of change of regime occurs.

Following the results of Figure 10, we have measured the kinematic parameter, that is the mean rotation angle using a method introduced by Contopoulos and Vog-

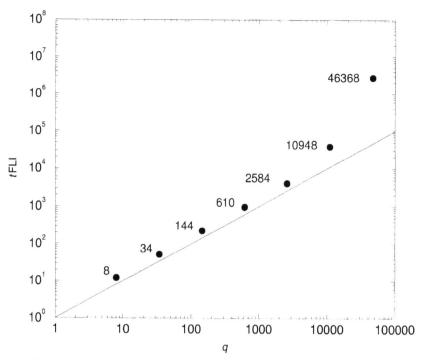

Figure 12. Time necessary for the FLI (*t*FLI) to become constant against the order q of the periodic orbits for the set of Fibonacci periodic orbits of the standard map with $\epsilon = 0.9715$ analyzed in Figure 8. The straight line is $t\text{FLI} = q$.

lis (1997) and called twist angle. In fact, the twist angle has been shown (Froeschlé and Lega, 1998) to give directly, in the case of regular resonant orbits, the frequency, or rotation angle measured by an observer in the center of the island. Let us call theoretical time, the time necessary for a vector to turn of an angle of 90°, that is, the ratio $tt = 90/\alpha$, α being the mean rotation angle. Figure 13 shows a surprisingly good agreement between the theoretical time and the experimental time of Figure 8 necessary for the FLI to reach its maximum value. The rotation angle turns out to be a quantity invariant to the change of regime observed for the size of the islands.

The model of linear elliptic rotation seems therefore to fit very well for the comprehension of the behavior of the FLI on periodic orbits.

6. Conclusion

The FLI (Froeschlé et al., 1997) was shown to be particularly sensitive for the distinction between regular and chaotic, even slow chaotic orbits. Recently, it has been shown (Froeschlé and Lega, 2000) that this indicator is also powerful for distinguish between resonant regular orbits and invariant tori. In the same art-

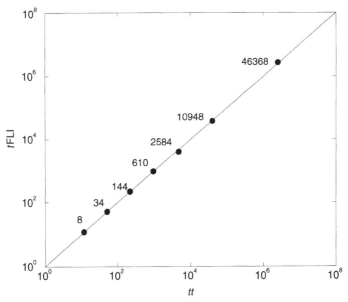

Figure 13. Time necessary for the FLI to become constant (tFLI) against the time expected (tt) for the vector $\vec{v}(t)$ to reach its maximum allowed value for the periodic Fibonacci orbits analyzed in Figure 8. The straight line is $t\text{FLI} = tt$

icle an explication for this behavior is given based on an argument related to the differential rotation.

Thanks to this property Froeschlé et al. (2000) have obtained a detailed representation of the Arnold web on a suitable Hamiltonian system with 3° of freedom and on a four-dimensional symplectic mapping (Froeschlé and Lega, 2000). The crucial question of the global stability of the system is also discussed, and an interval of transition between the exponentially long stable Nekhoroshev regime and the diffusive Cihirkov regime is also provided.

In Froeschlé et al. (2000) it was shown that the FLI had minimum values near the center of libration islands, that is seemed to be well suited also for the detection of periodic orbits. In this paper we have studied the FLI for periodic orbits showing that the FLI behaves like $\log q$, q being the order of periodic orbits. The application on a particular set of periodic orbits, the Fibonacci sequence which approximate the golden torus, has confirmed this result up to a given order q; for higher orders the size of the islands starts to shrinks exponentially to zero. The order q observed for the change of regime on the FLI is in perfect agreement with the one previously obtained (Lega and Froeschlé, 1996; Locatelli et al., 2000) on the size of the Fibonacci islands.

We have observed that the FLI for periodic orbits reaches a maximum value after a given time t, and we have shown that this time corresponds very well to the time necessary for a vector to 'visit' the elliptic orbits of the phase space from its semi-minor to its semi-major axis.

146 ELENA LEGA AND CLAUDE FROESCHLÉ

We have interpreted our results introducing a simple area-preserving model of linear elliptic rotation obtaining a good agreement between the FLI and both the geometric and the kinematic parameters introduced to explain both the maximum FLI value and the time necessary to reach it.

We have obtained encouraging results up to now on the standard mapping. Since this mapping can be considered as a paradigm model for the study of dynamical systems, in particular because of the richness of its structures, we intend to use the method for studies of physical interest like for example the three-body problem or the dynamic of stars inside a galaxy or even for problems of stability in celestial mechanics. Moreover, we also intend to use the method, thanks to its sensitivity, for a better comprehension of the properties of the standard map itself. In fact, the analysis of the Fibonacci sequence made in the present paper allowed us to find in a much simpler way, results in agreement with previous fine studies of the mapping (Lega and Froeschlé, 1996; Locatelli et al., 2000) and we think it worthwhile trying to get an even finer exploration of the neighborhood of an invariant torus.

Acknowledgement

We are indebted to Prof. Antonio Giorgilli for many discussions and suggestions.

References

Benettin, G., Galgani, L., Giorgilli, A. and Strelcyn, J. M.: 1980, 'Lyapunov characteristic exponents for smooth dynamical systems; a method for computing all of them', *Meccanica* **15**: Part I : Theory, 9–20 – Part 2: Numerical applications, 21–30.

Chirikov, B. V.: 1960, *Plasma Phys.* **1**, 253.

Contopoulos, G. and Voglis, N.: 1997, 'A fast method for distinguishing between order and chaotic orbits', *Astron. Astrophys.* **317**, 73–82.

Froeschlé, C.: 1970, 'A numerical study of the stochasticity of dynamical systems with two degrees of freedom', *Astron. Astrophys.* **9**, 15–23.

Froeschlé, C.: 1984, 'The Lyapunov characteristic exponents and applications', *J. de Méc. théor. et apll.* Numero spécial, 101–132.

Froeschlé, C. and Lega, E.: 1998, 'Twist angles: a fast method for distinguishing islands, tori and weak chaotic orbits. Comparison with other methods of analysis', *AA* **334**, 355–362.

Froeschlé, C., Lega, E. and Gonczi, R.: 1997, 'Fast Lyapunov indicators. Application to asteroidal motion', *Celest. Mech. & Dyn. Astr.* **67**, 41–62.

Froeschlé, C. and Lega, E.: 2000, 'On the structure of symplectic mappings. The Fast Lyapunov indicator: a very sensitive tool', *Celest. Mech. & Dyn. Astr.* **78**, 167–195.

Froeschlé, C., Guzzo, M. and Lega, E.: 2000, 'Graphical evolution of the Arnold's web: from order to chaos', *Science* **289-N.5487**, 2108–2110.

Greene, J. M.: 1979, 'A method for determining a stochastic transition', *J. Math. Phys.* **20**, 1183 pp.

Guzzo, M., Lega, E. and Froeschlé, C.: 2001, 'On the numerical detection of the effective stability of chaotic motions in quasi-integrable systems', (in press).

Laskar, J.: 1993, 'Frequency analysis for multi-dimensional systems. Global dynamics and diffusion', *Physica D* **67**, 257–281.

Laskar, J.: 1994, 'Frequency map analysis of an Hamiltonian system', *Workshop on Non-Linear Dynamics in Particle Accelerators*, September 1994, Arcidosso.

Laskar, J., Froeschlé, C. and Celletti, A.: 1992, 'The measure of chaos by the numerical analysis of the fundamental frequencies', Application to the standard mapping' *Physica D* **56**, 253.

Lega, E. and Froeschlé, C.: 1997, 'Fast Lyapunov Indicators. Comparison with other chaos indicators. Application to two and four dimensional maps', In: J. Henrard and R. Dvorak, (eds) *The Dynamical Behaviour of our Planetary System*, Kluwer Academic Publishers.

Lega, E. and Froeschlé, C.: 1996, 'Numerical investigations of the structure around an invariant KAM torus using the frequency map analysis', *Physica D* **95**, 97–106.

LeVeque, W. J.: 1977, *Fundamentals of Number Theory*, Addison–Wesley Publishing Company.

Lichtenberg, A. J. and Lieberman, M. A.: 1983, *Regular and Stochastic Motion*, Springer, Berlin, Heidelberg, New York.

Locatelli, U., Froeschlé, C., Lega, E. and Morbidelli, A.: 2000, 'On the relationship between the Bruno function and the breakdown of invariant tori', *Physica D* **139**, 48–71.

MacKay, R. S.: 1993, *Renormalisation in Area Preserving Maps*, World Scientific.

Mahler, K.: 1957, *Lectures on Diophantine Approximations*.

Morbidelli, A. and Giorgilli, A.: 1995, 'Superexponential stability of KAM tori', *J. Stat. Phys.* **78**, 1607 pp.

Nekhoroshev, N. N.: 1977, 'Exponential estimates of the stability time of near-integrable Hamiltonian systems', *Russ. Math. Surveys* **32**, 1–65.

Olivera, A. and Simó, C.: 1987, 'An obstruction method for the destruction of invariant curves', *Physica D* **26**, 181 pp.

ON THE OBSERVABILITY OF RADIATION FORCES ACTING ON NEAR-EARTH ASTEROIDS

DAVID VOKROUHLICKÝ[1], STEVEN R. CHESLEY[2] and ANDREA MILANI[3]

[1] *Institute of Astronomy, Charles University, V Holešovičkách 2, CZ-180 00 Prague 8, Czech Republic, e-mail: vokrouhl@mbox.cesnet.cz*
[2] *Jet Propulsion Laboratory, Pasadena, CA 91109, USA, e-mail: steven.chesley@jpl.nasa.gov*
[3] *Dipartimento di Matematica, Università di Pisa Via Buonarotti 2, I-56127 Pisa, Italy, e-mail: milani@dm.unipi.it*

Abstract. We consider the perturbations on near-earth asteroid orbits due to various forces stemming from solar radiation. We find that the existence of precise radar astrometric observations at multiple apparitions, spanning periods on the order of 10 years, allows the detection of such forces on bodies as large as kilometer across. Indeed, the perturbations are so substantial that certain of the forces can be essential to fit an orbit to the observations. In particular, we show that the recoil force of thermal radiation from the asteroid, known as the Yarkovsky effect, is the most important of these unmodeled perturbations. We also show that the effect of reflected light can be important if even moderate albedo variations are present, while moderate changes in oblateness appear to have a far smaller effect. An unexpected result is that the Poynting–Robertson effect, typically only considered for submillimeter dust particles, could be observable on smaller asteroids with high eccentricity, such as 1566 Icarus. Finally, we also study the possibility of improving the orbit uncertainty through well-timed optical observations which might help in better detection of these nongravitational perturbations.

Key words: Asteroids, near-earth, nongravitational effects

1. Introduction

The fact that nongravitational forces act on the natural bodies of the solar system is obvious, but the question of whether these forces are relevant in modeling the motion of these bodies is far more subtle. That question revolves on how accurate the model should be and on the time span over which the motion will be studied, as well as on the size of the body in question. The importance of outgassing on comets or electromagnetic and radiative effects (including the Poynting–Robertson drag) on dust particles is well documented, but for larger or passive bodies like asteroids, the prevailing understanding among solar system dynamicists to date has been that relativistic N-body dynamics are adequate to describe precisely their motion and that any other forces are well below the level of noise in the observations and orbital solutions. However, we show concretely with the present results that certain nongravitational forces, notably those related to the interaction with solar radiation, will no longer be ignorable for some asteroids within the next 5–10 years.

The availability of precise radar astrometry observations (with uncertainties on the order of a few tens of meters for ranging and several millimeters per second for

Celestial Mechanics and Dynamical Astronomy **81:** 149–165, 2001.
© 2001 *Kluwer Academic Publishers.*

Doppler observations) has allowed very precise orbit determination. If high quality radar observations are available at multiple apparitions covering a sufficiently long period, then the effect of very small but cumulative forces could be observable. Due to the limitations of radar technology, only asteroids which pass close to the Earth can be measured with the requisite precision.

Besides having precise measurements over a long interval, we also need the perturbations to be either cumulative, or large amplitude if periodic. It is well known that the long-term effect of solar radiation on a nonrotating spherical body with uniform albedo has no effect on average, only providing short periodic effects on the orbital motion since the forces are exactly aligned with the solar gravity. But these assumptions are rarely satisfied, even to a crude level, among the asteroid population. Rotating bodies reradiate solar energy in the thermal band in a direction offset from the sun by an angle depending on the rotation parameters and the surface thermal conductivity. If an object has albedo variations across its surface, in particular if it is darker in one hemisphere than the other, the net recoil force from reflected light will be greater on the brighter hemisphere, thus shifting the force away from the sun vector. We also consider a similar effect acting on oblate ellipsoids of revolution. Finally, the Poynting–Robertson (PR) effect, which is the (v/c) correction of the simple radiation pressure formula for a body moving with velocity v in the radiation source reference frame (c is the light velocity), needs to be considered for orbits with particularly high eccentricity.

In this paper we first collect and summarize results from two previous papers (Vokrouhlický et al., 2000; Vokrouhlický and Milani, 2000), hereafter Paper I and Paper II, respectively. Paper I provided an exhaustive analysis of the impact of the Yarkovsky effect on several asteroids, and Paper II considered in detail the other effects discussed here. Second, we also discuss importance of the precise astrometric observations that might be eventually taken before the next close approaches to the Earth of the selected asteroids in Papers I and II, and improve thus possibility to test the radiation perturbations.

2. Solar Radiation Forces

Let us consider a body of an arbitrary shape illuminated by solar radiation. The sun is assumed to be infinitely remote, so that the radiation field consists of parallel light rays characterized by a unit vector $\mathbf{n}$ ('outward from the sun'). Denote the geometric cross section of the body along the direction $\mathbf{n}$ by $P_\perp$, $\mathcal{E}$ the solar radiation flux, m the body's mass and c the light velocity. Then the radiation striking the body's surface imparts an acceleration $\mathbf{a}$ given by

$$\mathbf{a} = \frac{P_\perp \mathcal{E}}{mc} \mathbf{n} . \tag{1}$$

Part of the absorbed radiation is physically reprocessed in the body and later reemitted in the infrared band ('thermal radiation of the body'). The recoil acceleration due to this radiation field is the so-called Yarkovsky effect, which we discuss in Section 2.1. The complementary part of the absorbed radiation, characterized by the albedo coefficient A, is immediately reflected from the surface of the body. We discuss the recoil acceleration due to this reflected radiation in Section 2.2. The formula (1) is entirely correct only when the body is stationary with respect to the source of the radiation field. If it moves with a characteristic velocity v in the source frame, corrections of the order (v/c) (typically $\simeq 10^{-4}$ in the solar system) may be important. This is the PR effect, and we discuss its dynamical consequences for NEA motion in Section 2.3.

2.1. YARKOVSKY EFFECT

A portion of the solar radiation received by the body is thermally absorbed and reradiated with some time delay. The amount of energy transformed into heat is governed by the absorptivity (complementary to the albedo) of the asteroid's surface, and the delay depends on the thermal properties of the surface material. This delay causes the net force from the recoil acceleration of the thermal emission to shift away from the solar direction by an amount depending upon the angular rates of the body, resulting in some acceleration component along the asteroid's velocity vector. This in turn causes a more or less constant drift of the semi-major axis, which is the significant aspect of the Yarkovsky effect.

The phrase 'angular rates' used above is deliberately vague because the Yarkovsky effect can be neatly broken down into two variants, depending upon the fundamental frequencies involved. The Yarkovsky *diurnal* acceleration originates from the rotation rate of the asteroid, and the Yarkovsky *seasonal* acceleration results from the asteroid's mean motion. We describe these separately in the following paragraphs. An interested reader may find the rather involved quantitative formulation of the Yarkovsky acceleration, both diurnal and seasonal variants, in Paper I. Instead, we give here the results in a graphical form (Figure 1).

2.1.1. *Diurnal mode*

For the results in this paper we have used the diurnal model described by Vokrouhlický (1998a) and also outlined in Paper I. The most important simplifying assumptions are that (i) the temperature exhibits small variations around a mean value (so that linearization of the heat diffusion problem is allowed), (ii) the body is spherical with radius R, and (iii) the albedo parameter A is constant on the body's surface. While assumption (i) is relatively reasonable, assumptions (ii) and (iii) may be poorly satisfied for real objects. Vokrouhlický (1998b) considered the Yarkovsky effect on spheroidal objects, at the limit when the penetration depth of the diurnal thermal wave is much smaller then the body's size, and concluded that by using spherical model one introduces an error up to a factor 2. Given also the uncertainty in the surface thermal properties we cannot hope to determine the Yarkovsky force with much better precision at this stage of investigation. We feel

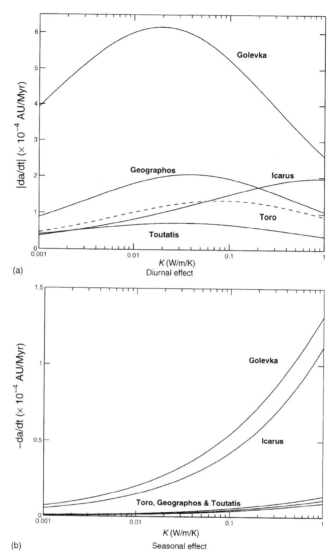

Figure 1. The estimated secular drift rate of the semi-major axis due to the (a) diurnal and (b) seasonal variants of the Yarkovsky effect versus the surface conductivity K. The asteroids considered here are 1566 Icarus, 1620 Geographos, 1685 Toro, 4179 Toutatis, and 6489 Golevka. All drift rates are negative, with the exception of the diurnal case of 1685 Toro (dashed line on panel (a)). The physical and spin parameters used for these plots are as specified in Paper I.

that thermal models tailored to the shape of a particular object will eventually be needed to compute these forces at the necessary precision in the future. It may be shown that the Yarkovsky force component related to the variable albedo on the surface partially, but not entirely, compensates the optical effect discussed in Section 2.2. For the sake of simplicity we omit discussion of the fine interplay of the Yarkovsky and optical effects in the present text (while we prepare a detailed study of this issue).

The diurnal Yarkovsky acceleration acts in the equatorial plane of the asteroid. In general, it depends strongly and nonlinearly on the orientation of the spin axis, the spin rate, and the surface thermal conductivity. This last parameter is the hardest to characterize from ground based observations. The lowest reasonable conductivities, of the order 0.001–0.01 W/m/K, correspond to a substantial regolith layer, while the highest considered conductivity, 1 W/m/K, corresponds to a fairly smooth solid rock surface. Most NEAs probably have K values in the range 0.01–0.1 W/m/K. The diurnal effect is maximized on average when the spin axis is orthogonal to the orbital plane.

The semi-major axis drift rate da/dt is the most important manifestation of the Yarkovsky acceleration because a linear drift in semi-major axis leads to a quadratic drift in mean anomaly over time. This fast accumulating variation in position is what allows the detection of this perturbation. Figure 1(a) depicts the diurnal semi-major axis drift rate for several interesting objects as a function of the surface thermal conductivity K. Note that for most of these objects da/dt depends only weakly on K, implying in these cases that our lack of knowledge of this parameter does not markedly affect our ability to predict the observability of this effect. On the other hand, in a case like 6489 Golevka where the dependency on K is substantial, we should be able to estimate better the surface conductivity if the effect is observable.

2.1.2. *Seasonal mode*

The seasonal mode of the Yarkovsky effect, which is always aligned with the body's spin axis, was first pointed out by Rubincam (1995). This force is somewhat more challenging from a computational perspective, especially for orbits with large orbital eccentricity. In such cases, the variations of temperature along the orbit, in particular over the thermal relaxation time scale of the seasonal effect, are large enough so that the basic assumptions of linearization of the heat diffusion problem are violated. Avoiding the linearization approach yields a precise result, but requires a completely numerical solution. In the present work we have used the model of Vokrouhlický and Farinella (1998) that solves the thermal state of the body along an arbitrarily eccentric orbit in the 'large body' approximation (penetration depth of the seasonal thermal wave much smaller than the geometric size of the body). These results have been also checked by a more involved numerical solution of Vokrouhlický and Farinella (1999), where the restricting large body assumption has been relaxed.

Figure 1(b) depicts the seasonal semi-major axis drift rate as a function of K for the same objects considered in Figure 1(a). Note that the seasonal effect tends to be about an order of magnitude smaller than the diurnal effect for small values of the surface conductivity, while for higher values ($K \simeq 1$ W/m/K) the seasonal effect becomes comparable to the diurnal effect. The seasonal effect is maximum when the spin axis lies in the orbital plane, and it is zero when the spin axis is perpendicular to the orbital plane.

2.2. REFLECTED RADIATION PRESSURE

In this section we consider separately two special cases for the recoil acceleration due to reflected radiation. Specifically we shall consider objects with axially symmetric but nonuniform albedo distribution and oblate ellipsoids of revolution. This allows us to determine the extent that nonuniform reflectivity or shape can perturb the orbit. In each case we assume the body has an axis of symmetry, and this is reasonable, at least in an average sense, for objects with principal axis rotation. For simplicity we also assume a Lambert (isotropic) law of reflection of the body's surface, although it is well known that this assumption is not exactly satisfied for cosmic bodies.

2.2.1. *Objects with albedo variations*

We shall assume albedo modes that have an odd symmetry with respect to the equator of the body since these have more important orbital effects than the even-symmetry terms. Such an albedo distribution has the form $A(\theta) = a_0 + a_k \cos^k \theta$ with an arbitrary odd number k. Here a_0 and a_k are constants and θ is the co-latitude measured from the symmetry axis $\mathbf{s}$. A linear combination of such terms in $A(\theta)$ would just mean a linear combination of the perturbations in a first order perturbation theory such as the one we are using, thus we shall assume the above description of the albedo distribution with k an odd integer and a_k an arbitrary constant.

In the case of the simple albedo distribution $A(\theta)$ introduced above, the reflected radiation pressure acceleration may be obtained analytically. For the sake of illustration we give here the total radiation acceleration in the dipole case ($k = 1$) which reads

$$ \mathbf{a} = \kappa \left\{ \left(1 + \frac{4}{9} a_0 \right) \mathbf{n} + \frac{a_1}{6} \left(\cos \theta_0 \, \mathbf{n} - \mathbf{s} \right) \right\} , \qquad (2) $$

where θ_0 is the colatitude of the sun measured from the symmetry axis ($\cos \theta_0 = -\mathbf{n} \cdot \mathbf{s}$) and $\kappa = \pi R^2 \mathcal{E}/mc$. The somewhat more complicated case of the arbitrary odd value of k is given in Paper II. Note that the first term in Equation (2) is the usual acceleration of a spherical body with homogeneous albedo a_0. This term has little importance since it can only result in short-periodic orbital effects. However, the second term in Equation (2) is due to the axially symmetric albedo term $\propto a_1 \cos \theta$ and holds potentially important orbital effects (even though a_1 is expected to be significantly smaller than a_0). Most importantly, it results in a long-term drift in semi-major axis that is associated with significant along-track orbital displacement (that may be directly observable in the NEA case). It is easy to check that the even-symmetry zonal albedo terms do not yield this effect. We also note that the semi-major axis effect disappears for near-circular orbits ($e = 0$) but this is rarely the case for NEAs.

2.2.2. *Ellipsoids of revolution*

Unlike the previous paragraph, we assume a constant albedo $A = a_0$, but we shall consider bodies of an ellipsoidal shape, with two equal axes (to maintain axial symmetry). The ratio of the polar (R_p) and the equatorial (R_e) radii will be denoted $\epsilon = R_p/R_e$.

In what follows we use the mathematical formulation of Vokrouhlický (1998b, Appendix). The radiation pressure due to the absorbed sunlight is still given by Equation (1); however, the geometric cross section $P_\perp$ is now given by $P_\perp(\theta_0) = \pi R_e^2 J_2(\theta_0)$. Here we have again assumed the sun direction at an angle θ_0 from the symmetry axis, and following Vokrouhlický (1998b) we introduce the auxiliary functions $J_n(x) = \sqrt{\epsilon^n \sin^2 x + \cos^2 x}$ ($n = 1, 2, 3, \ldots$).

As far as the recoil force due to the reflected radiation is concerned, we can still obtain an analytic result for the total radiation acceleration of the ellipsoidal body in the form

$$\mathbf{a} = \kappa \left\{ \left[J_2(\theta_0) + \frac{4}{9} a_0 \psi_x(\epsilon) \right] \mathbf{n} - \frac{4}{9} a_0 \psi_{zx}(\epsilon) \cos \theta_0 \, \mathbf{s} \right\}, \tag{3}$$

where for reasons of brevity we refer the reader to Vokrouhlický (1998b) or Paper II for the definitions of the auxiliary functions $\psi_x(\epsilon)$ and $\psi_{zx}(\epsilon)$. The parameter κ is defined as before, but with $R = R_e$. As expected, when $\epsilon \neq 0$ a symmetry-axis-aligned ($\propto \mathbf{s}$) acceleration components occur with 'seasonal' modulation due to the $\cos \theta_0$ ($\psi_{zx}(\epsilon) \neq 0$ for $\epsilon \neq 1$). However, the acceleration in Equations (3) does not lead to a secular drift in the semi-major axis, and thus there is no rapid accumulation of the displacement, in contrast to the Yarkovsky and albedo effects described above. Only the eccentricity and inclination of the orbit undergo long-term variations, which produce quasi-linear perturbations of the asteroid position with time. This implies that such a perturbation is unlikely to be observable, as we indeed confirm in Section 3.

2.3. POYNTING–ROBERTSON EFFECT

Details on this classical effect can be found in textbooks or journal reviews (e.g. Burns et al., 1979). Assuming diffuse reflection on a spherical body with constant surface albedo ($A = a_0$) one obtains

$$\mathbf{a} = -\frac{\kappa}{c} \left(1 + \frac{4}{9} a_0 \right) \left[\mathbf{v} + (\mathbf{n} \cdot \mathbf{v}) \mathbf{n} \right], \tag{4}$$

for the corresponding acceleration. It should be noted that little attention has been paid so far to the more general case of the PR effect when the body reflects/scatters the incoming radiation in an nonsymmetric way (see e.g. Kocifaj and Klačka, 1999). At this stage of investigation we omit this generalization, but future work might require a more precise model of the PR effect. One can easily show that the PR acceleration (4) results in a secular inward drift of the semi-major axis that appreciably increases in magnitude when the eccentricity of the orbit is high.

3. Examples

Hereafter, we focus our discussion to the case of two asteroids: Icarus and Golevka. An interested reader may find information about other potentially interesting orbits in Paper I.

3.1. 1566 ICARUS

Icarus has been observed with radar in June 1968 and in June 1996. The next opportunity for radar observations will occur when it comes within 0.05 AU of the earth in June 2015. For the computations in this section we use a radius of 450 m and a surface albedo of $a_0 = 0.4$ (Veeder et al., 1989). The spin axis direction and rate are taken from De Angelis (1995).

In Figure 2(a) we depict the observability of the total Yarkovsky effect at the 2015 encounter. For this experiment we have first computed the predicted positions and 3σ uncertainty ellipses without the Yarkovsky effect in our model. We plot this observation and the corresponding confidence ellipse (dashed lines) at the origin of the radar observable (range/range-rate) plane. Next we make the same prediction, but this time include Yarkovsky accelerations ($K = 0.01$ W/m/K), both in the orbit determination and propagation of the solution, obtaining a displaced observation and uncertainty ellipse, which are plotted (solid lines) relative to the non-Yarkovsky solution. The separation between the observations indicates the observable Yarkovsky-induced displacement, while the separation between the ellipses indicates the extent to which the effect is detectable. If the ellipses are entirely or nearly disjoint then the effect should be easily observable, but if they overlap to a large extent then the orbital uncertainty is too great to guarantee an unambiguous detection. We plot these ellipse pairs for several epochs surrounding the time of closest approach. The figure indicates that the effect will be large (tens of kilometers and several kilometers per day); however, there is a moderate overlap between the ellipses in each pair. The predicted observations themselves are separated at the 3σ level, which will generally allow a clear detection, but not in every case. Note that this plot reflects a worst case scenario since we can expect numerous optical observations up to the point of the 2015 encounter, which may significantly reduce the depicted uncertainty regions. Moreover, we have also studied the case with a higher surface conductivity ($K = 0.1$ W/m/K) and found that this leads to uncertainty ellipses with no overlap whatsoever.

Figure 2(b) shows the results of a similar experiment, but indicating the observability due to surface albedo variations as discussed in Section 2.2. The predictions with albedo variation included (solid curves) use $a_0 + a_1 \cos\theta$ where $a_1 = 0.01$, which is a relatively minor variation. The displacements are smaller than those

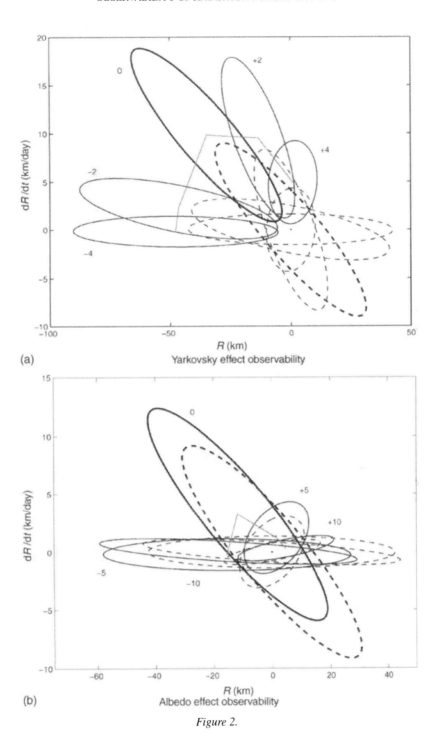

Figure 2.

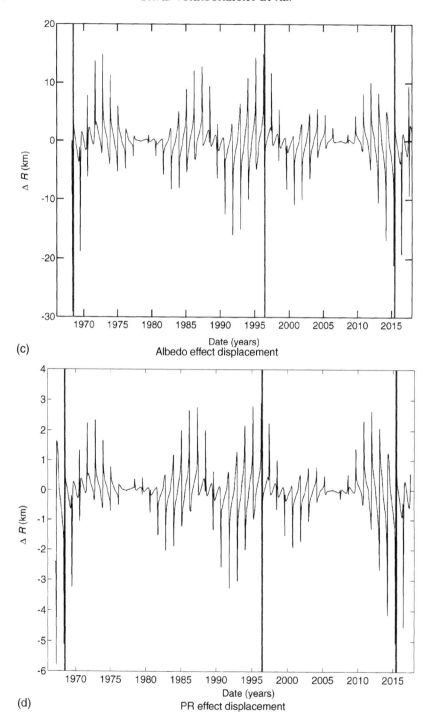

Figure 2. (continued)

Figure 2. The significance of some nongravitational forces on the orbit of asteroid 1566 Icarus. Parts (a) and (b) indicate the observability in the space of radar observables of the Yarkovsky and nonuniform albedo accelerations, respectively at the close encounter in June 2015. Dashed ellipses indicate the extent of 3σ uncertainty regions for the unperturbed case, and the effect of the perturbations is given by the solid curves. Several pairs of ellipses are plotted at various epochs, labeled with the number of days after the point of closest approach. Parts (c) and (d) depict the displacement ΔR along the earth line-of-sight caused by reflected solar radiation, assuming nonuniform albedo and the PR effect, respectively. The three close earth approaches are marked by the shaded strips.

from the Yarkovsky experiment above. The earth line-of-sight displacements from this effect are presented in Figure 2(c). At the 2015 encounter, the radar range perturbation could be as large as 25 km with the above albedo model. In reality the actual albedo will be difficult to ascertain, so, in contrast to the Yarkovsky experiment, this experiment shows only that albedo variation could be important in some cases, and not whether it will be important. As it has been also pointed out above, part of the effect might be compensated by the modification of the Yarkovsky effect due to variation of the albedo coefficient. The extent of this compensation, depending on the surface conductivity value and the rotation state, will be studied in a forthcoming work.

The earth line-of-sight displacements ΔR from PR accelerations on the orbit of Icarus are potentially at the multikilometer level, as shown in Figure 2(d). In large part, this is due to the unusually high eccentricity (0.83) for this body. Recently, Métris and Vokrouhlický (2001, in preparation) have shown that the PR effect probably produces observable effects on the long-term evolution of the semi-major axis of the artificial satellites Etalon 1 and 2. The perturbations of highly eccentric NEA orbits represent a second case of 'macroscopic objects' which orbit determination requires radiation pressure modeling at the PR (i.e. v/c) level. These two cases demonstrate the dramatic increase of the orbital data precision recently achieved.

Finally we note in passing that, as expected, the effect of oblateness described in Section 2.2 is much smaller than the other perturbations discussed above and it is unlikely to be of any importance.

3.2. 6489 GOLEVKA

We present in Figure 3 the results for 6489 Golevka in the same manner as above for 1566 Icarus. Golevka has been observed by radar in June 1991, 1995 and 1999. Unfortunately, astrometry has so far not been obtained from the 1999 data; thus the 1991–1995 baseline is rather short to detect subtle nongravitational phenomena in Golevka's orbit. Nevertheless, the next close approach to the earth occurs on May 20, 2003.

We consider the physical parameters of Golevka as derived by Hudson et al. (2000), notably a surface albedo (a_0) of 0.15, mean radius of 265 m and spin axis orientation **s** with ecliptic longitude and latitude $\ell = 202°$ and $b = -45°$. The shape model of Golevka, as derived from the radar observations, is very complex

160 DAVID VOKROUHLICKÝ ET AL.

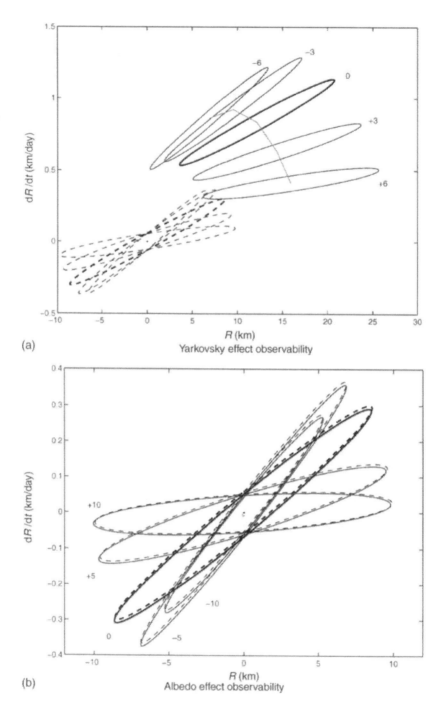

Figure 3. The impact of some nongravitational forces on the orbit of 6489 Golevka. The presentation is the same as in Figure 2. The radar observability predictions in parts (a) and (b) are for the May 2003 observing opportunity.

OBSERVABILITY OF RADIATION FORCES ON NEAS 161

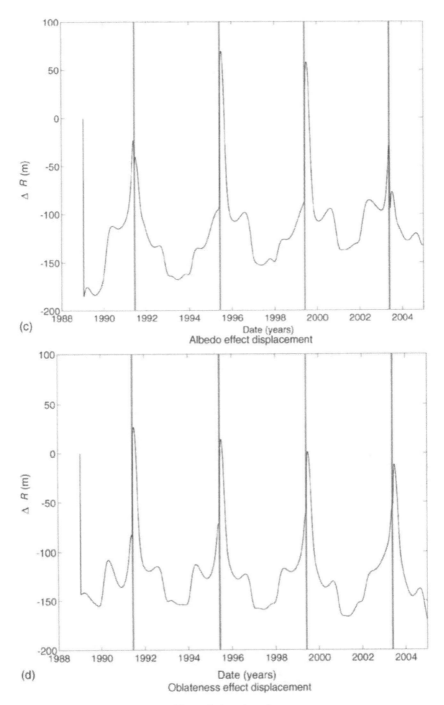

Figure 3. (continued)

and impossible to fit with an ellipsoidal model (to which our theory is limited). We can only obtain an order of magnitude of the nonsphericity effect by adopting $\epsilon \simeq 0.8$, a rather conservative value since the estimate of the longest to shortest geometric axes of Golevka is about 1.4 (Hudson et al., 2000). For the simulations we have again assumed $K = 0.01$ W/m/K.

The most important aspect is the observability of the Yarkovsky effect at the encounter in 2003. This encounter is relatively soon, and the radar observation predictions in Figure 3(a) indicate that the effect will almost certainly be detectable if accurate radar astrometry is obtained during the encounter. In fact the separation between the confidence regions with and without the Yarkovsky effect is large enough that a good estimate of the surface conductivity of Golevka may be available after 2003.

Figures 3(b)–(d) indicate that, at least for Golevka, the effect of anisotropic albedo or nonsphericity are unlikely to be observable/important in 2003. In fact, most of the perturbation observed in Figures 3(c)–(d) is of the short-periodic character, since the observational data do not cover a long enough time span to allow accumulation of the long-term effects.

3.3. DISCUSSION

The results presented in Figures 2 and 3 are based on propagation of the present knowledge of the orbit (derived from currently available data) to the next close encounters of Icarus and Golevka with the earth, when presumably the radar observations could be taken. The main obstacle for clear detection of the radiation effects studied in this paper (in particular the Yarkovsky effect) is due to the orbit uncertainty, which is much larger than the observation uncertainty. A possible approach to this problem might be to further constrain the orbit uncertainty through additional optical observations before the next radar observing opportunity. In the case of objects where all of the optical and radar observations are at a single apparition any further optical observations at subsequent oppositions will generally lead to a significant reduction in the orbit uncertainty. We have tested this idea in the case of the asteroid 1998 KY26 (see Paper I). When the orbit is known better, as in the case of Golevka or Icarus, the situation may be different and a careful study is needed. In what follows we shall focus mainly on Icarus, trying to predict if any observations before June 2015 may help to reduce the rather large orbital uncertainty depicted in Figure 2(a).

In Table I we have selected epochs suitable for future observations of both Golevka and Icarus until 2015. Interestingly, the more involved observations of Golevka have little influence on the orbit uncertainty in May 2003, since the orbit is already very tightly constrained by the two very precise radar observation runs in 1991 and 1995. We have checked that none of the potential optical observations until 2015 at the $\simeq 0.25$ arcsec level can significantly reduce the orbit uncertainty. On the other hand if the present efforts to obtain useful astrometry from the 1999

TABLE I

Observability conditions of Golevka and Icarus. We list epochs through 2015 when elongation of the asteroids is larger than 85° (Golevka) and 90° (Icarus)

Golevka		Icarus	
Time span	Min/max magnitude	Time span	Min/max magnitude
Sep 2000–Mar 2001	24.4/25.8	May–Sep 2001	18.6/20.9
Oct 2001–Apr 2002	24.5/25.9	May–Sep 2002	19.1/20.8
Dec 2002–Jan 2004[a]	16.0/23.9	May–Sep 2003	19.1/20.7
Sep 2004–Mar 2005	24.5/25.8	May–Sep 2004	18.5/20.9
Oct 2005–Apr 2006	24.3/25.8	May–Sep 2005[b]	16.4/20.8
Jan 2007–Jan 2008[b]	19.8/24.4	Jul–Aug 2006[b]	17.6/19.9
Sep 2008–Mar 2009	24.6/25.9	Jun–Aug 2009[a]	17.9/20.2
Oct 2009–Apr 2010	24.3/25.8	May–Sep 2010	18.2/20.8
Aug 2011–Jan 2012	21.2/24.4	May–Sep 2011	19.0/20.9
Sep 2012–Mar 2013	24.6/25.9	May–Sep 2012	19.2/20.6
Oct 2013–Apr 2014	24.2/25.8	May–Sep 2013	18.8/20.9
		May–Sep 2014	17.6/20.9
		Jun–Sep 2015[a]	13.6/20.4

[a] Approach to within 0.1 AU of the earth.
[b] Approach to within 0.5 AU of the earth.

Golevka radar campaign (Steve Ostro, private communication) are successful, the orbital uncertainty in 2003 would improve considerably.

The situation is somewhat different for Icarus. Figure 4 shows the sky-plane uncertainty of the current orbit till the next close approach in June 2015. If an observational campaign is organized close to the maxima of this uncertainty plot, so that the resulting 'normal point' precision drops below $\simeq 0.1$–0.15 arcsec, we may hope to constrain better the orbit in 2015. As a test, we have simulated results of such campaigns in September 2001, August 2002, August 2009 and September 2010 by representing each with a single observation with an effective precision of 0.1 arcsec. The resulting 3σ uncertainty in range during the 2015 close approach (about 30.8 km with the current orbit solution; see Figure 2(a)) becomes 28.5 km. If the effective level of 0.05 arcsec for the normal points might be reached, the orbit uncertainty in 2015 further constrains to 25.4 km. Moreover, at this level of precision, the sky-plane displacement due to the Yarkovsky effect might start to play important role given the very long time span of the observations. We should point out that none of the assumed campaigns requires exceptional equipment, since the Icarus visual magnitude ranges from 17.5 to 19 for the selected dates, but some of the observations must be done from the southern hemisphere because of low declination. Finally, we note that we have disregarded the possibility of radar

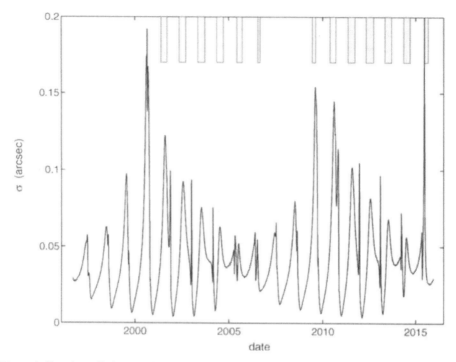

Figure 4. Sky-plane (1σ) uncertainty of the Icarus orbit from the currently available data. Shaded areas show possible observational windows as described in Table I. Notice that maximum observational effort should be concentrated in 2001, 2009 and 2010, since these data may eventually better constrain the currently available solution, and thus shrink the orbit uncertainty at the next close approach to the earth in 2015.

ranging in June 2006, when Icarus gets at 0.29 AU from the earth. Obviously, this observation would be also very helpful.

References

Burns, J. A., Lamy, P. L. and Soter, S.: 1979, 'Radiation forces on small particles in the solar system', *Icarus* **40**, 1–48.

De Angelis, G.: 1995, 'Asteroid spin, pole and shape determinations', *Planet. Sp. Sci.* **43**, 649–682.

Hudson, R. S., Ostro, S. J., Jurgens, R. F. et al.: 2000, 'Radar observations and physical model of asteroid 6489 Golevka', *Icarus* **148**, 37–51.

Kocifaj, M. and Klačka, J.: 1999, 'Real dust particles and unimportance of the Poynting–Robertson effect', *http://xxx.lanl.gov/astro-ph/9910042*

Rubincam, D. P.: 1995, 'Asteroid orbit evolution due to thermal drag', *J. Geophys. Res.* **100**, 1585–1594.

Veeder, G. J., Hanner, M. S., Matson, D. L., Tedesco, E. F., Lebofsky, L. A. and Tokunaga, A. T.: 1989, 'Radiometry of near-earth asteroids', *Astron. J.* **97**, 1211–1219.

Vokrouhlický, D.: 1998a, 'Diurnal Yarkovsky effect as a source of mobility of meter-sized asteroidal fragments. I. Linear theory', *A&A* **335**, 1093–1100.

Vokrouhlický, D.: 1998b, 'Diurnal Yarkovsky effect as a source of mobility of meter-sized asteroidal fragments. II. Non-sphericity effects', *A&A* **338**, 353–363.

Vokrouhlický, D.: 1999, 'A complete linear model for the Yarkovsky thermal force on spherical asteroid fragments', *A&A* **344**, 362–366.

Vokrouhlický, D. and Farinella, P.: 1998, 'The Yarkovsky seasonal effect on asteroidal fragments: a nonlinearized theory for the plane-parallel case', *Astron. J.* **116**, 2032–2041.

Vokrouhlický, D. and Farinella, P.: 1999, 'The Yarkovsky seasonal effect on asteroidal fragments: a nonlinearized theory for spherical bodies', *Astron. J.* **118**, 3049–3060.

Vokrouhlický, D., Milani, A. and Chesley, S. R.: 2000, 'Yarkovsky effect on small near-earth asteroids: mathematical formulation and examples', *Icarus* **148**, 118–138 (Paper I).

Vokrouhlický, D. and Milani, A.: 2000, 'Direct solar radiation pressure on orbits of small near-earth asteroids: observable effects?', *A&A* **362**, 746–755 (Paper II).

STELLAR MOTIONS IN GALACTIC SATELLITES

JUAN C. MUZZIO, M. MARCELA VERGNE, FELIPE C. WACHLIN and DANIEL D. CARPINTERO

Facultad de Ciencias Astronómicas y Geofísicas de la Universidad Nacional de La Plata and Instituto Astrofísico La Plata (CONICET), La Plata, Argentina

Abstract. The study of the motions of the stars that belong to a galactic satellite (i.e. a globular cluster or a dwarf galaxy orbiting a larger one) has some similarities, as well as significant differences, with that of the restricted three-body problem of celestial mechanics. The high percentage of chaotic orbits present in some models is of particular interest because it rises, on the one hand, the question of the origin of those chaotic motions and, on the other hand, the question of whether an equilibrium stellar system can be built when the bulk of the stars that make it up behave chaotically.

Key words: galactic satellites, stellar orbits, chaotic motion

1. Introduction

Many stellar systems are satellites of larger ones: the globular clusters, spheroidal systems, and the Magellanic clouds that orbit our own Galaxy, the Milky Way, are typical examples. It has long been recognized that the tidal field of the main galaxy affects the size of the satellite, imposing a limiting *tidal radius*, but aside from that effect, satellite models like the ones due to King, tend to ignore the tidal influence inside the body of the satellite (see, e.g. Binney and Tremaine, 1987). Nevertheless, Carpintero et al. (1999), Muzzio et al. (2000a,b) have shown that the motions of the stars that belong to the satellite are strongly affected, even those that pertain to the innermost regions of the satellite which are usually regarded as the ones best shielded from tidal effects.

The problem of the stellar motions inside a galactic satellite has some obvious similarities with the restricted three-body problem of celestial mechanics, as one can identify the main galaxy with the Sun, the satellite with a planet and the star with a minor body. A little thought shows, however, that there are also important differences. First, while the age of the Solar System is of the order of hundreds of millions of orbital periods, the age of a small stellar system is of the order of thousands of orbital periods only. Second, while the attractive force goes to infinity as one approaches the planet, that force goes to zero as one approaches the center of the galactic satellite. Third, while planets are usually taken as spherical (at least in the first approximation), galactic satellites are triaxially shaped by the tidal forces (and can have even more complicated forms, as the Magellanic clouds themselves show). Finally, planetary orbits do not depart much from circles, while

Celestial Mechanics and Dynamical Astronomy **81**: 167–176, 2001.
© 2001 *Kluwer Academic Publishers.*

galactic satellites tend to follow strongly elongated orbits (nevertheless, as in our previous investigations, here we will consider only the case of satellites on circular orbits). The similarities and differences between both problems were discussed by Muzzio et al. (2000b); here we want only to point out that this research may be of some interest to celestial mechanicists and to encourage them to undertake similar investigations.

Having that aim in mind, in the next section we explain the importance of stellar orbits for the construction of models of stellar systems and the problems posed by the presence of chaotic orbits. We then summarize the results of our previous investigations on this matter and present some new results.

2. The Importance of Stellar Orbits

Models of stellar systems must be *self–consistent*: the distribution of stars (i.e. mass) produces a gravitational potential; that potential determines the kind of orbits that the stars can follow in the system and the orbits, in turn, determine the distribution of the stars, which must be precisely the initial one to have self-consistency. The regions of space visited by the stars as they move on their orbits should, therefore, be consistent with the form of the stellar system we want to model: if we want to have a stellar system elongated in the, say, x direction, the bulk of the orbits has to be elongated in that direction too.

Now, since we usually want to build stationary models, all the orbits will obey the energy integral and they will be confined within the zero-velocity surface that corresponds to their energy. In three-dimensional space, chaotic orbits will obey, at most, one additional isolating integral, while regular orbits have two additional isolating integrals. Nevertheless, if the distribution function depends only on the energy, the stellar system must be spherical and, moreover, an equipotential surface tends to be more spherical than the equidensity surface through the same point (see, e.g. Binney and Tremaine, 1987). Therefore, chaotic orbits are not very good building blocks for stellar systems that depart much from sphericity (see Merritt, 1999 for further details).

3. Previous Results

Recognizing the importance of orbits to build a self-consistent stellar system, Carpintero et al. (1999) investigated what types of orbits might be present in a galactic satellite. They represented the satellite with a potential that was spherical at its center and increased its triaxiality outwards, just as it would happen with a satellite deformed by tidal effects. The satellite was placed on a circular orbit and stellar encounters among the satellite's stars were neglected. They reasoned that, if there were chaotic orbits under these conditions, even more could be expected

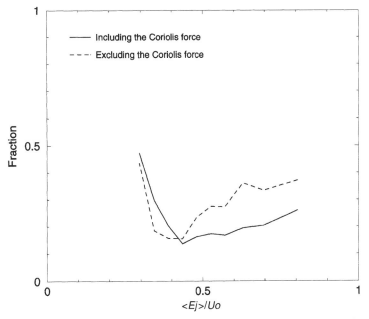

Figure 1. Fraction of chaotic orbits versus the value of the Jacobi integral normalized to the value of the central potential. We reproduce the values of the $b = 0.2290$ model of Muzzio et al. (2000b) and we add the new values obtained for the same model after eliminating the Coriolis force from the equations of motion.

under more realistic conditions (i.e. elongated orbit, stellar encounters, and so on). They investigated sets of initial conditions with fixed values of the Jacobi integral and starting, either with zero initial velocity, or from the main planes of symmetry with the velocity needed to yield the adopted value of the integral. The resulting orbits were classified using the Carpintero and Aguilar (1998) code of frequency analysis, and the Liapunov exponents were also obtained for a small number of orbits. The latter method was much slower than the former and it only allowed to distinguish chaotic from regular orbits, without providing a classification of the regular ones, but the results of both methods were very similar (see Figures 2 through 5 of Carpintero et al.), and the authors concluded that stellar orbits within galactic satellites were highly chaotic, affecting even the innermost regions of the satellite. Moreover, Liapunov times turned out to be surprisingly short, both in terms of orbital periods and satellite age.

Later on, Muzzio et al. (2000a) used the Carpintero and Aguilar (1998) code to classify the stellar orbits in satellites modelled according to the recipe of Heggie and Ramamani (1995). The satellites were again on circular orbits, but the big advantage offered by these models is that they are self-consistent (albeit involving certain approximations) and provide a distribution function. Muzzio et al., could then obtain the fractions of chaotic orbits, which turned out to be very high, indeed: almost one-fourth of the orbits of a satellite modeled after a dwarf galaxy

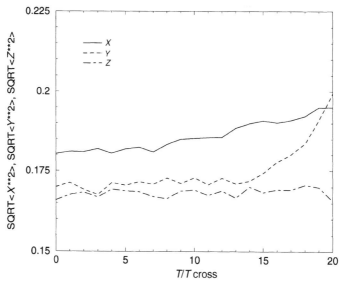

Figure 2. Evolution of the Heggie–Ramamani model with $Wo = 2.5$ of Muzzio et al. (2000a). The square roots of the mean square values of x, y and z are given as a function of the time (in units of crossing time). A total number of 6,944 bodies and a softening parameter of 0.0025 were used in the simulations.

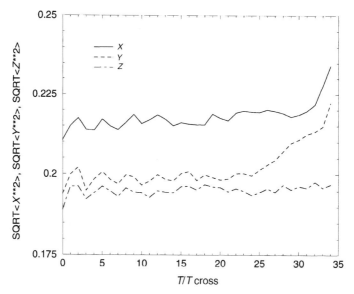

Figure 3. Same as Figure 2 for a Heggie–Ramamani model with $Wo = 0.5$. A total number of 20 000 bodies and a softening parameter of 0.0050 were used in the simulations.

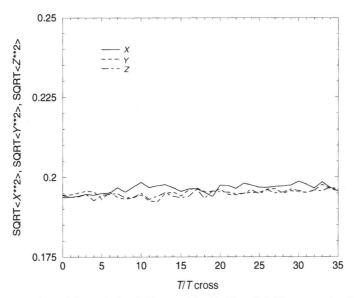

Figure 4. Same as Figure 2 for an isolated King model with $Wo = 0.5$. The x, y and z directions are purely arbitrary in this case. A total number of 20 000 bodies and a softening parameter of 0.0025 were used in the simulations.

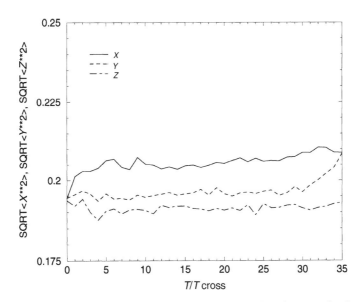

Figure 5. Same as Figure 2 for a King model with $Wo = 0.5$, placed on the same circular orbit as the Heggie–Ramamani models. A total number of 20 000 bodies and a softening parameter of 0.0050 were used in the simulations.

and between half and two-thirds of the orbits in models of globular clusters were found to be chaotic. They could also investigate the dependence of the fractions of chaotic orbits on the values of the Jacobi integral of the orbits, which they took as proof that the chaoticity arises from the interplay of the three forces that are present: the attraction from the cluster, the centripetal–centrifugal differential force and the Coriolis force. The role of those three forces was confirmed by Muzzio et al. (2000b) in their investigation of spherical satellites modeled according to the Schuster (or Plummer) law. About one-fourth of the stellar orbits in these models turned out to be chaotic. Summarizing all their results the authors concluded that, while mild triaxiality alone yielded very little chaos and a spherical satellite had a significant fraction of chaotic orbits, it was the combination of both triaxiality and tidal forces that yielded the highest fractions of chaotic orbits.

4. New Results

4.1. THE EFFECT OF THE CORIOLIS FORCE

Since the Coriolis force does not alter the values of the Jacobi integral, we repeated the analysis of the Schuster (or Plummer) models of Muzzio et al. (2000b) eliminating that force from the equations of motion. Figure 1 shows the fractions of chaotic orbits for the $b = 0.2290$ model. Clearly, the Coriolis force has an important effect on chaoticity, but it seems to be opposite to the one envisaged by Muzzio et al. (2000a): the presence of the Coriolis force marginally increases the chaoticity where that force is smallest (low $< Ej > /Uo$ values) and significantly decreases it where the Coriolis force is largest (large $< Ej > /Uo$ values). Curiously, however, for the $b = 0.0458$ model we got the opposite result. Therefore, while it is clear that the Coriolis force has an important role in the chaotization (or regularization) of the orbits of galactic satellites, that role seems to be much more complex than we had originally surmised and further studies are needed to decide which is its precise contribution.

4.2. STABILITY OF THE HEGGIE–RAMAMANI MODELS

Much of our previous work was based on the models of Heggie and Ramamani (1995), which have the great advantage of providing a known distribution function. Although these models are self-consistent, because they satisfy both the Boltzmann and Poisson equations, the solution is only a first order approximation. It is important, therefore, to check whether these are actually equilibrium models and, furthermore, whether they are stable in the long run. Heggie and Ramamani themselves did such a check, using an N-body simulation, but their results are far from convincing because their models only remained stable for a little less than two crossing times.

Two problems are immediately evident in the Heggie and Ramamani tests: they used only 1,000 bodies in their simulations, and their particle–particle interactions were purely Newtonian, with no softening included. The relaxation time of their models was then only about 15 crossing times, and even their isolated King model shows clear signs of evolution. Therefore, we decided to make our own N-body simulations, including larger numbers of bodies and using softened particle–particle forces. The NBODY2 code, kindly provided by S. Aarseth, was adopted for our experiments. We used the same constants and parameters as in our previous work, that is, gravitational constant and satellite mass equal to 1, orbit radius equal to 100 and angular velocity equal to 0.5.

Our first check was done with the $Wo = 2.5$ model of Muzzio et al. (2000a), with $N = 6,944$. We adopted softening parameters ranging between 0.0025 and 0.0050, which yielded relaxation times of the order of 100 crossing times (see, e.g. Huang et al., 1993). Figure 2 shows the evolution of the square root of the mean square values of the x, y and z coordinates (longest, intermediate and shortest axes, respectively, in the notation of Heggie and Ramamani) and it is clear that the model remains stable for at least 12 or 13 crossing times, that is, much longer than in the experiment of Heggie and Ramamani. Another $Wo = 2.5$ model, including 10 000 bodies, remained stable even longer, about 17 crossing times.

In order to test more triaxial cases, we also investigated models with $Wo = 0.5$. Simulations including 10 000 bodies remained stable for about 17 crossing times and this result encouraged us to perform another experiment, this time with 20 000 bodies. Figure 3 presents our results, in the same way as Figure 2, and it is clear that the model remains stable for at least 24 crossing times. Figure 4 provides the corresponding results for an isolated King model with $Wo = 0.5$, and Figure 5 those corresponding to that King model when it is subject to the same tidal field as the Heggie–Ramamani model of Figure 3. While the isolated model is very stable, indeed, displaying only a very mild expansion, the behavior of the King model suffering the action of the tidal forces resembles that of the Heggie–Ramamani model: it becomes unstable after 30 crossing times and the longer endurance can be attributed to the fact that, while the Heggie–Ramamani model fills in its Roche lobe from the start, the King model takes a while to reach a similar size in the x direction. The relaxation times for these models are very long, of the order of 300 crossing times.

Our experiments clearly show that increasing the number of bodies in the simulations extends the interval over which the Heggie–Ramamani models remain stable. It seems very likely that what is taking place in the experiments is completely analogous to what happens in a real globular cluster: particle–particle interactions increase the energy of some particles causing an expansion of the outermost regions and the tidal forces end up destabilizing the system. We feel that we can safely conclude that, in the continuous limit envisaged by the collisionless Boltzmann equation (i.e. for an infinite number of particles), the Heggie–Ramamani are indeed stable equilibrium models.

4.3. THE LACK OF CHAOTIC DIFUSSION

Merritt and Fridman (1996) have discussed the problems posed by chaotic orbits to the building of stationary stellar models. In their triaxial models, where orbits that fill in elongated regions of space (i.e. mainly boxes) are needed for self-consistency, chaotic orbits that tend to fill in all the region allowed to them by the energy integral are troublesome.

The high percentages of chaotic orbits found by Muzzio et al. (2000a) in the Heggie–Ramamani models seem, therefore, to pose a curious puzzle. A little reflection shows, however, that there is no problem at all, because the distribution function of those models depends only on the Jacobi integral and the stellar density results from integrating that function over all the velocities. Therefore, the isodensity surfaces coincide with the (effective) equipotential surfaces, so that chaotic orbits that fill in those equipotential (or zero-velocity) surfaces are perfect building blocks for the Heggie–Ramamani models.

As a check, we used a Heggie–Ramamani model with $Wo = 0.5$ (i.e. the most triaxial one of those investigated by us). We generated 10 000 initial conditions for that model and we followed the orbits in the (fixed) corresponding potential. In other words, now we are not using interacting particles, but letting the particles evolve in the Heggie–Ramamani potential with $Wo = 0.5$. Particle–particle interactions and self-consistency are thus suppressed, and the orbits can fill in all the space allowed to them by the isolating integrals that they obey. Orbit classification with the Carpintero and Aguilar (1998) code, as well as the Liapunov exponents computed for a small sample, showed that almost 40% of the orbits were chaotic.

Figure 6 presents our results; note that now the interval covered is 200 crossing times. The range of the Liapunov times for the chaotic orbits was approximately 15–75 crossing times, with a rough average of about 35 crossing times, so that the model remains perfectly stationary for several Liapunov times. To check whether the scatter was purely random, we also run 20 000 initial conditions for an interval of 20 crossing times. The dispersions computed for the 10 000 and 20 000 runs turned out to be proportional to the inverses of the square roots of the numbers involved, so that only noise seems to be present here.

5. Conclusion

Our previous work had shown that chaotic motions play an important role in galactic satellites: not only there are large fractions of chaotic orbits in the satellites, but the Liapunov times involved are short. The confirmation that the Coriolis force has a significant role on the (regular or chaotic) behavior of the stellar orbits inside the satellite emphasizes the need to take into account the tidal effects when building satellite models: clearly, those effects are not limited to imposing a tidal radius, but are present well inside that tidal radius too.

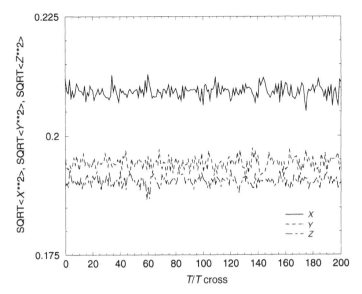

Figure 6. Same as Figure 2 for a Heggie–Ramamani model with $Wo = 0.5$. In this case the simulation is not self-consistent. Instead, 10 000 orbits whose initial conditions were randomly selected from the distribution function of that model were followed on the corresponding fixed potential of that same model.

We have also presented here a much needed confirmation that the Heggie–Ramamani models, on which much of our previous work was based, are truly stable. Not even the presence of large numbers of chaotic orbits prevents them from being stationary.

Acknowledgements

The technical assistance of R.E. Martínez is gratefully acknowledged. This investigation was supported with grants from the Universidad Nacional de La Plata and from the Consejo Nacional de Investigaciones Científicas y Técnicas de la República Argentina.

References

Binney, J. and Tremaine, S.: 1987, *Galactic Dynamics*, Princeton University Press.
Carpintero, D. D. and Aguilar, L. A.: 1998, 'Orbit classification in arbitrary 2D and 3D potentials', *Monthly Notices Royal Astron. Soc.* **298**(1), 1–21.
Carpintero, D. D., Muzzio, J. C. and Wachlin, F. C. : 1999, 'Regular and chaotic motion in globular clusters', *Celest. Mech. Dyn. Astr.* **73**(1), 159–168.
Heggie, D. C. and Ramamani, N. : 1995, 'Approximate self-consistent models for tidally truncated star clusters', *Monthly Notices Royal Astron. Soc.* **272**(2), 317–322.

Huang, S., Dubinski, J. and Carlberg, R. G.: 1993, 'Orbital deflections in N-body systems', *Astrophys. J.*, **404** (1), 73–80.

Merritt, D.: 1999, 'Elliptical galaxy dynamics', *Publ. Astron. Soc. Pacific* **111**(756), 129–168.

Merritt, D. and Fridman, T.: 1996, 'Triaxial galaxies with Cusps', *Astrophys. J.*, **460**(1), pp. 136–162.

Muzzio, J. C., Carpintero, D. D. and Wachlin, F. C.: 2000a, 'Regular and chaotic motion in galactic satellites', In: V. G. Gurzadyan and R. Ruffini (eds), *The Chaotic Universe, Proceedings of the Second ICRA Network Workshop*, Rome, Pescara, Italy, February 1999, World Scientific, Singapore, pp. 107–114.

Muzzio, J. C., Wachlin, F. C. and Carpintero, D. D.: 2000b, 'Regular and chaotic motion in a restricted three-body problem of astrophysical interest', In: M. Valtonen and C. Flynn (eds), *Small Galaxy Groups, IAU Colloquium 174*, Turku, Finland, June 1999, Astronomical Society of the Pacific, United States of America, pp. 281–285.

ATTITUDE DYNAMICS OF ORBITING GYROSTATS

CHRISTOPHER D. HALL

Aerospace and Ocean Engineering, Virginia Polytechnic Institute and State University, Blacksburg, Virginia

Abstract. Equilibrium attitudes of a rigid satellite with N rotors in a central gravitational field are investigated. The equations of motion are written as a noncanonical Hamiltonian system, where the Hamiltonian includes the potential, a volume integral over the body of the gyrostat. In practice, the Hamiltonian is approximated to partially decouple the position and attitude equations. The equilibria of this system of equations represent the steady motions of the body as seen in the body frame, and correspond to stationary points of the Hamiltonian constrained by the Casimir functions. This defines an algorithm for computing equilibria. In contrast to other approaches, this algorithm provides stability information directly, since the calculations required to solve the constrained minimization problem are also involved in computing the positive definiteness of the Hamiltonian as a Lyapunov function.

Key words: attitude dynamics, gyrostat, Hamiltonian

1. Introduction

In this paper we study a subset of the equilibrium attitudes of a rigid satellite with N rotors in a central gravitational field. The work presented herein is an extension of similar results for a rigid body (Beck and Hall, 1998), with the added complexity of the flywheels or rotors representing reaction or momentum wheels. The model of a rigid body with axisymmetric wheels is termed a gyrostat. As a result of numerous studies [*e.g.*, (Volterra, 1899; Krishnaprasad and Berenstein, 1984)], the global torque-free motion of a gyrostat is understood in cases with freely spinning rotors or with rotors constrained to spin at a constant speed relative to the platform. The basic results are well-known (Hughes, 1986).

There are also many reports relevant to orbiting gyrostats, where the gravity gradient torque is included (Kane and Mingori, 1965; Roberson and Sarychev, 1985; Anchev, 1973). These papers characterize the relative equilibrium motions of gyrostats in circular orbits; as a result, the steady motions of orbiting gyrostats, the subject of this paper, are fairly well understood. The main results are typically given as a set of special cases (Roberson and Sarychev, 1985; Hughes, 1986). Note that the gravitational moment used in all these studies is obtained by truncating the gravitational potential in a way that has been shown to be inconsistent (Wang et al., 1991). The significance of the inconsistency has been shown to be negligible for "ordinary" asymmetric, rigid, gravity-gradient spacecraft (Beck and Hall, 1998). The gyrostat case has been investigated and some stability criteria have been

Pretka-Ziomek et al./Dynamics of Natural and Artificial Celestial Bodies, 177–186, 2001.
© 2001 *Kluwer Academic Publishers.*

obtained for a variety of cases (Wang et al., 1995). However, their analysis was based on the constant-speed rotor case, and is not directly applicable to the problem of performing rotational maneuvers.

Most papers parameterize a rotor's motion by its constant angular velocity relative to the body. However, the rotor's absolute angular velocity about its spin axis is more important in developing control torques to perform rotational maneuvers. The problem of performing rotational maneuvers using flywheels has been investigated by numerous authors, and a brief literature review has been presented by the present author (Hall, 1995). Only one article (Anchev, 1973) has used information about equilibrium motions to develop reorientation control laws.

Here we develop the general equations of motion for an N-rotor gyrostat in a central gravitational field. We specialize the equations of motion to the problem of a Keplerian circular orbit, and develop a new noncanonical Hamiltonian formulation similar to that developed for a rigid body (Beck and Hall, 1998). This formulation is equivalent to the equations that have been used by others to study equilibria of orbiting gyrostats, but has the advantage that standard methods can be used to obtain equilibria and to characterize their stability. We then develop the stability criteria for the simplest case of the cylindrical equilibria.

2. System and Equations of Motion

The system model is a gyrostat ($\mathcal{G}$): a rigid body ($\mathcal{B}$) with N axisymmetric flywheels ($\mathcal{R}_j$), with spin axes fixed in the body frame, $\mathcal{F}_b$ (see Fig. 1). The system inertia tensor is $\mathbf{I}$, expressed in $\mathcal{F}_b$. The wheels have axial moments of inertia, I_{sj}, $j = 1, ..., N$, which are collected into a diagonal matrix $\mathbf{I}_s = \text{diag}[I_{s1}...I_{sN}]$, and their spin axes are defined by the vectors $\mathbf{a}_j$, $j = 1, ..., N$, expressed in $\mathcal{F}_b$, and collected into a $3 \times N$ matrix as $\mathbf{A} = [\mathbf{a}_1...\mathbf{a}_N]$. The inertia-like matrix $\mathbf{J} = \mathbf{I} - \mathbf{A}\mathbf{I}_s\mathbf{A}^\mathsf{T}$ is symmetric and positive definite.

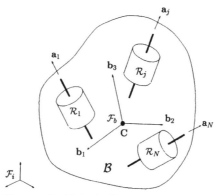

Figure 1. Gyrostat $\mathcal{G}$ with body $\mathcal{B}$ and N momentum wheels $\mathcal{R}_j$.

The gyrostat rotational equations of motion may be written as

$$\dot{\mathbf{h}} = \mathbf{h}^\times \mathbf{J}^{-1}(\mathbf{h} - \mathbf{A}\mathbf{h}_a) + \mathbf{g}_e \tag{1}$$
$$\dot{\mathbf{h}}_a = \mathbf{g}_a \tag{2}$$

where $\mathbf{h}$ is the 3×1 angular momentum vector, $\mathbf{h}_a$ is the $N \times 1$ vector of the absolute axial momenta of the wheels, "$\times$" denotes the skew-symmetric matrix form of a vector, $\mathbf{g}_e$ is the 3×1 vector of external torques, and $\mathbf{g}_a$ is the $N \times 1$ matrix containing the axial torques applied by the platform on the rotors. The term $\mathbf{J}^{-1}(\mathbf{h} - \mathbf{A}\mathbf{h}_a)$ is recognized as the angular velocity ω of $\mathcal{F}_b$ with respect to $\mathcal{F}_i$. To study equilibrium motions, we set $\mathbf{g}_a = \mathbf{0}$, and to study rotational maneuvers, we choose a suitable control law for the rotor torques. In this paper, we deal exclusively with the $\mathbf{g}_a = \mathbf{0}$ case.

The dynamics of Eqs. (1–2) must be augmented if the external or internal torques depend on the position or orientation of the gyrostat. For example, the gravity gradient torque is (Hughes, 1986):

$$\mathbf{g}_e = \mathbf{r}^\times \nabla_{\mathbf{r}} V(\mathbf{r}), \quad V(\mathbf{r}) = -\int_\mathcal{G} \frac{\mu}{\|\mathbf{r} + \boldsymbol{\rho}\|} dm \tag{3}$$

and the force acting on the body is $-\nabla_{\mathbf{r}} V(\mathbf{r})$. Here μ is the gravitational constant for the central body, $\mathbf{r}$ is the position vector from the center of attraction to the gyrostat mass center, and $\boldsymbol{\rho}$ is the position vector from the mass center to a mass element dm. This integral depends on the orientation of the gyrostat. A standard approximation, assuming $\|\boldsymbol{\rho}\| \ll \|\mathbf{r}\|$ is

$$\mathbf{g}_e = 3 \frac{\mu}{\|\mathbf{r}\|^3} \mathbf{o}_3^\times \mathbf{I}\mathbf{o}_3 \tag{4}$$

where $\mathbf{o}_3$ is the nadir vector; i.e. $\mathbf{o}_3 = -\mathbf{r}/\|\mathbf{r}\|$ (see Fig. 2). Here $\mathbf{o}_3$ denotes the third column of the rotation matrix $\mathbf{R}^{bo}$ that takes vectors from the orbital frame, $\mathcal{F}_o$, to $\mathcal{F}_b$. The orbital frame's remaining unit vectors are arranged so that $\mathbf{o}_2$ is in the negative orbit normal direction, and $\mathbf{o}_1 = \mathbf{o}_2^\times \mathbf{o}_3$. For circular orbits, $\mathbf{o}_1$ is in the velocity direction. Furthermore, for circular orbits, the term $\mu/\|\mathbf{r}\|^3$ is constant, and is denoted by ω_c^2. In the circular case, we append $\dot{\mathbf{o}}_3 = \mathbf{o}_3^\times \mathbf{J}^{-1}(\mathbf{h} - \mathbf{A}\mathbf{h}_a)$ so that the current state of $\mathbf{o}_3$ is available for computing the gravity gradient torque; otherwise,

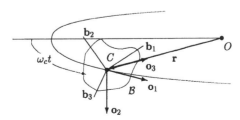

Figure 2. Configuration of Keplerian Orbit and Orbital Frame.

180   CHRISTOPHER D. HALL

the translational equations of motion are required to describe the variable radius vector $\mathbf{r}$.

The full equations can be recognized as a noncanonical Hamiltonian system (Maddocks, 1991; Beck and Hall, 1998):

$$\dot{\mathbf{z}} = \mathcal{J}(\mathbf{z})\nabla H(\mathbf{z}) \tag{5}$$

where $\mathbf{z}$ is the vector of states, $\mathcal{J}(\mathbf{z})$ is the skew-symmetric Poisson tensor or structure matrix, $H(\mathbf{z})$ is the Hamiltonian, and ∇ represents the gradient of H with respect to $\mathbf{z}$.

In systems of the form of Eq. (5), there are special first integrals, known as Casimir functions, whose gradients span the nullspace of the structure matrix, $\mathcal{J}(\mathbf{z})$. For the general equations of motion for an orbiting gyrostat, the nullspace of $\mathcal{J}(\mathbf{z})$ is one-dimensional and is spanned by the vector ∇C, where C is the total angular momentum about the center of the attracting body. This is the only Casimir-type first integral for this system. A second integral of the motion is the Hamiltonian, which because of the skew symmetry of the structure matrix, is easily seen to be constant:

$$\dot{H} = \nabla H^{\mathsf{T}} \dot{\mathbf{z}} = \nabla H^{\mathsf{T}} \mathcal{J}(\mathbf{z})\nabla H = 0$$

Thus the system of equations admits two first integrals. If the internal wheel torques, $\mathbf{g}_a$, are not all zero, then Eq. (2) must be used with the general equations. In this case, the Casimir function is still a first integral, since it is independent of the particular form of the Hamiltonian; however, the Hamiltonian is not conserved, but satisfies the following differential equation:

$$\dot{H} = [\partial H/\partial \mathbf{h}_a]\,\dot{\mathbf{h}}_a = -\mathbf{h}^{\mathsf{T}} \mathbf{J}^{-1} \mathbf{A} \mathbf{g}_a \tag{6}$$

Elsewhere, we have used this relationship with the method of averaging to reduce the "spinup" problem of torque-free gyrostats from five dimensions to two (Hall, 1995).

The general form of the equations of motion for a gyrostat in a central gravitational field is usually approximated in some way. Three types of approximations are usually necessary to obtain useful results (Beck and Hall, 1998): a $3 \times 3 \times \infty$ "matrix" of approximations is possible, where the three dimensions are approximation of the potential, restriction of the mass center motion, and material symmetry of the body. Here we study the second-order potential approximation for an arbitrary body moving in a circular Keplerian orbit.

3. Second-order Keplerian Approximation

We assume the gyrostat is in a circular Keplerian orbit, and approximate the potential with a second-order expansion, obtaining results that are equivalent to work previously reported. However, our approach is distinct from the classical approach in that we develop a new noncanonical formulation similar to that developed for the rigid body problem (Beck and Hall, 1998). We define the relative angular velocity and angular momentum of the gyrostat with respect to the rotating orbital reference frame, $\mathcal{F}_o$:

$$\omega_r = \omega + \omega_c \mathbf{o}_2 = \mathbf{J}^{-1}(\mathbf{h}_r - \mathbf{A}\mathbf{h}_a) \tag{7}$$

$$\mathbf{h}_r = \mathbf{h} + \omega_c \mathbf{J}\mathbf{o}_2 = \mathbf{J}\omega_r + \mathbf{A}\mathbf{h}_a \tag{8}$$

We also use a second-order approximation of the potential:

$$V_2(\mathbf{o}_3) = \frac{3}{2}\omega_c^2 \mathbf{o}_3{}^\mathsf{T} \mathbf{I} \mathbf{o}_3 \tag{9}$$

Using these definitions, and the mass, length, and time scales

$$m = \overline{m} = \int_{\mathcal{G}} d\overline{m} \qquad l = \left(\frac{\operatorname{tr} \overline{\mathbf{J}}}{\overline{m}}\right)^{\frac{1}{2}} \qquad t = \overline{\omega}_c{}^{-1}. \tag{10}$$

the dimensionless equations of motion become

$$\begin{bmatrix} \dot{\mathbf{h}}_r \\ \dot{\mathbf{o}}_2 \\ \dot{\mathbf{o}}_3 \end{bmatrix} = \begin{bmatrix} [\mathbf{h}_r + \{(1 - 2\mathbf{J})\mathbf{o}_2\}]^\times & \mathbf{o}_2^\times & \mathbf{o}_3^\times \\ \mathbf{o}_2^\times & 0 & 0 \\ \mathbf{o}_3^\times & 0 & 0 \end{bmatrix} \begin{bmatrix} \mathbf{J}^{-1}(\mathbf{h}_r - \mathbf{A}\mathbf{h}_a) \\ -\mathbf{J}\mathbf{o}_2 + \mathbf{A}\mathbf{h}_a \\ 3\mathbf{I}\mathbf{o}_3 \end{bmatrix} \tag{11}$$

This system is in the form of Eq. (5), with $\mathbf{z} = (\mathbf{h}_r, \mathbf{o}_2, \mathbf{o}_3)^\mathsf{T}$, and

$$H = \frac{1}{2}\mathbf{h}_r{}^\mathsf{T}\mathbf{J}^{-1}\mathbf{h}_r - \mathbf{h}_r{}^\mathsf{T}\mathbf{J}^{-1}\mathbf{A}\mathbf{h}_a - \frac{1}{2}\mathbf{o}_2{}^\mathsf{T}\mathbf{J}\mathbf{o}_2 + \mathbf{o}_2{}^\mathsf{T}\mathbf{A}\mathbf{h}_a + V(\mathbf{o}_3) \tag{12}$$

This system admits three Casimir functions, as the nullspace of the structure matrix is spanned by the three vectors:

$$\mathcal{N}\left[\mathcal{J}(\mathbf{z})\right] = \operatorname{span}\left\{ \begin{pmatrix} \mathbf{0} \\ \mathbf{o}_2 \\ \mathbf{0} \end{pmatrix}, \begin{pmatrix} \mathbf{0} \\ \mathbf{0} \\ \mathbf{o}_3 \end{pmatrix}, \begin{pmatrix} \mathbf{0} \\ \mathbf{o}_3 \\ \mathbf{o}_2 \end{pmatrix} \right\} \tag{13}$$

From the spanning vectors, we identify three independent Casimir functions:

$$C_1(\mathbf{z}) = \mathbf{o}_2{}^\mathsf{T}\mathbf{o}_2 \qquad C_2(\mathbf{z}) = \mathbf{o}_3{}^\mathsf{T}\mathbf{o}_3 \qquad C_3(\mathbf{z}) = \mathbf{o}_2{}^\mathsf{T}\mathbf{o}_3 \tag{14}$$

In this problem, all three Casimir functions are trivial since $\mathbf{o}_2$ and $\mathbf{o}_3$ are columns of the rotation matrix $\mathbf{R}^{bo}$ so that

$$\mathbf{o}_2{}^\mathsf{T}\mathbf{o}_2 \equiv 1 \qquad \mathbf{o}_3{}^\mathsf{T}\mathbf{o}_3 \equiv 1 \qquad \mathbf{o}_2{}^\mathsf{T}\mathbf{o}_3 \equiv 0 \tag{15}$$

The Hamiltonian is also constant (if $\mathbf{g}_a = \mathbf{0}$). We thus have a ninth-order system with four known first integrals.

182   CHRISTOPHER D. HALL

4. Equilibria

In canonical Hamiltonian systems, equilibria are found as the critical points of the Hamiltonian; *i.e.*, by setting $\nabla H = \mathbf{0}$, and computing $\mathbf{q}_e$ and $\mathbf{p}_e$. In the noncanonical case, the structure matrix can be singular, so that equilibria can also satisfy $\nabla H \in \mathcal{N}[\mathcal{J}(\mathbf{z})]$. Since the gradients of the Casimir functions lie in $\mathcal{N}[\mathcal{J}(\mathbf{z})]$, equilibria may be expressed as the critical points of a "variational Lagrangian:"

$$F(\mathbf{z}, \boldsymbol{\mu}) = H(\mathbf{z}) - \mu_1(C_1(\mathbf{z}) - 1) - \mu_2(C_2(\mathbf{z}) - 1) - \mu_3 C_3(\mathbf{z}) \tag{16}$$

subject to the constraints that the Casimir functions are constant. For the system represented by Eqs. (11), setting $\nabla F = \mathbf{0}$ leads to a nonlinear algebraic system with 12 unknowns: $(\mathbf{z}, \boldsymbol{\mu}) = (\mathbf{h}_r, \mathbf{0}_2, \mathbf{0}_3, \mu_1, \mu_2, \mu_3)$.

A typical problem involves fixing the wheel momenta $\mathbf{h}_a$, and computing the associated equilibria. Using Newton's method (Seydel, 1988) requires the Hessian $\nabla^2 F(\mathbf{z}_e, \boldsymbol{\mu}_e)$. The full 12×12 Hessian is required for numerical computation of equilibria, but as we will see below, the Hessian also plays a role in computing the stability of equilibria, and only the upper left 9×9 block associated with the states is required for the stability calculations.

5. Linearization, Linear Stability, and Nonlinear Stability

There are two approaches to computing stability in Hamiltonian systems (Beck and Hall, 1998): linear or spectral stability, and nonlinear stability. The former is based on linearizing the equations of motion about equilibrium, and the latter is based on establishing a suitable Lyapunov function.

The linearization of Eq. (5) about an equilibrium, $\mathbf{z}_e$ leads to

$$\mathbf{A}_e(\mathbf{z}_e) = \mathcal{J}(\mathbf{z}_e) \nabla^2 F(\mathbf{z}_e) \tag{17}$$

Thus, one checks the linear stability of an equilibrium by computing the eigenvalues of $\mathbf{A}_e(\mathbf{z}_e)$. Because the eigenvalues of a Hamiltonian system occur in pairs that are symmetric about both the real and imaginary axes, this approach only provides conditions for instability.

Since the Hamiltonian and Casimir functions are constants, the variational Lagrangian, $F(\mathbf{z})$, is a candidate Lyapunov function. Thus, the eigenvalues of $\nabla^2 F(\mathbf{z}_e)$ can be used to determine nonlinear stability.

Considering only the upper left 9×9 block of $\nabla^2 F$, one finds that the matrix is block diagonal. The upper left 3×3 block, $\mathbf{J}^{-1}$, is positive definite. If the eigenvalues of the remaining 6×6 block are positive, then the equilibrium is stable. The Lagrange multipliers appear in these terms, and are important in determining stability.

Possibly $\nabla^2 F$ is indefinite and hence F is not useful as a Lyapunov function. However, the equilibrium may be viewed as a constrained extremum of the

Hamiltonian subject to the constant values of the C_i (Beck and Hall, 1998). We introduce the orthogonal projection matrix, $\mathbf{P}(\mathbf{z})$ onto the range of $\mathbf{A}(\mathbf{z})$ as follows. Define

$$\mathbf{K}(\mathbf{z}) = [\nabla C_1(\mathbf{z}) \ \nabla C_2(\mathbf{z}) \ \nabla C_3(\mathbf{z})] \tag{18}$$

and let $\mathbf{Q}(\mathbf{z})$ be the projection onto $\mathcal{N}\left[\mathbf{A}^{\mathsf{T}}(\mathbf{z})\right]$:

$$\mathbf{Q}(\mathbf{z}) = \mathbf{K}(\mathbf{z}) \left(\mathbf{K}^{\mathsf{T}}(\mathbf{z})\mathbf{K}(\mathbf{z})\right)^{-1} \mathbf{K}^{\mathsf{T}}(\mathbf{z}) \tag{19}$$

Then the desired projection operator is

$$\mathbf{P}(\mathbf{z}) = 1 - \mathbf{Q}(\mathbf{z}) \tag{20}$$

The projected Hessian is then given by $\mathbf{P}(\mathbf{z}_e)\nabla^2 F(\mathbf{z}_e)\mathbf{P}(\mathbf{z}_e)$. This matrix has three zero eigenvalues associated with the nullspace of $\mathbf{A}(\mathbf{z}_e)$, and with the three Casimir functions. If its remaining eigenvalues are all positive, then the equilibrium is a constrained minimum and the relative equilibrium is nonlinearly stable.

6. An Example: Cylindrical Equilibria

Most presentations consider the equilibria of orbiting gyrostats as a variety of cases, including cylindrical, conical, hyperbolic, and offset hyperbolic. We examine the existence of and stability of the cylindrical case, with a single wheel aligned with the $\mathbf{b}_2$ axis. Thus $\mathbf{I}$ and $\mathbf{J}$ are diagonal, and the diagonal elements of $\mathbf{J}$ are $\{I_1, I_2 - I_s, I_3\}$. Using the principal reference frame, the standard gravity gradient equilibria have $\mathcal{F}_b$ aligned with $\mathcal{F}_o$, whence $\omega_r = \mathbf{0}$. We assume without loss of generality that $\mathbf{o}_2 = [0, 1, 0]^{\mathsf{T}}$ and $\mathbf{o}_3 = [0, 0, 1]^{\mathsf{T}}$. The equilibrium conditions lead to restrictions on the first and third elements of the wheel momenta in the body frame:

$$[\mathbf{Ah}_a]_1 = 0 \qquad [\mathbf{Ah}_a]_3 = 0 \tag{21}$$

The second element, $[\mathbf{Ah}_a]_2 = h_{r2}$, is arbitrary, but affects the stability of the steady attitude. The Lagrange multipliers are

$$\mu_1 = -J_2 + h_{r2} \qquad \mu_2 = 3I_3 \qquad \mu_3 = 0 \tag{22}$$

Thus the 9×9 block of $\nabla^2 F$ simplifies to

$$\nabla^2 F = \begin{bmatrix} \mathbf{J}^{-1} & \mathbf{0} & \mathbf{0} \\ \mathbf{0} & -\mathbf{J} - \mu_1 \mathbf{1} & \mathbf{0} \\ \mathbf{0} & \mathbf{0} & 3\mathbf{I} - \mu_2 \mathbf{1} \end{bmatrix} \tag{23}$$

which is diagonal, but is not positive definite, since the last diagonal element is zero. Using the values of the Lagrange multipliers from Eq. (22), the Hessian has

184 CHRISTOPHER D. HALL

a zero eigenvalue, 3 positive eigenvalues $(1/I_1, 1/(I_2 - I_s), 1/I_3)$, two eigenvalues that place constraints on the moments of inertia $(3(I_1 - I_3), 3(I_2 - I_3))$, and 3 eigenvalues that involve the rotor momentum $(J_2 - I_1 - h_{r2}, -h_{r2}, J_2 - I_3 - h_{r2})$. Introducing the Smelt parameters $k_1 = (I_2 - I_3)/I_1$ and $k_3 = (I_1 - I_2)/I_3$, these conditions can be written as $k_1 > 0, k_1 > k_3$, and

$$\frac{(1 - k_1)k_3}{3 - k_3 - k_1(1 + k_3)} - h > 0 \quad \text{and} \quad \frac{(1 - k_3)k_1}{3 - k_3 - k_1(1 + k_3)} - h > 0 \qquad (24)$$

where $h = I_s + h_{r2}$. When $h = 0$, the third of these four conditions is equivalent to $k_3 > 0$, and the fourth is equivalent to $k_1 > k_3$. When $h \neq 0$, the third condition becomes $k_3 > h(3 - k_1)/(1 - k_1 + h(1 + k_1))$, and the fourth becomes $k_3 > (h(3 - k_1) - k_1)/(h(1 - k_1) - k_1)$. Notice that the well-known DeBra-Delp region is not predicted by these stability conditions. The criteria provided here are based on using F as a Lyapunov function, and are therefore sufficient conditions only. (However, since $\nabla^2 F$ is only positive semidefinite, strictly speaking, we cannot even make these weak conclusions.)

The projected Hessian, $\mathbf{P}\nabla^2 F\mathbf{P}$ yields sharper stability criteria. For the single-rotor, cylindrical equilibria case, the important eigenvalues of the projected Hessian are:

$$\{I_2 - I_1 - h, 2(I_2 - I_3) - h/2, 3(I_1 - I_3)\} \qquad (25)$$

There are three zero eigenvalues associated with the Casimir functions, and 3 eigenvalues that are always positive. If three stated eigenvalues are positive, then the equilibrium is stable (this is a sufficient condition). This leads to the following sufficient conditions: $k_1 > k_3$, $(k_3(1 - k_1))/(3 - k_3 - k_1(1 + k_3)) - h > 0$, and $(4k_1(1 - k_3))/(3 - k_3 - k_1(1 + k_3)) - h > 0$. As stated above, the eigenvalues of the projected Hessian give sharper stability conditions than those of the Hessian. The corresponding stability regions are shown in Fig. 3. The principal benefit is that the $k_1 > 0$ condition is replaced with a criterion that provides an additional region of nonlinear stability. Furthermore, the conditions derived from the projected Hessian permit stability even in the case of $h_{r2} > 0$, which immediately leads to a negative eigenvalue of the unprojected Hessian. The stability region is shown in Fig. 3, with the unshaded region indicating the region for which the projected Hessian guarantees nonlinear stability. The stable region in the lower left quadrant is the additional region of stability provided by the sharper conditions obtained with the projected Hessian.

7. Conclusions

A noncanonical Hamiltonian formulation of the equations of motion for orbiting gyrostats leads to straightforward algorithms for computing relative equilibria and

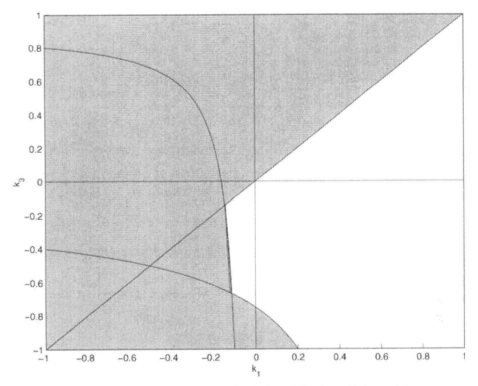

Figure 3. Smelt Parameter Plane Using Projected Hessian, with $h = -0.2$.

determining their stability. Results that have been obtained previously using a variety of manipulations of the equations of motion and their conserved quantities are obtained in a more straightforward fashion when the equations are put into a noncanonical form. Although we have treated only the simplest case of gyrostat equilibria in this paper, this approach should provide the means to unify the various cases that are usually treated separately. Our goal is to apply this approach to unify the treatment of both rigid body and gyrostat relative equilibria.

Acknowledgements

This work was supported by Arje Nachman of the Air Force Office of Scientific Research and Alison Flatau of the National Science Foundation.

References

Anchev, A. A.: 1973, 'Equilibrium Attitude Transitions of a Three-Rotor Gyrostat in a Circular Orbit', *AIAA Journal* **11**(4), pp. 467–472.

186 CHRISTOPHER D. HALL

Beck, J. A. and Hall, C. D.: 1998, 'Relative Equilibria of a Rigid Satellite in a Circular Keplerian Orbit', *Journal of the Astronautical Sciences* **46**(3), pp. 215–247.

Hall, C. D.: 1995, 'Spinup Dynamics of Gyrostats', *Journal of Guidance, Control and Dynamics* **18**(5), pp. 1177–1183.

Hughes, P. C.: 1986, *Spacecraft Attitude Dynamics*, New York: John Wiley & Sons.

Kane, T. R. and Mingori, D. L.: 1965, 'Effect of a Rotor on the Attitude Stability of a Satellite in a Circular Orbit', *AIAA Journal* **3**(5), pp. 936–940.

Krishnaprasad, P. S. and Berenstein, C. A.: 1984, 'On the Equilibria of Rigid Spacecraft with Rotors', *Systems and Control Letters* **4**, pp. 157–163.

Maddocks, J. H.: 1991, 'On the Stability of Relative Equilibria', *IMA Journal of Applied Mathematics* **46**(1–2), pp. 71–99.

Roberson, R. E. and Sarychev, V. A.: 1985, 'Equilibria and Stability of Satellite Gyrostats: A Comprehensive Review', Berlin, Springer-Verlag. Symposium Udine/Italy, September 16–20, 1985, pp. 227–236.

Seydel, R.: 1988, *From Equilibrium to Chaos: Practical Bifurcation and Stability Analysis*, New York: Elsevier.

Volterra, V.: Dec 1898–Jun 1899, 'Sur la Théorie des Variations des Latitudes', *Acta Mathematica* **22**(3–4), pp. 201–357.

Wang, L.-S., Krishnaprasad, P. S. and Maddocks, J. H. : 1991, 'Hamiltonian Dynamics of a Rigid Body in a Central Gravitational Field', *Celestial Mechanics and Dynamical Astronomy* **50**, pp. 349–386.

Wang, L.-S., Lian, K.-Y. and Chen, P.-T.: 1995, 'Steady Motions of Gyrostat Satellites and Their Stability', *IEEE Transactions on Automatic Control* **40**(10), pp. 1732–1743.

NON-INTEGRABILITY OF A CERTAIN PROBLEM OF ROTATIONAL MOTION OF A RIGID SATELLITE

ANDRZEJ J. MACIEJEWSKI

Toruń Centre for Astronomy, N.Copernicus University, 87-100 Toruń, Gagarina 11, Poland

Abstract. Rotational motion of a rigid satellite moving in a circular orbit around an oblate planet is considered. For a case of a symmetric satellite, we prove non-integrability of equations of motion applying Morales-Ramis extension of Ziglin theory.

Key words: Rotational Motion, Non-integrability, Ziglin Theory, Morales-Ramis Theory

To Félix Mondéjar—in memoriam

1. Introduction

A wide class of problems in celestial mechanics is connected with the rotational motion of planets, moons, asteroids and comets which can be considered as rigid objects. There exist several model problems for the rotational motion of a body. They give beautiful examples of dynamical systems.

In a study of a dynamical system the fundamental question is whether the system is integrable or not. When the answer is affirmative then we are able to describe the behavior of the system completely, if it is not, we are forced to use many advanced mathematical tools to partially describe the motion. However, having a dynamical system we do not know in advance if it is integrable or not. In fact to find an answer to this question is a hard task.

It is really astonishing that for the mentioned model problems for rotational motion there are no strict statements about their non-integrability. An exception are models of planar rotational motion in different configurations where the Melnikov method of separatrices splitting can be used for an effective proof of non-integrability, see [3] for overview.

In our first attempt to prove the non-integrability of a model with more that one and half degrees of freedom—the problem of the rotational motion of a satellite in a circular orbit—although we applied strong theories, they lead to difficulties which were solved numerically, see e.g. [6, 7, 5]. The final success [8] in a fully analytic proof of non-integrability was possible thanks to the application of Morales-Ramis extension [9] of Ziglin theory [11, 12].

In this paper we use this approach for proving the non-integrability of rotational motion of a rigid symmetric satellite moving in a circular orbit around an oblate planet.

Pretka-Ziomek et al./Dynamics of Natural and Artificial Celestial Bodies, 187–192, 2001.
© 2001 *Kluwer Academic Publishers.*

2. Equations of Motion and Particular Solutions

Let us consider a rigid body whose mass center moves in a circular equatorial orbit around an axially symmetric planet. Equations of the rotational motion of the body have the following form

$$
\left.
\begin{aligned}
\frac{d}{dt}\mathbf{M} &= \mathbf{M} \times \mathbf{\Omega} + 3\omega_K^2(1 + 5\epsilon)\mathbf{\Gamma} \times \mathbf{I}\mathbf{\Gamma} - 6\omega_K^2\epsilon\mathbf{N} \times \mathbf{I}\mathbf{N}, \\
\frac{d}{dt}\mathbf{\Gamma} &= \mathbf{\Gamma} \times (\mathbf{\Omega} - \omega_O\mathbf{N}), \qquad \frac{d}{dt}\mathbf{N} = \mathbf{N} \times \mathbf{\Omega},
\end{aligned}
\right\}
\tag{1}
$$

where $\mathbf{\Omega}$, $\mathbf{I} = \mathrm{diag}(A, B, C)$, $\mathbf{M} = \mathbf{I}\mathbf{\Omega}$ are the angular velocity, the inertia tensor and the angular momentum of the body, respectively; $\mathbf{\Gamma}$ and $\mathbf{N}$ are unit vector in direction of the radius vector of the center of mass of the body and the normal to the orbit, respectively. Components of all vectors are taken with respect to the principal axes frame of the body. In (1) ω_K and ω_O denote the Keplerian and the orbital angular velocity of the center of mass of the body and parameter ϵ depends on the flattening of the planet. For derivation and details see [4].

Equations (1) have the following Jacobi type first integral

$$
H = H_1 = \frac{1}{2}\langle\mathbf{M}, \mathbf{I}^{-1}\mathbf{M}\rangle - \omega_O\langle\mathbf{M}, \mathbf{N}\rangle + V(\mathbf{\Gamma}, \mathbf{N}),
\tag{2}
$$

where

$$
V(\mathbf{\Gamma}, \mathbf{N}) = \frac{3}{2}\omega_K^2(1 + 5\epsilon)\langle\mathbf{\Gamma}, \mathbf{I}\mathbf{\Gamma}\rangle - 3\omega_K^2\epsilon\langle\mathbf{N}, \mathbf{I}\mathbf{N}\rangle,
$$

and three geometric first integrals

$$
H_2 = \langle\mathbf{\Gamma}, \mathbf{\Gamma}\rangle, \qquad H_3 = \langle\mathbf{N}, \mathbf{N}\rangle, \qquad H_4 = \langle\mathbf{\Gamma}, \mathbf{N}\rangle.
\tag{3}
$$

As it was shown in [4] equations (1) are Hamiltonian ones and they generalize the Beletskii system [1]. For their integrability we need two additional first integrals.

In order to apply Ziglin or Morales-Ramis theory it is necessary to know a particular solution of the system. We show how to find it for system (1). Manifold

$$
\mathcal{N} = \{(\mathbf{M}, \mathbf{\Gamma}, \mathbf{N}) | M_1 = M_2 = N_1 = N_2 = \Gamma_3 = 0, N_3 = 1\},
\tag{4}
$$

is invariant with respect to the flow generated by (1). System (1) restricted to $\mathcal{N}$ can be written in the form

$$
\dot{\Omega}_3 = 3\frac{B - A}{C}\Gamma_1\Gamma_2, \quad \dot{\Gamma}_1 = (\Omega_3 - 1)\Gamma_2, \quad \dot{\Gamma}_2 = -(\Omega_3 - 1)\Gamma_1,
\tag{5}
$$

and it possesses two first integrals

$$
H_{|\mathcal{N}} = \frac{1}{2}C\Omega_3^2 - \Omega_3 + \frac{3}{2}(A\Gamma_1^2 + B\Gamma_2^2), \qquad H_{2|\mathcal{N}} = \Gamma_1^2 + \Gamma_2^2,
$$

and can be solved analytically. In fact, for $H_{2|\mathcal{N}} = 1$ they describe planar oscillation of the body when it rotates around the third principal axis which is permanently perpendicular to the orbital plane. It is convenient to consider system (5) on a cylinder

$$\Pi = \mathbb{R} \times \mathbb{B}S^1 = \{(\Omega_3, \Gamma_1, \Gamma_2) \in \mathbb{R}^3 \mid \Gamma_1^2 + \Gamma_2^2 = 1\} = \mathcal{M}^6 \cap \mathcal{N}.$$

Introducing variable $\varphi \mod 2\pi$ such that $\Gamma_1 = \cos\varphi$, and $\Gamma_2 = \sin\varphi$ we obtain the following system

$$\dot{\Omega}_3 = 3\frac{B-A}{C}\cos\varphi\sin\varphi, \qquad \dot{\varphi} = 1 - \Omega_3. \tag{6}$$

Integration of this system reduces to an integration of the equation of a physical pendulum.

Although Morales-Ramis theory can be applied for a system with many degrees of freedom, due to lack of appropriate results concerning linear equations of order greater than two, it is most effective for a system with two degrees of freedom. This is why we assume that the body is symmetric, e.g. $B = C$. Then M_1 is a first integral and for the fixed value of M_1 system (1) has two degrees of freedom. In what follows we investigate this case. To obtain the Hamiltonian of the reduced system we take the Euler angles $\mathbf{q} = [q_1, q_2, q_3]^T$ of the type 3-2-1 as the canonical coordinates. These angles describe the orientation of the principal axes reference frame with respect to the orbital frame. Omitting details, we can write the Hamiltonian for a symmetric satellite in the form

$$H = \frac{1}{2}\left[\frac{p_1 + \gamma \sin q_2}{\cos q_2}\right]^2 + \frac{p_2^2}{2} - p_1 + \frac{3}{2}\alpha\left(\cos^2 q_1 - b\right)\cos^2 q_2, \tag{7}$$

where $\alpha = A - 1$, $\gamma = M_1$ is a fixed value of the first integral M_1, parameter b describes the flattening of the planet. Above we choose units in such a way that $\omega_O = 1$ and $B = C = 1$. A family of particular solutions of Hamiltonian system given by (7) exists when $\gamma = 0$ and lies on (q_1, p_1)-plane. For further considerations we take one solution from this family given by

$$\Gamma = \begin{cases} \sin q_1(t) = \dfrac{1}{\cosh\omega t} & p_1(t) = 1 + \dfrac{\omega}{\cosh\omega t} & \text{for} \quad A > 1, \\[3mm] \sin q_1(t) = \tanh\omega t & p_1(t) = 1 + \dfrac{\omega}{\cosh\omega t} & \text{for} \quad A < 1, \end{cases} \tag{8}$$

where $\omega = \sqrt{3|A-1|} \in (0, \sqrt{3})$. This is a heteroclinic solution joining two unstable equilibria.

3. Non-integrability Theorem

Having a Hamiltonian system (on a complex symplectic manifold) and its particular solution we can write down the variational equations corresponding to this

190   A. J. MACIEJEWSKI

solution. From them we can derive the Normal Variational Equations (NVE) taking only variations transversal to the considered phase curve. The basic statement of Morales-Ramis theory claims that if the system is integrable in the Liouville sense (with meromorphic first integrals) in a neighborhood of the considered phase curve then the identity component of differential Galois group of NVE is Abelian, see [9] and references therein.

Parameter b in (7) for real bodies is positive and small. Our main result is the following.

THEOREM 1. *Hamiltonian system given by* (7) *with* $\gamma = 0$, $A \neq 1$ *and* $b \in (0, 1/3)$ *does not possess an additional meromorphic first integral.*

To prove this theorem first we have to determine NVE for solution (8). In our case it is easy—NVE correspond to variations of q_2 and p_2 and they can be written as one equation of the second order

$$\ddot{\xi} = [3\alpha(\cos^2 q_1(t) - b) - p_1(t)^2]\xi, \tag{9}$$

where $(q_1(t), p_1(t))$ are given by (8). To determine the identity component of differential Galois group G^0 of this equation we apply the Kovacic algorithm, see [9]. To this end we make the following covering

$$t \longrightarrow z := \tanh \frac{\omega t}{2},$$

and we change the dependent variable $\xi \to w$, $\xi = w \exp[-\frac{1}{2} \int_{z_0}^{z} p(s)\, ds]$. Then equation (9) transforms to the invariant form

$$w'' = r(z)w, \tag{10}$$

where $r(z)$ is a rational function. All these transformations do not change G^0. We have to analyze this equation for $A < 1$ and $A > 1$ separately. In both cases equation (10) is a Fuchsian equation with four singularities $z_{1,2} = \pm i$, $z_{3,4} = \pm 1$ on Riemannian sphere $\mathbb{P}^1 = \mathbb{C} \cup \{\infty\}$. Points $z_{1,2}$ correspond to poles of the heteroclinic solution located at $t_{1,2} = \pm i\frac{\pi}{2\omega}$, while points $z_{3,4}$ correspond to $t_{3,4} = \pm\infty$ in the original time parameterization. For (10) G^0 is contained in SL(2, $\mathbb{C}$).

Theoretically the Kovacic algorithm [2] always solves the problem, however in our case its direct application leads to very complicated computations. Because of this, we first restrict admissible values of b (although it is interesting to analyze integrability of the system for $b \in \mathbb{R}$), and then we are able to exclude some complicated cases performing the following analysis of the local monodromy.

Exponents of equation (10) at regular singular point $z_1 = i$ are equal $\alpha_1 = 2$ and $\alpha_2 = -1$. Their difference $s = \alpha_1 - \alpha_2 = 3$ is an integer and, according to general theory, see [10], in a neighbourhood of z_1 (10) one solution has the form

$$w_1(z) = (z - z_1)^{\alpha_1} f(z), \qquad f(z) = 1 + \sum_{k=}^{\infty} f_k(z - z_1)^k, \tag{11}$$

where series defining $f(z)$ converges in the considered region. The second solution, independent of $w_1(z)$ is defined by integral

$$w_2(z) = w_1(z) \int^z \frac{d\zeta}{w_1(\zeta)^2} = w_1(z) \int^z (\zeta - z_1)^{-s-1} \frac{d\zeta}{f(\zeta)^2}. \tag{12}$$

If we denote $f(z)^{-2} = 1 + \sum_{k=}^{\infty} g_k(z - z_1)^k$, then solution $w_2(z)$ can be written in the form

$$w_2(z) = w_1(z)g_s \ln(z - z_1) + (z - z_1)^{\alpha_2} v(z), \tag{13}$$

where $v(z)$ is holomorphic in a neighbourhood of z_1. The form of local monodromy depends on whether logarithmic term is present or not in the solution. Direct calculations shows that when $b \neq 1/2$ then $g_3 \neq 0$. It follows that the monodromy matrix corresponding to a small loop C encircling singular point z_1 counterclockwise is following

$$\begin{bmatrix} 1 & 2\pi i \\ 0 & 1 \end{bmatrix}. \tag{14}$$

A subgroup of $SL(2, \mathbb{C})$ generated by a triangular matrix cannot be finite and thus the identity component of differential Galois group G^0 cannot be finite as it contains the triangular matrix given above. Similarly we show that G^0 cannot be an imprimitive subgroup of $SL(2, \mathbb{C})$. Finally, using the Case I of the Kovacic algorithm [2, p. 11] we prove that G^0 cannot be triangulisabe subgroup of $SL(2, \mathbb{C})$. From all these considerations follows that $G^0 = SL(2, \mathbb{C})$, and thus is not Abelian. This finishes the proof of our theorem.

The local monodromy plays the fundamental role in the above prove. It is trivial only when $b = 1/2$. For this case application of the Kovacic algorithm is not trivial. Detailed discussion of non-integrability of the system studied in this paper with arbitrary $b \in \mathbb{R}$ will be published elsewhere.

References

1. Beletskii, V. V.: 1965, *Motion of a Satellite about its Mass Center*. Moscow: Nauka (In Russian).
2. Kovacic, J. J.: 1986, 'An algorithm for solving second order linear homogeneous differential equations', *J. Symbolic Comput.* **2**(1), 3–43.
3. Maciejewski, A. J.: 1995, 'Non-Integrability of the Planar Oscillations of a Satellite', *Acta Astronomica* **232**, 167–184.
4. Maciejewski, A. J.: 1997, 'A Simple Model of the Rotational Motion of a Rigid Satellite Around an Oblate Planet', *Acta Astronomica* **47**, 387–398.
5. Maciejewski, A. J.: 1999, 'Problems Connected with the Dynamics of Celestial Bodies Rotational Motion', In: : *Dynamics of Small Bodies in the Solar System* (B. Steves and A. E. Roy, Eds.), Dordrecht, 315–320.
6. Maciejewski, A. J. and Goździewski, K.: 1995, 'Solutions homoclinic to regular precessions of a symmetric satellite', *Celestial Mech.* **61**, 347–368.

7. Maciejewski, A. J. and Goździewski, K. : 1999, 'Numerical evidence of non-integrability of certain Lie-Poisson equations', *Rep. Math. Phys.* **44**(1/2), 133–142.
8. Maciejewski, A. J. and Simo, C.: 2000, 'Non-integrability of Rotational Motion of Symmetric Satellite in Circular Orbit', *C. R. Acad. Sci. Paris Sér. I Math.* in preparation.
9. Morales Ruiz, J. J.: 1999, *Differential Galois theory and non-integrability of Hamiltonian systems*, Basel: Birkhäuser Verlag.
10. Whittaker, E. T. and Watson, G. N.: 1935, *A Course of Modern Analysis*, London: Cambridge University Press.
11. Ziglin, S. L.: 1982, 'Branching of Solutions and non-existence of first integrals in Hamiltonian mechanics. I', *Functional Anal. Appl.* **16**, 181–189.
12. Ziglin, S. L.: 1983, 'Branching of Solutions and non-existence of first integrals in Hamiltonian mechanics. II', *Functional Anal. Appl.* **17**, 6–17.

MUTUAL EVENTS OF THE SATURNIAN SATELLITES: A TEST OF THE DYNAMICAL MODELS

J.E. ARLOT, W. THUILLOT and CH. RUATTI

Institut de mécanique céleste et de calcul deséphémérides IMCCE–Unité Mixte de Recherche 8028
du CNRS, Paris Observatory, 77 avenue Denfert-Rocherau, 75014 , Paris (France)
e-mail: arlot@bdl.fr or thuillot@bdl.fr

Abstract. In this paper, we use very accurate astrometric observations of the major satellites of Saturn in order to evaluate the quality of the theoretical models of their motions. The use of the mutual events allows this analysis and it appears that the most recent and complete dynamical models have similar external precision than the old ones. This may be due to the method of reduction of the data used in the fit of the constants of these models.

Key words: Natural satellites, Saturnian system, eclipses, occultations

1. Introduction

Among the different types of observations leading to astrometric positions, the mutual events of the natural planetary satellites provide the most accurate data. Therefore, it appears that their use may help to evaluate the precision of the ephemerides issued from the theoretical models of the motion of the satellites. Since the internal precision of the theoretical models was recently improved (Vienne and Duriez, 1995), we should find an improvement for the external precision determined from high precision mutual events observations.

2. The Mutual Events and the Eclipses by Saturn

Every 15 years, the Earth and the Sun pass through the equatorial plane of Saturn and, for a terrestrial observer, the planet Saturn seems to have lost its rings. Since the major satellites of Saturn are orbiting in this equatorial plane, they occult and eclipse each other during a period of one year. Furthermore, during this period, occultations and eclipses of the satellites by the planet itself are also observable. The observations of these events can be made as photometric observations: we measure the light flux received from one or two satellites when it decreases during an occultation or an eclipse.

The absence of atmosphere on the satellites (except Titan) leads to sharp lightcurves easy to interpretate. From the lightcurve of a mutual event observation, we are able to infer the relative positions of the involved satellites with an accuracy

Pretka-Ziomek et al./Dynamics of Natural and Artificial Celestial Bodies, 193–196, 2001.
© 2001 *Kluwer Academic Publishers.*

10 times better than the accuracy of a positional observation of the satellites. On the other hand, the eclipses by Saturn itself allow to directly determine a shift in longitude in the theoretical model for each observation.

In 1995, the occurrence of mutual events led us to organize a campaign of observations and we got 66 observations published in a catalogue (Thuillot et al., 2001). From these observations, it is possible to deduce a relationship between the orbital longitudes of the two involved satellites. In this work, we reduced 25 good observations of these events observations in order to get results from a preliminary analysis.

3. Comparison of the Observations with Ephemerides

The observation of an eclipse by Saturn provides a direct evaluation of the error of the theoretical model in describing the motion of the satellite in its orbit. A shift in the orbital longitude may be observed as an advance or a delay in the time of the event. Contrarily, the observation of a mutual event involving two satellites provides a combination of the two possible shifts in longitude of both satellites. However, the observation of several uncorrelated mutual events permit to determine individual shifts in longitude for each satellite.

Thanks to this method, we estimate the external precision for two ephemerides: the first one based upon the theoretical model by (Dourneau, 1993), the second one based upon the TASS model (Vienne and Duriez, 1995). The model by Dourneau is given to have an internal precision of 100 km (0.010 geocentric arcsec) and the model by Duriez and Vienne 10 km (0.001 arcsec). Table I provides the shifts in longitude deduced from the observed mutual events for the involved satellites and for Dourneau's and TASS theory (Arlot and Thuillot, in preparation).

This observational accuracy is given by the errors on the determination of the shifts in longitude, and, as supposed, the observations of mutual events appear to be very accurate. The external precision of the theory is provided by the values of these shifts in longitude (O-C's) and it is curious to see that the TASS theory does not provide the large improvement that we could expect regarding the improvement of the internal error.

In order to verify the reality of these shifts in longitude, we can use the eclipses observations. The Table II presents a comparison of the shifts deduced from the mutual events (with Dourneau's ephemerides) with the ones deduced from the eclipses by Saturn. These values are coherent and this comparison shows the reality of the results.

4. Interpretation of the Results

How to explain that a theory with a better internal precision leads to ephemerides of similar precision than the former one? The explanation is probably in the reduction

TABLE I

Shifts in orbital longitude, expressed in km, deduced from the time residuals measured during the mutual events

satellites	Dourneau's theory O–C		TASS theory O–C	
Mimas	−57	± 198	351	± 107
Enceladus	211	± 135	−42	± 73
Tethys	172	± 62	−116	± 34
Dione	7	± 49	50	± 26
Rhea	58	± 27	−99	± 15
Titan	−410	± 32	−121	± 17

TABLE II

Comparison of the shifts in orbital longitudes, expressed in km, issued from the observations of mutual events and eclipses by Saturn. [1]: Arlot and Thuillot (in preparation), [2]: Nicholson et al., 1999

satellites	O–C Eclipses[1]	O–C Eclipses[2]	O–C Mutual events
Mimas	−187	–	−57
Rhea	–	0	+58
Titan	−796	–	−410

of the data used in the fit of these theories. The errors are probably not random errors. Some systematic errors appeared, either from reference frame or from the reduction. An explanation may be in the reduction of small field observations. The inner satellites of Saturn are observed using small field receptors and no reference stars are present in the field. In that case two ways are possible: the first one is recommended by Pascu (Pascu, 1996) and proposes to calibrate the field using nearby clusters of reference stars. The scale and the orientation (if the cluster is not too far from the field of Saturn) are determined independently from the Saturnian system. The second way for the reduction is to use the satellites themselves to determine the scale and the orientation of the field. In that case, the observations are biased by the ephemerides used for the satellites and the theoretical models fitted on these observations are, in fact, adjusted on the old ephemerides used for the calibration. Unfortunately, it appears that a large number of observations have been reduced using this second way. Such observations should be published in units of the receptor in order to allow an iteration for the determination of the scale and of the orientation. A new reduction of these data should be made and the modern theoretical models should not be fitted on this type of data but only on data correctly reduced independently of the former model of motion of the satellites.

5. Conclusion

We have estimated the external precision of two ephemerides of the Saturnian satellites thanks to the observation of their mutual events. We did not find the large expected improvement between the Dourneau's theoretical model and the recent one of Duriez and Vienne. The explanation may be in the reduction method of some observations used in the adjustment of the models (calibration with an old ephemeris of the satellites). The new theoretical models should be fitted only on well-calibrated observations using reference stars.

References

Thuillot, W., Arlot, J. E., Ruatti et al.: 2001, 'The PHESAT95 catalogue of observations of the mutual events of the Saturnian satellites', *Astron. Astrophys.*, in press.

Dourneau, G.: 1993, 'Orbital elements of the eight major satellites of Saturn determined from a fit of their theories of motion to observations from 1886 to 1985', *Astron. Astrophys.*, **267**, 292–299.

Vienne, A. and Duriez, L.: 1995, 'TASS1.6: Ephemerides of the major Saturnian satellites', *Astron. Astrophys.*, **297**, 588–605.

Nicholson, P., French, R.G. and Mattews, K.: 1999, 'Eclipses and occultations of Saturn's satellites observed at Palomar in 1995', In J. -E. Arlot and Blanco (Eds.), *Proceedings of the PHEMU97 workshop*, Catania, Italy.

Pascu, D. : 1996, 'Long-focus CCD astrometry of planetary satellites', In S. Ferraz-Mello, B. Morando, and J. -E. Arlot (Eds.), *Dynamics, ephemerides, and astrometry of the solar system* (Proceedings of the 172nd Symposium of the IAU) Paris, France.

AN ANALYTICAL THEORY OF MOTION OF NEREID

ABDEL-NABY S. SAAD
National Astronomical Observatory, Mitaka, Tokyo 181-8588, Japan
E-mail : Kinoshita@nao.ac.jp

HIROSHI KINOSHITA
Dept. of Astronomical Science, School of Mathematical and Physical Science,
The Graduate Univ. For Advanced Studies, Mitaka, Tokyo 181-8588, Japan
E-mail : saad@pluto.mtk.nao.ac.jp

Abstract. In this paper, an analytical theory of motion of the second Neptunian satellite Nereid is constructed using Lie transformation approach. The main perturbing forces which come from the solar influence are only taken into account. The disturbing function is developed in powers of the ratio of the semimajor axes of the satellite and the Sun and put in a closed form with respect to the eccentricity. The theory includes secular perturbations up to the fourth order, short, intermediate and long period perturbations up to the third order. The osculating orbital elements which describe the orbital motion of Nereid are evaluated analytically. The comparison with the numerical integration of the equations of motion gives an accuracy on the level of 0.2 km in the semimajor axis, 10^{-7} in the eccentricity and 10^{-4} degree in the angular variables over a period of several hundred years. The results of the present theory satisfy the required accuracy for the observations.

Key words: Celestial mechanics, satellite theory, osculating elements, ephemerides, Nereid.

1. Introduction

Neptune has 8 satellites. Triton was discovered by Lassel in 1846 and is a massive retrograde highly inclined satellite, Nereid was discovered by Kuiper in 1949 and is a small satellite in a highly eccentric orbit, and the other 6 satellites were discovered in 1989 during the Voyager 2 encounter with the Neptunian system (Stone and Miner 1989) and are located in the vicinity of their mother planet Neptune.

Mignard (1975) constructed a satellite theory disturbed by the Sun with use of the Von Zeipel method and applied this theory to Nereid (1981). He adopted an eccentric anomaly for the expressions of the solution, of which method was proposed by Hori (1963). In the elimination of the long periodic terms he assumed the inclination is small and the solution was obtained up to the second order of the inclination. Oberti (1990) extended the theory of Mignard by including the perturbation from Triton with use of Deprit method and Segerman and Richardson (1997) included the oblateness perturbation of Neptune with use of Deprit method. Several authors have dealt with the orbital determinations of Nereid (Rose 1974; Mignard 1975, 1981; Veillet 1982, 1988; Jacobson 1990, 1991). Rose fit van Biesbroeck's (1951, 1957) observations, while Veillet used the theory of Mignard.

Pretka-Ziomek et al./Dynamics of Natural and Artificial Celestial Bodies, 197–204, 2001.
© 2001 *Kluwer Academic Publishers.*

198 A. SAAD AND H. KINOSHITA

Jacobson fit the numerically integrated Neptunian satellite orbits (Nereid and Triton) to Earth-based astrometric observations and Voyager spacecraft observations.

In this paper we study the dynamical motion of Nereid using both analytical and numerical methods. We construct a third order analytical theory on the motion of Nereid which is mainly perturbed by the Sun. The theory is elaborated by the use of Lie transforms approach advanced by Hori (1966). The small parameter is the ratio of the mean motion of the Sun and Nereid $\sim 6 \times 10^{-3}$. The disturbing function is developed in this small parameter and put in a closed form with respect to the eccentricity. The present theory can be applied to any inclination. The ephemerides evaluated by the analytic expressions of the present theory are compared with those computed by the numerical integration of the equations of motion. The accuracy and the amplitudes of the osculating orbital elements of Nereid are shown by tables and figures.

2. Hamiltonian of the Motion

The Hamiltonian equation of the nonplanar case is given by

$$F = F_0 + F_1 + F_2, \tag{1}$$

where

$$F_0 = \frac{\mu^2}{2L^2}, F_1 = \nu G, \tag{2}$$

$$F_2 = F_{21} + F_{22} + F_{23}, \tag{3}$$

$$\left. \begin{aligned} F_{21} &= \nu^2 a^2 \left(\frac{r}{a}\right)^2 \left\{ (-\frac{1}{8} + \frac{3}{8}\theta^2) + \frac{3}{16}(1+\theta)^2 \cos(2f + 2y_2) \right\}, \\ F_{22} &= \nu^2 a^2 \left(\frac{r}{a}\right)^2 \left\{ \frac{3}{8}(1-\theta^2) [\cos(2f + 2g) + \cos(2g - 2y_2)] \right\}, \\ F_{23} &= \nu^2 a^2 \left(\frac{r}{a}\right)^2 \left\{ \frac{3}{16}(1-\theta)^2 \cos(2f + 4g - 2y_2) \right\}, \end{aligned} \right\} \tag{4}$$

$\mu = n^2 a^3$, ν and n are the mean motions of the Sun and the satellite, $\theta = \cos i = H/G$, $y_2 = g + h - k$, $(L, G, H; l, g, h)$ are Delaunay's elements and k defines the longitude of the Sun (which is moving in a Keplerian orbit). To reach our goal we applied a succession of canonical transformations on the Hamiltonian equation. At first stage we eliminate the short-periodic terms and take into account the eccentric anomaly of Nereid u as independent variable, since the high eccentric orbit of Nereid precludes replacing functions of the true anomaly by expansions involving the mean anomaly. The analytical expressions of the new Hamiltonian F_i^* ($i = 1, 2, 3, 4$) and the determining function S_j ($j = 1, 2, 3$) are given in (Saad

& Kinoshita 2000). In the present study, the short-period is $l = 360$ days, which describes the orbital revolution of Nereid around Neptune.

Removal the intermediate term k will be achieved by building another transformation. Here the intermediate period k is about 165 years which defines the orbital revolution of Neptune around the Sun. The Hamiltonian is also free from the node h^* since the disturbing potential becomes axial symmetric. After eliminating the long (intermediate) terms, the orbital elements a^{**}, e^{**}, n^{**} and η^{**} are computed from

$$a^{**} = \frac{L^{**2}}{\mu}, e^{**} = \sqrt{1 - \left(\frac{G^{**}}{L^{**}}\right)^2}, n^{**} = \frac{\mu^2}{L^{**3}}, \eta^{**} = \frac{G^{**}}{L^{**}}. \tag{5}$$

The analytical expressions of the new Hamiltonian F_i^{**} ($i = 1, 2, 3, 4$) and determining function S_j^* ($j = 1, 2, 3$) are given in Appendix. By two ways we have checked the validity of these expressions from the analytical point of view. The first one is satisfying d'Alembert characteristics. The second way we put $i = 0$ (inclination of Nereid) in the general formulae and got the the analytical expressions of a fictitious Nereid (Saad & Kinoshita 1999). Up to this stage, the Hamiltonian system is still including the long terms g ($\sim$ 13000 years). We overcome these terms by using Jacobian elliptic functions (Kinoshita and Nakai 1991, 1999) and got the mean elements of Nereid.

3. The Osculating Orbital Elements

The osculating orbital elements are a combination of secular, long-periodic and short-periodic variations in elements. In order to get the osculating orbital elements, and hence, evaluate ephemerides for Nereid, we have started with the mean elements (by solving the final Hamiltonian system in Jacobian elliptic functions). Then we substituted reversely in the formulae of long-periodic and short-periodic perturbations. The above analytical expressions are implemented for digital computations by constructing a computational algorithm described by its purpose, input and its computational sequence. The final equations of the osculating elements have the forms

$$\left. \begin{aligned}
a_{osc} &= a_0'' + \delta a_{sho}, \\
e_{osc} &= e_{long}' + \delta e_{sho}, \\
G_{osc} &= G_{long}' + \delta G_{sho}, \\
H_{osc} &= H_{long}' + \delta H_{sho}, \\
I_{osc} &= \arccos\left(H_{osc}/G_{osc}\right), \\
\omega_{osc} &= \omega_{long}' + \delta \omega_{sho}, \\
\Omega_{osc} &= \Omega_{long}' + \delta \Omega_{sho}, \\
l_{osc} &= l_{long}' + \delta l_{sho},
\end{aligned} \right\} \tag{6}$$

where the subscripts *long* and *sho* have the meanings long-period and short-period perturbations respectively. For any element x, $x'_{long} = x'' + \delta x'$, x'' defines the mean elements and the symbol δ refers to the variations of the elements. Tables I and II show the amplitudes and the accuracy in the osculating orbital elements for both short and long periodic perturbations respectively.

TABLE I

Amplitudes of the osculating elements

Elements	Short-period	Long-period	Units
semi-major axis	747.989	1196.78	km
eccentricity	0.0004	0.0115	rad
arg. of pericenter	0.006	0.7	deg
inclination	0.0025	0.16	deg
long. of asc. node	0.01	0.17	deg
mean anomaly	0.0325	2.0	deg

TABLE II

Accuracy of the osculating elements

Elements	Short-period	Long-period	Units
semi-major axis	0.2	0.2	km
eccentricity	3×10^{-8}	1×10^{-7}	rad
arg. of pericenter	3×10^{-6}	4×10^{-4}	deg
inclination	1.5×10^{-6}	6×10^{-5}	deg
long. of asc. node	3×10^{-6}	4×10^{-4}	deg
mean anomaly	2.5×10^{-5}	4×10^{-5}	deg

4. Discussion and Conclusion

Figure 1 shows the residuals in the osculating elements after making some corrections in mean motions of the elements ℓ, g and h using least-squares method. The global internal accuracy of the present theory is obtained by direct comparison with the numerical integration of the equations of motion. The maximum discrepancies reached 0.2 km in the semi-major axis, $\sim 10^{-7}$ in the eccentricity and 4×10^{-4} degree in the angular variables over several hundred years. The theory has not been fitted to the observations. We intend to do that after including the perturbations of Triton and the oblateness of Neptune although the latter's effect is very small. Then

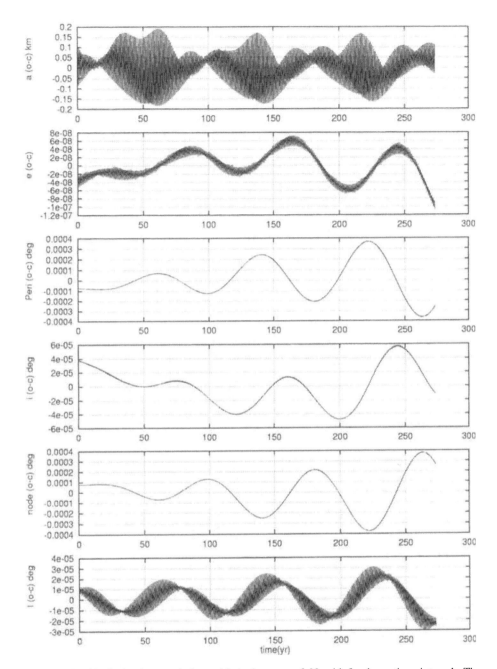

Figure 1. Residuals in the osculating orbital elements of Nereid for long time interval. The semi-major axis is given in *km*, eccentricity in radians and the rest of elements are in degree.

the integrations constants can have the real meaning. However, the accuracy we have got is much better than the present accuracy for Earth-based observations.

Acknowledgements

The first author acknowledges the support of a MONBUSHO Doctoral Fellowship at the Graduate University For Advanced Studies, Tokyo. He is also thankful to the staff of the National Astronomical observatory, Tokyo for their generous hospitality and excellent working conditions.

Appendix

In this section right now, the superscript ** will be omitted from the orbital elements. We follow the algorithm concerned the long-period terms to get the new Hamiltonian and determining functions as follows

$$F_0^{**} = \frac{\mu^2}{2L^2}, \quad F_1^{**} = \nu G, \tag{7}$$

$$F_2^{**} = \nu^2 a^2 \left\{ \left(1 + \frac{3}{2}e^2\right) + \left(\frac{-1}{8} + \frac{3}{8}\theta^2\right) + \frac{15}{16}e^2(1 - \theta^2)\cos(2g) \right\}, \tag{8}$$

$$S_1^* = \nu a^2 \left\{ \frac{3}{16}\left(1 + \frac{3}{2}e^2\right)(1 - \theta^2)\sin(2k - 2h) + \frac{15}{64}e^2(1 + \theta)^2 \right.$$
$$\left. * \sin(2k - 2g - 2h) + \frac{15}{64}e^2(1 - \theta)^2 \sin(2k + 2g - 2h) \right\}, \tag{9}$$

$$F_3^{**} = \frac{\nu^3 a^2}{n}\eta \left\{ \frac{9}{128}\theta(2 - 2\theta^2 + e^2(33 + 17\theta^2)) + \right.$$
$$\left. + \frac{135}{128}e^2\theta(1 - \theta^2)\cos(2g) \right\}, \tag{10}$$

$$S_2^* = \frac{\nu^2 a^2}{n}\eta \left\{ \frac{9}{128}(-2 + 17e^2)\theta(1 - \theta^2)\sin(2k - 2h) \right.$$
$$+ \frac{45}{256}e^2(1 + \theta)^2(-2 + 3\theta)\sin(2k - 2g - 2h)$$
$$\left. + \frac{45}{256}e^2(1 - \theta)^2(2 + 3\theta)\sin(2k + 2g - 2h) \right\}, \tag{11}$$

$$F_4^{**} = \frac{v^4 a^2}{n^2} \left\{ \frac{1}{8192} (8(-67 - 726\theta^2 + 9\theta^4) + 144e^2(329 + 253\theta^2 \right.$$
$$+344\theta^4) - 9e^4(2527 + 2794\theta^2 + 5407\theta^4))$$
$$+\frac{3}{2048} e^2(-1 + \theta^2)(-12(68 + 345\theta^2)$$
$$+e^2(307 + 4035\theta^2)) \cos(2g)$$
$$\left. +\frac{315}{8192} e^4(-1 + \theta^2)^2 \cos(4g) \right\}, \tag{12}$$

while S_3^* has the analytical expression

$$S_3^* = \frac{v^3 a^2}{16n^2} \left\{ (-1 + \theta^2)^2 \left(-\frac{69}{256} + 3e^2 + \frac{123}{2048} e^4 \right) \sin(4k - 4h) \right.$$
$$+ \frac{1905}{4096} e^4(1 + \theta)^4 \sin(4k - 4g - 4h)$$
$$+ (-1 + \theta)(1 + \theta)^3 \left(-\frac{315}{256} e^2 - \frac{645}{1024} e^4 \right) \sin(4k - 2g - 4h)$$
$$+ (1 + \theta)(-1 + \theta)^3 \left(-\frac{315}{256} e^2 - \frac{645}{1024} e^4 \right) \sin(4k + 2g - 4h)$$
$$+ \frac{1905}{4096} e^4(-1 + \theta)^4 \sin(4k + 4g - 4h) + \frac{3}{256}(-1 + \theta^2)$$
$$* \left(456 - 1720e^2 + 1355e^4 + \theta^2(-104 - 7232e^2 + 8131e^4) \right)$$
$$* \sin(2k - 2h) + \frac{795}{512} e^4(-1 + \theta)(1 + \theta)^3 \sin(2k - 4g - 2h)$$
$$-\frac{1}{128}(1 + \theta)^2 \left(3e^2(-474 + 563e^2) \right.$$
$$+15e^2\theta(210 - 372\theta + e^2(-263 + 425\theta)))$$
$$* \sin(2k - 2g - 2h) - \frac{1}{128}(-1 + \theta)^2 \left(3e^2(-474 + 563e^2) \right.$$
$$+ 15e^2\theta(-6(35 + 62\theta) + e^2(263 + 245\theta))) \sin(2k + 2g - 2h)$$
$$\left. + \frac{795}{512} e^4(-1 + \theta)^3(1 + \theta) \sin(2k + 4g - 2h) \right\}. \tag{13}$$

Notice that all the above analytical expressions satisfy d'Alembert characteristics. This may prove the validity of these expressions from the analytical point of view. Up to this stage, the Hamiltonian system is still include the long terms g. Omitting these terms has been done by using Jacobian elliptic functions (Kinoshita and Nakai 1991, 1999) in solving the Hamiltonian equations system.

References

Hori, G.: 1963, 'A new approach to the solution of the main problem of the lunar theory', *Astron. J.* **68**, 125–146.

Hori, G.: 1966, 'Theory of general perturbations with unspecified canonical variables', *Publ. Astron. Soc. Japan* **18**, 287–296.

Jacobson, R. A.: 1990, 'The orbits of the satellites of Neptune', *Astron. Astrophys.* **231**, 241– 250.

Jacobson, R. A., Riedel, J. E. and Taylor, A. H.: 1991, 'The orbits of Triton and Nereid from spacecraft and Earthbased observations', *Astron. Astrophys.* **247**, 565–575.

Kinoshita, H. and Nakai, H.: 1991, 'Secular perturbations of fictitious satellites of Uranus', *Celest. Mech. & Dyn. Astron.* **52**, 293–303.

Kinoshita, H. and Nakai, H.: 1999, 'Analytical solution of the Kozai resonance and its application', *Celest. Mech. & Dyn. Astron.* **75**, 125–147.

Mignard, F.: 1975, 'Satellite with high eccentricity. Application to Nereid', *Astron. Astrophys.* **43**, 359–379.

Mignard, F.: 1981, 'The mean elements of Nereid', *Astron. J.* **86**, 1728–1729.

Oberti, P.: 1990, 'An accurate solution for Nereid's motion. I. Analytic modeling', *Astron. Astrophys.* **239**, 381–386.

Rose, L. E.: 1974, 'Orbit of Nereid and the mass of Neptune', *Astron. J.* **79**, 489–490.

Saad, A. S. and Kinoshita, H.: 1999, 'An analytical theory on a satellite motion with highly eccentric orbit', *Proceedings of the 31st Symposium on Celestial Mechanics*, Kashima, Ibaraki, Japan, March 3-5, (H. Umehara, Ed.), pp. 249–268.

Saad, A. S. and H. Kinoshita : 2000, 'The Theory of Motion of Nereid. II. Non-planar Case', *Proceedings of the 32nd Symposium on Celestial Mechanics*, Hayama, Kanagawa, Japan, March 15-17, (T. Fukushima, T. Ito, H. Arakida and M. Yoshimitsu, Eds.), in press.

Segerman, A. M. and Richardson, D. L.: 1997, 'An analytical theory for the orbit of Nereid', *Celest. Mech. & Dyn. Astron.* **66**, 321–344.

Stone, E. C. and Miner, E. D.: 1989, 'The Voyager 2 encounter with the Neptunian system', *Science* **246**, 1417–1421.

Veillet, C.: 1982, 'Orbital elements of Nereid from new observations', *Astron. Astrophys.* **112**, 277–280.

Van Biesbroeck, G.: 1951, 'The orbit of Nereid, Neptune's second satellite', *Astron. J.* **56**, 110–111.

MPP01, A NEW SOLUTION FOR MOON'S PLANETARY PERTURBATIONS. COMPARISONS TO NUMERICAL INTEGRATIONS

P. BIDART (bidart@danof.obspm.fr)

DANOF/UMR8630 - Observatoire de Paris 61, avenue de l'Observatoire, 75014 Paris, France

Abstract. A new solution for planetary perturbations in the orbital motion of the Moon has been constructed, and is intended to be substitued to the previous one, computed twenty years ago. We present here the results of our computations and comparisons to ELP2000-82B planetary perturbations solution and to the JPL numerical integrations.

Key words: celestial mechanics, lunar theory, ephemerides

Abbreviations: JPL – Jet Propulsion Laboratory; IMCCE – Institut de Mécanique Céleste et de Calcul des Ephémérides; CERGA – Centre d'Etude et de Recherche en Géodynamique et Astronomie; LLR – Lunar Laser Ranging

1. Introduction

Since 1996, the LLR at CERGA provides observations with a precision under the centimeter level.

In another hand, the r.m.s. residuals computed with the ELP2000-82B solution are estimated roughly at 3 centimeters. The main deficiencies of the ELP solution come from planetary perturbations whose precision is around a few meters. Numerical complements are introduced to improve the solution and thus to allow any comparison with observations. The aim of this work is to reduce the contribution of these complements and to improve the validity of the solution over a large time span. It has been motivated by the construction of a new planetary solution VSOP2000 by (Moisson, 2000) at the IMCCE. Besides, progresses in numerical tools allow us to manipulate large series with a great precision.

The main features of the computation of the new solution MPP01 has been explained in (Bidart, 2000). In the first part of this paper, we present differencies beetween MPP01 and the ELP2000-82B solution for planetary perturbations, and comparisons with external results obtained by Moshier in his paper (Moshier, 1992). In the second part, we present the comparisons after fitting the solution ELP/MPP01 to numerical integrations of the JPL, DE403, DE405, and DE406 on a 6000-years time span.

Pretka-Ziomek et al./Dynamics of Natural and Artificial Celestial Bodies, 205–210, 2001.
© 2001 *Kluwer Academic Publishers.*

2. Analytical Analysis of MPP01

2.1. DIFFERENCES WITH ELP2000-82B SOLUTION FOR PLANETARY PERTURBATIONS

The construction of planetary perturbations takes its inspiration from Brown's method. In a first step, we consider the problem including the Earth, the Moon and a keplerian Sun restricted to their point masses. The solution provided is actually of a high precision (Chapront-Touzé, 1980). In a second step, the neglected effects (planetary perturbations, figures of Earth and Moon, tidal and relativistic effects) are considered as perturbations to the main problem.

Planetary perturbations in longitude ΔV, latitude ΔU and radius vector ΔR are expressed with Poisson series :

$$\sum_{n=0}^{5} t^n \sum_{i_1,\dots,i_p} C_{i_1,\dots,i_p}^{(n)} \sin(i_1\lambda_1 + \cdots + i_p\lambda_p + \Phi_{i_1,\dots,i_p}^{(n)}) \tag{1}$$

where $C_{i_1,\dots,i_p}^{(n)}$ are numerical coefficients, $\Phi_{i_1,\dots,i_p}^{(n)}$ are numerical phases, and $i_1\lambda_1 + \cdots + i_p\lambda_p$ are combinations of the four Delaunay arguments (D, F, l, l') and of the mean longitudes of the planets (except Pluto). The numerical values of orbital parameters have been fitted to DE403 (Chapront, Chapront-Touzé, 1996).

The planetary solution VSOP2000 that we used here has been set up by Moisson at the IMCCE (Moisson, 2000). The IERS92 set of masses was used for computations. The VSOP2000 constants have been fitted to DE403. We neglected the effects of asteroids and Pluto on the planets and we truncated the series at the fifth power of the time. The accuracy of VSOP2000, has roughly been improved with respect to VSOP82, by a factor 10 to 20 for inner planets and by a factor 100 for the outers.

The computations lead to the determination of series with and internal precision $P = 2.10^{-12}$ radians $\simeq 4.10^{-7}$ arcsecondes. Table I summarizes the main differences in the construction between ELP2000-82B and MPP01. Comparing the series of both solutions, the main differences are produced by long periodic terms, *i.e.* terms whose frequency is small compared to the mean motion of the Moon. Those differences are explained by the improvement of the planetary solution in one hand, and by a more accurate internal precision in another hand.

2.2. COMPARISON OF LONG PERIODIC TERMS

We consider now the results presented by Moshier on the comparison of ELP2000-82B with a lunar ephemeris over a 7000 years time span (Moshier, 1992).

To improve the agreement of the ELP solution with the numerical integration on a long time span, Moshier adds a quadratic expansion of three long periodic terms and secular contributions in lunar arguments. The effect of these adjustments is to

A NEW SOLUTION FOR MOON'S PLANETARY PERTURBATIONS

TABLE I

Summary of main differences in the construction of ELP2000-82B and MPP01

	ELP2000-82B	MPP01
Indirect perturbations method	increaments with respect to elements	complete motion for EMB – keplerian contribution
Poisson Series	developped up to t^2 (solar eccentricity perturbations only)	developped up to t^5
Planetary solution	VSOP82	VSOP2000
Set of masses	IAU76	IERS92
Internal precision	10^{-5}"	4.10^{-7}"
Size of series	10000 terms	100000 terms

TABLE II

Comparison of long periodic terms determined by Moshier with those present in MPP01. Periods are in julian years, amplitudes are in arcseconds per century^2, and phases are in degrees

terms in $t^2 \sin(\omega + \Phi)$		Moshier		MPP01	
ω	period (y)	A $("/cy^2)$	Φ $(°)$	A $("/cy^2)$	Φ $(°)$
$18Ve - 16Te - l$	273.045	$-230.3\ 10^{-5}$	0	$-228.0\ 10^{-5}$	$+23.6$
$8Ve - 13Te$	238.922	$-17.7\ 10^{-5}$	$+22.5$	$-4.1\ 10^{-5}$	$+58.9$
$10Ve - 3Te - l$	1911.772	$-30.1\ 10^{-5}$	-60.8	$-25.4\ 10^{-5}$	-17.3

reduce the residuals in longitude to about 1.5" r.m.s. over the 7000 years time span. Table II presents the comparison on these terms. We observe a good agreement concerning the amplitude of the term $18Ve - 16Te - l$. The differences in the amplitudes of the terms $10Ve - 3Te - l$ and $8Ve - 13Te$ are reduced when we consider the development of these terms in upper powers of the time and by taking into account the other long periodic terms ($2Ju - 5Sa$ and $4Te - 8Ma + 3Ju$) which were not mentioned in (Moshier, 1992). The comparisons are also in good agreement with the 25° shift in phases observed by Moshier, produced by the restriction of lunar arguments in planetary terms involving l to their linear development.

3. Comparison to Numerical Integrations

3.1. ELP/MPP01 FITTED ON DE403 AND DE405

A complete solution for the orbital motion of the Moon is then constructed adding to the planetary perturbations, the main problem, the Earth and Moon figures perturbations and tidal and relativistic effects from ELP2000-82B solution. We name

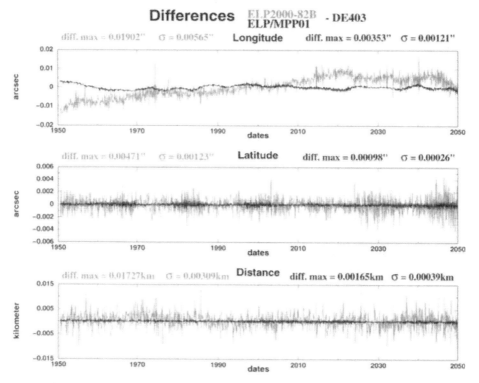

Figure 1. Longitude, latitude and distance of ELP2000-82B and ELP/MPP01 fitted and compared to DE403.

it ELP/MPP01. We have then computed the differences between the two solutions on the time span [January 1st 1950 - January 1st 2050]. The corrections to the orbital parameters

$$w_1^0,\ w_2^0,\ w_3^0,\ Te^0,\ \varpi'^0,\ \nu,\ n',\ \Gamma,\ e,\ e'$$

are evaluated, applying the least squares method to the residuals. w_i^0 are respectively the phases of the mean longitude of the Moon, the longitude of the perigee and of the node of the Moon. Te^0 and ϖ'^0 are the phases of the heliocentric mean longitudes of the Earth and of the Earth's perihelion. ν, n', e, e' are respectively the sideral mean motions and eccentricities of the Moon and the Sun. Γ being the inclination of the Moon's orbital plane on the inertial mean ecliptic.

Once the parameters described above have been fitted, we iterate the computation of differences. The Figure 1 presents the results of ELP2000-82B and ELP/MPP01 both fitted and compared to DE403. The comparison shows that our solution is improved in longitude and latitude by half an order of magnitude, but the most significant change concerns the r.m.s. in distance, roughly 10 times smaller than with ELP2000-82B. This point is of great importance considering that LLR observations give the determination of the Earth-Moon distance.

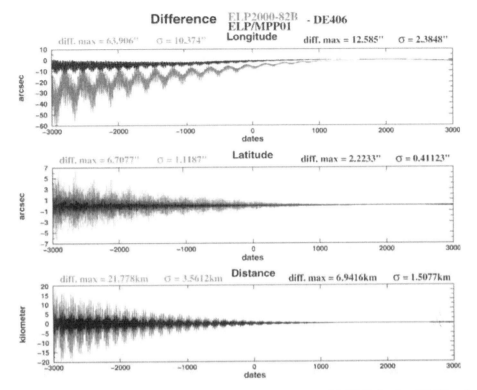

Figure 2. Longitude, latitude and distance of ELP2000-82B and ELP/MPP01 fitted to DE405 and compared to DE406.

3.2. COMPARISON ON A 6000-YEARS TIME SPAN

We have carried out the same analysis as described for DE403 with the numerical integration DE405 of the JPL (Standish, 1997). Both models providing the lunar ephemeris are similar except for the tides (Love number) which provide a drift in the quadratic term of the lunar mean longitude $\delta w_1^2 = 0".11333/\text{century}^2$. As a consequence, the corrections to orbital parameters are roughly the same as those obtained comparing ELP/MPP01 to DE403.

Besides the fit to DE405, this allows us to compare our solution to the numerical integration DE406 which is an extended ephemeris of DE405 on a 6000-years time span. Nevertheless, further corrections should be brought to t^4 coefficients of the polynomials for the longitude, by taking into account the results found by Moshier. The figure 2 shows the result of the comparisons of ELP2000-82B and ELP/MPP01 to DE406 on the time span [January 1st 3000 B.C. - January 1st 3000 A.C.]. We observe a reduction of the differences, mainly due to the better determination of long periodic terms and more particularly the 273 years' period term $18Ve - 16Te - l$. The differences become lower than 2.4" r.m.s..

4. Conclusion

The improved solution for planetary perturbations in the orbital motion of the Moon should contribute to reduce the contribution of numerical complements that still have to be brought for LLR observations reduction. Nevertheless, $O - C$ residuals are strongly affected by the atmospheric noise and the Earth rotation, so the improvement should not be important.

Acknowledgements

I would like to thank X. Moisson, M. Chapront-Touzé and specially J. Chapront for their many usefull informations, and for their help in numerical computations. J. Souchay is also thanked for his help in improving the English of this paper.

References

Bidart, P.: 2000, *Astron. & Astrophys.*, accepted for publication.
Chapront-Touzé, M.: 1980, *Astron. & Astrophys.* **83**, 86.
Chapront-Touzé, M. and Chapront, J.: 1980, *Astron. & Astrophys.* **91**, 233.
Chapront, J. and Chapront-Touzé, M.: 1996, *Celest. Mech.* **66**, 31.
Moisson, X.: 1999, *Astron. & Astrophys.* **341**, 318.
Moisson, X.: 2000, Private communication.
Moshier, S.L.: 1992, *Astron. & Astrophys.* **262**, 613.
Standish, E.M., Newhall, X.X., William, J.G. and Folkner, W.F.: 1995, JPL IOM 314.10–127.
Standish, E.M.: 1997, Magnetic tapes.

AN EFFICIENT ALGORITHM FOR COMPUTATION OF HANSEN COEFFICIENTS AND THEIR DERIVATIVES

AKMAL VAKHIDOV

Lohrmann Observatory, Technical University Dresden, Germany

Abstract. An efficient algorithm for approximate evaluation of Hansen coefficients and their derivatives is presented

Key words: Hansen coefficients, recurrence relations, polynomial approximation.

The problem of efficient computation of Hansen coefficients and their derivatives was studied in detail in many papers (see, for exam-pie, (Giacaglia, 1976), (Brumberg, 1995)). But this problem is really quite complicated and up to now we have no standard algorithms, which could allow to evaluate Hansen coefficients and their derivatives efficiently for all possible practical cases.

In our research we develop an efficient computational scheme for a proximate evaluation of Hansen coefficients and their derivatives. Our approach is based on the use of polynomial approximations and special recurrence relations, which allow to make the process of computing the Hansen coefficients much more convenient.

Because in theories of motion of celestial bodies we often compute a large set of Hansen coefficients many times for different (sometimes only a little bit different) values of eccentricity (for example, in the process of integration of averaged equations of motion in semianalytical theories), it seems to be much more efficient to use some simple polynomial approximations of Hansen coefficients instead of much more complicated direct methods (Brumberg, 1995). In our research we study in detail the properties of different approximating schemes: approximation by fragments of Taylor series, interpolation by Lagrange polynomials, Chebyshev approximation. Our investigations show, that the Taylor expansions are very efficient in the case of small eccentricities and can approximate the Hansen coefficients with very high relative accuracy. Lagrange interpolations are quite suitable for moderately large eccentricities, but the interval of approximation should be small. Chebyshev approximations are quite convenient for moderately large and even for large eccentricities, the interval of approximation can be also large, but the process of constructing the Chebyshev approximations needs much more computer resources than for Lagrange interpolations. Both Chebyshev and Lagrange approximations give high absolute accuracy of approximation, but relative accuracy can be low. More detailed analysis of numerical efficiency of these approximating schemes is presented in (Vakhidov, 2000a).

Pretka-Ziomek et al./Dynamics of Natural and Artificial Celestial Bodies, 213–214, 2001.
© 2001 *Kluwer Academic Publishers.*

Using polynomial approximations we can compute a basic set of Hansen coefficients and all other coefficients can be evaluated using recurrence algorithms. In our research we obtaln a new system of recurrence currence relations for Hansen coefficients, which gives a connection only between Hansen coefficients included into disturbing function of plan tary or satellite motion. For constructing this new system of recurrence formulae we use the approach developed in (Sokolsky et al., 1995), which is based on computer-algebraic generating the special systems of recurrence relations with prior determined properties. Our new recurrence relations allow to organize the recurrence process very efficiently both from high to low and from low to high harmonics of disturbing function both for the perturbations from internal and external bodies. These new recurrence formulae are presented in (Vakhidov, 2000b). They are available also from the author upon request.

Finally, we construct the following algorithm. At first, we compute some Hansen coefficients as initial values for recurrences using polynomial approximations. After this we use our new recurrence relations and compute the coefficients for several harmonics of disturbing function. After this we correct the values of Hansen coefficients using the polynomial approximations and continue the recurrence process for several next harmonics, after this correct again etc. Such a correction of values of Hansen coefficients inside recurrences with the help of polynomial approximations allows to avoid the accumulation of errors in the r currence process. Our test experiments show the high efficiency of our approach and the possibility to use it efficiently in solving different problems of applied celestial mechanics.

The author thanks DFG (Deutsche Forschungsgemeinschaft) for the financial support of this research.

References

Brumberg, V.: 1995, *Analytical techniques of celestial mechanics*, Springer, Berlin-Heideib erg.

Giacaglia, G.E.O.: 1976, 'A note on Hansen's coefficients in satellite theory', *Celest. Mech.* **14**, 515.

Sokoisky, A., Vakhidov, A. and Vasiliev, N.: 1995, 'Generating a new recurrence relations system for elliptic Hansen coefficients by means of computer algebra', *SIGSAM Bulletin*, June special issue, 16.

Vakhidov, A.: 2000a, 'Construction of polynomial approximation for Hansen coefficients', *Comp. Phys. Comm.* **124**, 40.

Vakhidov, A.: 2000b, 'Some recurrence relations between Hansen coefficients', *Celest. Mech. and Dynam. Astron.* (in press).

ON THE POYNTING-ROBERTSON EFFECT AND ANALYTICAL SOLUTIONS

JOZEF KLAČKA (klacka@fmph.uniba.sk)
Institute of Astronomy, Faculty for Mathematics and Physics, Comenius University Mlynská dolina, 842 48 Bratislava, Slovak Republic

Abstract. Analytical solutions of the equation of motion for the Poynting-Robertson effect obtained by Breiter and Jackson (1998) are discussed. Special attention is devoted to pseudo-circular orbits and terminal values of osculating elements – terminal values are of no physical sense since relativistic equation of motion containing only first order of $\vec{v}/c$ was used by the authors.

1. Introduction

Breiter and Jackson (1998; BJ-paper in the following text) have presented analytical solutions of the Poynting-Robertson effect (P-R effect). The aim of this paper is to discuss analytical solutions for the P-R effect in the two limiting cases: pseudo-circular orbits and terminal values of osculating elements.

2. Pseudo-circular Orbits

Klačka and Kaufmannová (1992, 1993 – typewriting error in Eq. (1)) numerically solved the P-R effect for nearly circular orbits – pseudo-circular orbits. Their results were confirmed in BJ-paper in an analytical way. Moreover, analytical solution yields a new result: eccentricity is still an increasing function of time for a special case of initial conditions. However, this very special mathematical result is caused by the fact that BJ-paper considers only first order in v/c.

3. Terminal Values of Osculating Elements

As for the terminal values of osculating elements presented in BJ-paper, the results may be collected in two important statements: $\lim_{x \to 0^+} f = \pi$ and $\lim_{x \to 0^+} e = 1$. Let us calculate other important quantities. The results are:

$$\lim_{x \to 0^+} r = 0, \quad \lim_{x \to 0^+} a = 0, \quad \lim_{x \to 0^+} H = 0, \quad \lim_{x \to 0^+} E = -\infty,$$
$$\lim_{x \to 0^+} v_T = 0, \quad \lim_{x \to 0^+} v_R = -(c/2)(1-\beta)/\beta, \qquad (1)$$

where r is particle's distance from the central point mass, a – semimajor axis, v_T – transversal component of the velocity vector, $v_R \equiv \dot{r}$ – radial component of the velocity vector, H – angular momentum, E – total energy of the particle with respect to the central point mass.

The obtained results yield inconsistencies. The osculating trajectory is parabola ($e = 1$), the particle is situated at apocenter ($f = \pi$) and the total energy is $E = -\infty$. Normal result is that $E = 0$ for the case $e = 1$. The limiting mathematical results correspond to non-physical situation: the particle spirals toward $r = 0$ in a finite time and its potential energy decreases in an unlimited value.

Eqs. (1) yield a hint how to put the discussed inconsistencies into a correct physics. Since the last of Eqs. (1) yields that $\lim_{x \to 0^+} v_R < -c$ for $0 < \beta < 1/3$, we have to use complete form of the P-R effect – relativistic effect (Klačka 1992a, Eq. (140)).

The last of Eqs. (1) yields that the form of the equation of motion containing only first order of v/c could be acceptable only for $0 \leqslant 1 - \beta \ll 1$ – only in this case the requirement $v \ll c$ holds. However, the third Kepler's law yields $T^2 = 4 \pi^2 a^3 \{\mu_{\beta=0} (1 - \beta)\}^{-1}$ and $\lim_{\beta \to 1^-} T = \infty$. Thus, the situation $0 \leqslant 1 - \beta \ll 1$ is not physically interesting. This situation is evident also from Eq. (30) in Klačka (1992b; error is in Eq. (10) of the paper – the right-hand side of Eq. (10) must contain $\mu (1 - \beta)$ instead of μ when used in the section 3): $\lim_{\beta \to 1^-} e_{in} > 1$ – no inspiralling toward the center occurs.

4. P-R Effect and Reality

As it was already discussed in Klačka (1992a – Eq. (122)), P-R effect holds only for a special form of interaction between electromagnetic radiation and dust particle. As for general interaction we refer to Klačka (2000). Application to real cosmic dust particle is presented in Klačka and Kocifaj (2000) – real orbital evolution does not correspond to the P-R effect.

5. Conclusions

We have shown that terminal values of osculating elements lead to serious inconsistencies caused by the fact that relativistic equations of motion only in first order in v/c were used in Breiter and Jackson (1998) – (general) relativity theory must be used. Also the case of still growing eccentricity does not occur for the complete form of the P-R effect.

Orbital evolution of real dust particle may not correspond to the P-R effect.

Acknowledgements

The author wants to thank to the Organizing Committees for the financial support which enabled his participation at the conference. The paper was supported by the Scientific Grant Agency VEGA (grant No. 1/7067/20).

References

Breiter, S. and Jackson, A. A.: 1998, 'Unified analytical solutions to two-body problems with drag', *MNRAS* **299**, 237–243.

Klačka, J.: 1992a, 'Poynting-Robertson effect: I. Equation of motion', *Earth, Moon and Planets* **59**, 41–59.

Klačka, J.: 1992b, 'Poynting-Robertson effect: II. Perturbation equations', *Earth, Moon and Planets* **59**, 211–218.

Klačka, J. and Kaufmannová, J.: 1992, 'Poynting-Robertson effect: 'circular' orbit', *Earth, Moon and Planets* **59**, 97–102.

Klačka, J. and Kaufmannová, J.: 1993, 'Poynting-Robertson effect and small eccentric orbits', *Earth, Moon and Planets* **63**, 271–274.

Klačka, J.: 2000, 'Electromagnetic radiation and motion of real particle', *Icarus* (submitted; astroph/0008510).

Klačka, J. and Kocifaj, M.: 2000, 'On the stability of the zodiacal cloud' (this conference).

DEVELOPMENT OF THE NUMERICAL THEORY OF THE RIGID EARTH ROTATION

V. V. PASHKEVICH (apeks@gao.spb.ru)
Central Astronomical Observatory of the Russian Academy of Sciences Pulkovskoe shosse, 65/1, 196140, St.Petersburg, Russia

Development of the numerical theory of the rigid Earth's rotation is carried out in the Rodrigues-Hamilton parameters, which define a position of the principle axes of inertia of the Earth with respect to the fixed ecliptic plane and equinox J2000.0. The rigid Earth perturbed rotation is a result of the gravitational interaction of the Earth's body with the point mass disturbing bodies (the Sun, Moon and major planets). The orbital motions of the disturbing bodies are defined by the DE403/LE403 ephemeris (Standish et al., 1995). In the paper (Pashkevich, 1999) the numerical solution of the rigid Earth rotation was received in the Rodrigues-Hamilton parameters. The comparison of the results of the numerical solution of the problem with the semi-analytical solution of the Earth's rotation SMART97 (Bretagnon et al., 1998) was carried out in Euler angles ψ, ω, ϕ over 2000-2199 time interval.

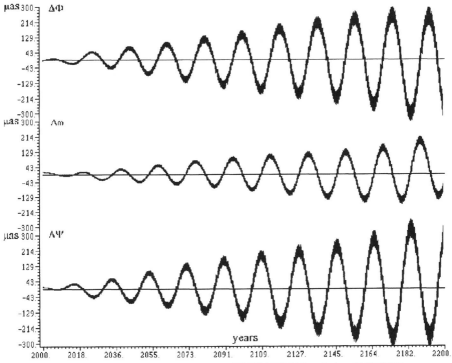

Figure 1. Numerical integration minus KINEMATICAL solution SMART97.

Good agreement (~10 microarcseconds) between the numerical integration and the semi-analytical solution SMART97 in the DYNAMICAL case (without taking into account the geodetic perturbations) and not good agreement (~ 300 microarcseconds) for the solutions in the KINEMATICAL case (Fig.1) (with accounting the geodetic perturbations) is a problem studied in this investigation.

The analysis of the semi-analytical theory SMART97 is carried out in order to study this phenomenon in more details. The authors of SMART97 "...have added the precession to the mean longitudes of the planets in order to have the same frequencies and the same periods as in the other solutions. Thus, the mean longitudes of the planets are reckoned from the equinox of date" (Bretagnon et al., 1998). Thus was found out, that added of the precession contains the value of the geodetic precession. Therefore in the DYNAMICAL solution SMART97 the arguments of the nutational harmonics already contain the geodetic precession. Consequently, in the result of the analytical solution of the corresponding differential equations the coefficients of the nutational harmonics in the DYNAMICAL solution SMART97 also contain the geodetic perturbations. Then from of the DYNAMICAL solution the authors of SMART97 constructed the KINEMATICAL solution SMART97 by means of the 3 geodetic rotations. It seems evident that the DYNAMICAL solution must not contain geodetic perturbations when constructing the KINEMATICAL solution in this way. The following experiment confirms this

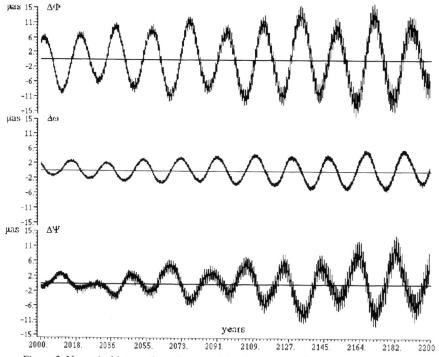

Figure 2. Numerical integration minus modified KINEMATICAL solution SMART97.

statement: in the KINEMATICAL solution SMART97 the geodetic precession is excluded from the mean longitudes of the planets, then the initial conditions are calculated and the numerical integration is performed. After such modification of the solution SMART97 the residuals over all the interval of comparison do not surpass ±15 of microarcseconds in the longitude of the ascending node of the dynamical Earth's equator and in the proper rotation angle, and ±6 of microarcseconds in the inclination of the dynamical Earth's equator to the fixed ecliptic J2000.0 (Fig.2).

The investigation was carried out at the Central (Pulkovo) Astronomical Observatory, under a financial support of the **Russian Foundation for Fundamental Research**, Grant No 98-02-18318.

References

Bretagnon,P., Francou, G., Rocher, P. and Simon, J. L.: 1998, 'SMART97: A new solution for the rotation of the rigid Earth', *Astronomy and Astrophysics*, **329, No.1**, pp. 329–338.

Pashkevich, V. V.: 1999, 'Results of the numerical investigation of the rotational motion of the Earth', In *Deposition No 2514-B99 30 July 1999*, pp. 1–30 (in Russian).

Standish, E. M., Newhall, X. X., Williams, J. G. and Folkner, W. M.: 1995, 'JPL Planetary and Lunar Ephemerides, DE403/LE403', JPL IOM 314.10-127.

ALGORITHMS OF THE NUMERICAL SIMULATION OF THE MOTION OF SATELLITES

VICTOR AVDYUSHEV and TATYANA BORDOVITSYNA
Applied Mathematics and Mechanics Institute, Tomsk, Russia

Key words: natural satellites, numerical methods, Kustaanheimo–Stiefel transformations, Encke-type algorithms

The paper presents efficient algorithms for the numerical simulation of motion as applied to natural satellites. These algorithms are based on using regularizing and stabilizing Kustaanheimo–Stiefel (KS) transformations (Stiefel and Scheifele, 1971).

For the examples of hypothetical and real satellites it is shown that the application of KS-transformations improves both accuracy and speed of numerical integration.

The principles of constructing Encke-type equations of motion in KS-coordinates are presented. Two intermediate solutions in Encke method are considered. The first one is Keplerian orbit (K) described by the differential equations (Avdyushev, 1999)

$$\frac{d^2 u_K}{dE^2} + \frac{1}{4} u_K = 0, \quad \frac{d\omega_K}{dE} = 0, \quad \frac{d\tau_K}{dE} = \frac{\mu}{8\omega_K^3};$$

and the second one is a circular orbit of an equatorial satellite of an oblate planet (J) described by the equations

$$\frac{d^2 u_J}{dE^2} + \frac{1}{4}(1 + 4\Phi)u_J = 0, \quad \frac{d\omega_J}{dE} = 0,$$

$$\frac{d\tau_J}{dE} = \frac{\mu}{8\omega_J^3}(1 - \Phi), \quad \Phi = \frac{(2\omega_J)^4 J_2 b^2}{2\mu^2}.$$

Here $u = u(E)$ stands for the 4-dimensional position vector of the satellite in KS-space, E stands for the eccentric anomaly, $\omega = \sqrt{h/2}$ stands for the frequency where h is the energy of the dynamical system, $\tau = t - (u, u')/\omega$ is the temporal element, t is physical time, and μ, J_2 and b are the gravitational parameter, the coefficient of the second zonal harmonic and the equatorial radius of the central planet respectively.

Since the eccentricities and inclinations for the orbits of most natural satellites are rather small the J-equations present their dynamics more accurate than the K-equations. Therefore, using J-solution as an intermediate one in Encke method can be more preferable for constructing the equations of the satellite motion in perturbations.

In the work several numerical experiments are carried out which allow making some follow conclusions.

First and foremost, the application of equations in increments $(\delta u_K, \delta u_J)$ improves the accuracy of numerical simulation in KS-space (u) by 2–3 orders without any additional expenses in computations. At the same time the direct integration in rectangular space (x) for the best accuracy takes computing operations in 2–6 times greater than in KS-space. The classical algorithm (in rectangular coordinates) cannot compete with KS-algorithms by speed at all. Under similar computations, the results of the classical algorithm become quite useless for the representation of solution (Table I).

The results of one experiment for some satellites as an example are given in Table I. Here ΔT is the integration interval from the discovery of a satellite till the present time and $L = \lg \Delta r$, where Δr is numerical error calculated in astronomical units by the well–known forward–and–backward integration method. It should be noted that the accuracy of the classical Encke method goes down when Φ increases. But the accuracy of the generalized one is quite insensitive to the variation of Φ.

TABLE I

The estimation of the numerical error of integration

	Satellite	$\Phi, 10^{-4}$	ΔT, rev.	$L(\delta u_J)$	$L(\delta u_K)$	$L(u)$	$L(x)$
J5	Amalthea	11	78900	-13	-11	-10	-3
S1	Mimas	9	81700	-13	-12	-11	-6
M1	Phobos	1	140400	-13	-13	-11	-6

In this way the experiments show the high efficiency of KS-algorithms in the problems of the satellite dynamics. In particular Encke KS-algorithms are distinguished both by the accuracy of numerical simulation and by the speed of computations.

References

Stiefel, E. L. and Scheifele, G.: 1971, 'Linear and Regular Celestial Mechanics', Springer–Verlag, Berlin, Heidelberg, New York.

Avdyushev, V. A.: 1999, 'A New Intermediate Orbit in the Problem on the Motion of an Inner Satellite of an Oblate Planet', *Research in Ballistics and Contiguous Problems of Dynamics*, Tomsk, **3**, pp. 126–127 (in Russian).

NUMERICAL SIMULATION OF THE MOTION OF MARTIAN SATELLITES

EKATERINA TITARENKO, TATYANA BORDOVITSYNA and VICTOR
AVDYUSHEV
Research Institute of Applied Mathematics and Mechanics, Tomsk, Russia

Key words: Martian satellites, motion, numerical simulation

The numerical long-term prediction of the satellite motion is very laborious. This is mainly because the high rate of the change of the right member functions of the classical equation of motion results in a small step in numerical integration and, in turn, in a rapid accumulation of the round-off errors. As is well known, the application of regularizing and stabilizing variables allows avoiding these difficulties. The authors have constructed the numerical model of the motion of Martian satellites using the equation of motion written in rectangular coordinates (x), in regularizing and stabilizing Kustaanheimo–Stiefel (KS) variables (u) and in increments of KS-coordinates (δu).

Initial conditions of the satellite motion have been calculated and fitted by Chapront–Touze analytical theory (Chapront–Touze, 1990) and the programming system ERA (Krasinsky and Vasilyev, 1996). Preliminary orbital parameters of Martian satellites have been calculated by Chapront-Touze theory on 1976 August 10 0h UT.

As a rule, in an analytical theory the accuracy of calculated velocities is worse then that of coordinates. So the initial parameters of the motion of the satellites have been fitted to a large number of the positions obtained by Chapront Touze theory. Isochronous derivatives have been computed by simultaneous integrating the equations of motion and the variation equations. The improvement of the initial parameters has been carried out over a time span of 400 days. The obtained parameters are given in Table I. The comparison of the results of the numerical ephemeris of the satellites with the analytical theory shows good fitness over a time span of 800 days. The maximal deviations are not greater than 12 km for Phobos and 20 km for Deimos. All the calculations have been executed by Everhart 15-order method.

To estimate the efficiency of the application of the algorithms a numerical experiment has been carried out. Three algorithms have been considered: in rectangular coordinates, in KS-coordinates and in increments of KS-coordinates (Bordovitsyna et al., 1998). The accuracy and speed of the algorithms for the example of Phobos over a time span 120 years have been estimated by means of varying the integrator parameter controlling the size of integrating step. The results of the experiment are given in Figure 1.

The figure shows that the speed of integration in rectangular coordinates is in 3 times slower than in KS-coordinates. At the same time, the accuracy of Encke

---

Pretka-Ziomek et al./Dynamics of Natural and Artificial Celestial Bodies, 225–226, 2001.
© 2001 *Kluwer Academic Publishers.*

TABLE I
Initial parameters of the motion of the satellites. 1976 August 10 0h UT

Satellite	Coordinates, km			Velocity, m/sec		
	x_1	x_2	x_3	$\dot{x}_1$	$\dot{x}_2$	$\dot{x}_3$
Phobos	7017.189	−2755.989	−5369.463	1055.780	1847.151	401.824
Deimos	−17632.313	−15418.667	1135.269	702.030	−860.295	−770.783

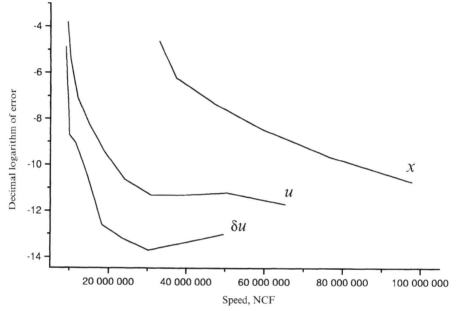

Figure 1. Estimations of efficiency algorithms. Phobos.

method is higher than that of the classical algorithm by 2 orders. The efficiency characteristics of the numerical algorithms NCF is a Number of Calling Functions of the right member of equations.

References

Bordovitsyna, T. V., Bykova, L. E. and Avdyushev, V. A.: 1998, 'Problems in Applications of Regularizing and Stabilizing KS–transformations tosks of Dynamics of Planets' Natural Satellites and Asteroids', *Astronom and Geodezy*, **16**, Tomsk, 33–57.

Chapront–Touze', M.: 1990, 'Orbits of Martial Satellites from ESAPHO and ESADE theories', *Astron. Astrophys.*, **240**, 159–172.

Krasinsky, G. A. and Vasilyev, M. V.: 1996, 'ERA: Knowledge Base for Ephemeris and Dynamical Astronomy', In *Proceeding of IAU Coloquium 165 "Dynamics and Astrometry of Natural and Artificial Celestial Bodies"*, Kluwer Acad. Publ., 239–244.

STABILITY OF LONG-PERIOD PLANAR SATELLITE MOTIONS IN A CIRCULAR ORBIT

A.I. NEISHTADT
Space Research Institute, Profsoyuznaya 84/32, Moscow 117810, Russia

V.V. SIDORENKO
Keldysh Institute of Applied Mathematics, Miusskaya Sq. 4, Moscow 125047, Russia

C. SIMÓ
*Dept. de Matemàtica Aplicada i Anàlisi, Universitat de Barcelona, Gran Via 585,
Barcelona 08007, Spain*

Abstract. We study, under the linear approximation, the stability of planar motions – a one parameter family of partial solutions of the equations describing the satellite attitude motion in a circular orbit.

Key words: attitude dynamics, stability of motion

1. Introduction

When a satellite moves in the central gravity field, its attitude motion can be a planar motion, in which case one of the principal central axes of inertia is perpendicular to the orbital plane (Beletsky, 1975). If the satellite orbit is elliptical, such planar motions have rather complicated behavior (in particular, chaotic motions are possible). The situation is simplified for a circular orbit: the equation describing the discussed class of motions represents the mathematical pendulum equation in this case. Properties of the planar satellite motions in a circular orbit were investigated previously by Kane and Shippy (1963), Markeev and Sokolsky (1977) and other specialists.

2. Planar Motions: Formal Definition and Classification

To describe the satellite attitude motion we introduce two right-handed Cartesian coordinate systems. Let $OXYZ$ be the orbital system (axis OZ is directed along the radius-vector of the satellite's center of mass O, the axis OX is aligned with the tangent to the orbit, unidirectionally with the satellite motion), $Oxyz$ is a body-fixed system, whose axes are the principal central axes of inertia of the satellite. The current position of the trihedron $Oxyz$ can be obtained from an initial orientation with the Ox, Oy, Oz axes coinciding with the axes OX, OY, OZ by the product of rotations $R_z(\beta) \circ R_x(\gamma) \circ R_y(\alpha)$ (Figure 1). Being interested in the study of motions

Pretka-Ziomek et al./Dynamics of Natural and Artificial Celestial Bodies, 227–233, 2001.
© 2001 *Kluwer Academic Publishers.*

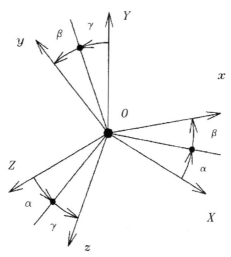

Figure 1. Reference frames used.

close to planar, the angles β and γ shall be assumed to be small. The true anomaly of the orbital motion τ will be used as independent variable.

The Hamiltonian that allows one to obtain the equations of attitude motion in canonical form has the following structure

$$\mathcal{H} = \sum_{k=0}^{\infty} \mathcal{H}_{2k}(p_\alpha, p_\beta, p_\gamma, \alpha, \beta, \gamma), \qquad (2.1)$$

where p_α, p_β and p_γ are the momenta conjugated to variables α, β and γ, $\mathcal{H}_k$ are homogeneous polynomials of degree k in p_β, p_γ, β and γ with coefficients being functions of p_α and α (Markeev and Sokolsky, 1977).

The planar motions correspond to the solutions of canonical equations with Hamiltonian (2.1) lying on the invariant manifold $p_\beta = p_\gamma = \beta = \gamma = 0$. The behavior of variables p_α and α in planar motions is described by the equations (Beletsky, 1975)

$$\frac{dp_\alpha}{d\tau} = -\frac{\partial \mathcal{H}_0}{\partial \alpha}, \quad \frac{d\alpha}{d\tau} = \frac{\partial \mathcal{H}_0}{\partial p_\alpha}, \qquad (2.2)$$

$$\mathcal{H}_0 = \frac{1}{2}\left[(p_\alpha - 1)^2 + 3(\Theta_A - \Theta_C)\sin^2\alpha\right].$$

The parameters Θ_A and Θ_C in the expression for $\mathcal{H}_0(p_\alpha, \alpha)$ characterize the geometry of masses of the satellite: $\Theta_A = A/B$, $\Theta_C = C/B$, where A, B and C are the principal central momenta of inertia. Without any loss of generality we can assume that $\Theta_A > \Theta_C$.

The planar motions form a one-parameter family of partial solutions. We will use the value h of the function $\mathcal{H}_0$ along the solution as a family parameter. If

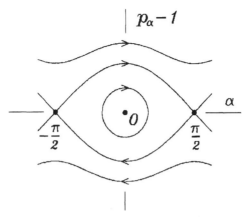

Figure 2. Phase portrait of system (2.2).

$h > h_\circ = \frac{3}{2}(\Theta_A - \Theta_C)$, then in the orbital reference frame the satellite will rotate around the normal to the orbital plane. The case $0 < h < h_\circ$ corresponds to motion of oscillatory type.

On the phase portrait of system (2.2) the regions of oscillatory and rotational motions are separated by separatrices (Figure 2). The separatrices represent the asymptotic motions $p_{\alpha_\pm}(\tau), \alpha_\pm(\tau)$ at $h = h_\circ$ ($\alpha_\pm(\tau) \to \pm\frac{\pi}{2}$ for $\tau \to +\infty$, $\alpha_\pm(\tau) \to \mp\frac{\pi}{2}$ for $\tau \to -\infty$).

The period of motions with initial data in the vicinity of the separatrices $T(h) \to \infty$ for $h \to h_\circ$ ($T(h) \sim \ln|h - h_\circ|$). Below the motions with period $T(h) \gg 1$ will be called long-periodic.

3. Stability Analysis of Planar Motions

In the linear approximation such an analysis is reduced to the investigation of the stability of a trivial solution of the system

$$\frac{d\mathbf{w}}{d\tau} = JH(p_\alpha(\tau, h), \alpha(\tau, h))\mathbf{w}, \tag{3.1}$$

where $\mathbf{w} = (p_\beta, p_\gamma, \beta, \gamma)^T$, $J = \begin{bmatrix} 0 & -E_2 \\ E_2 & 0 \end{bmatrix}$ is the standard symplectic matrix, $H(p_\alpha, \alpha) = [n_{jk}]_{j,k=1}^4$ is the matrix of the quadratic form $\mathcal{H}_2$ in (2.1): $\mathcal{H}_2 = \frac{1}{2}\langle \mathbf{w}, H\mathbf{w}\rangle$, $\langle \cdot, \cdot \rangle$ is the Euclidean scalar product in $\mathbb{R}^4$. The non identically zero elements of the matrix $H(p_\alpha, \alpha)$ have the form:

$$n_{11} = \frac{1+\lambda}{1-\mu}, \quad n_{22} = \frac{1+\lambda}{1+\lambda\mu}, \quad n_{33} = \lambda(1-\mu)\left(\frac{p_\alpha^2}{1+\lambda\mu} + \frac{3\sin^2\alpha}{1+\lambda}\right),$$

$$n_{44} = p_\alpha^2 + \frac{3(\lambda+\mu)}{1+\lambda}\cos^2\alpha, \quad n_{14} = n_{41} = p_\alpha, \quad n_{23} = n_{32} = \frac{\lambda(1-\mu)}{1+\lambda\mu}p_\alpha,$$

$$n_{34} = n_{43} = \frac{3}{2} \frac{\lambda(1-\mu)}{1+\lambda} \sin 2\alpha, \quad \lambda = \frac{1-\Theta_A}{\Theta_C}, \quad \mu = \Theta_A - \Theta_C,$$

$$(\mu, \lambda) \in D = \{0 < \mu \leqslant 1, \ -1 \leqslant \lambda \leqslant 1\}.$$

Since the solution $p_\alpha(\tau, h), \alpha(\tau, h)$ is invariant with respect to the shifts in τ, we will suppose below that $\alpha(0, h) = 0$.

Matrix $H(p_\alpha(\tau, h), \alpha(\tau, h))$ possesses certain symmetry with respect to the reversal of time:

$$H(p_\alpha(-\tau, h), \alpha(-\tau, h)) = QH(p_\alpha(\tau, h), \alpha(\tau, h))Q,$$

where $Q = \mathrm{diag}(1, -1, -1, 1)$. As a consequence of this symmetry we note the relationship used later on for the calculation the monodromy matrix $M(h) = \left[m_{jk}(h)\right]_{j,k=1}^4$ of system (3.1):

$$M(h) = QW^{-1}\left(\frac{T(h)}{2}, h\right) QW\left(\frac{T(h)}{2}, h\right).$$

Here $W(\tau, h)$ is the matrix of fundamental solutions of (3.1) ($W(0, h) = E_4$).

The solution $\mathbf{w} = 0$ is stable when the roots of characteristic equation

$$\rho^4 - a_1\rho^3 + a_2\rho^2 - a_1\rho + 1 = 0, \tag{3.2}$$

$$a_1 = \mathrm{tr}\, M(h), \quad a_2 = \sum_{j=1}^{3} \sum_{k=j+1}^{4} (m_{jj}m_{kk} - m_{jk}m_{kj}),$$

lie on the unit circle in the complex plane and are pairwise different. In this case the coefficients a_1 and a_2 in (3.2) should satisfy the inequalities:

$$|a_1| < \min\left\{\left(1 + \frac{a_2}{2}\right), 4\right\}, \quad a_2 < \frac{a_1^2}{4} + 2. \tag{3.3}$$

4. Asymptotic Expressions for the Monodromy Matrix

In motion along the phase trajectories of equation (2.2), corresponding to long-periodic solutions, the point $(p_\alpha(\tau, h), \alpha(\tau, h))$ stays for a long time in the vicinity of unstable stationary points $(p_\alpha^*, \alpha^*) = \left(1, \frac{\pi}{2}(2k+1)\right), k \in \mathbb{Z}$. Thus, on the major part of a period the qualitative behavior of $W(\tau, h)$ is described by the equation

$$\dot{W} = \Lambda W, \tag{4.1}$$

where $\Lambda = JH(p_\alpha^*, \alpha^*)$. The general solution of the equation (4.1) represents a product of the matrix function $G(\tau) = \exp(\Lambda\tau)$ and an arbitrary constant matrix. The properties of $G(\tau)$ depend on the values of the roots of the secular equation

$$\det(\Lambda - \chi E_4) = \chi^4 + d_1\chi^2 + d_2 = 0. \tag{4.2}$$

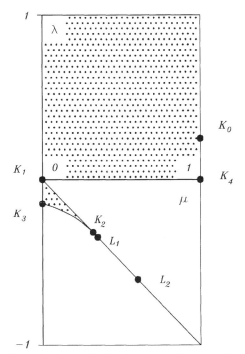

Figure 3. The set D. Shading shows the subset D_s.

If the conditions

$$d_1 > 0, \ d_2 > 0, \ d_1^2 > 4d_2 \tag{4.3}$$

are fulfilled, equation (4.2) has purely imaginary roots, which are pairwise different. The matrix $G(\tau)$ possesses invariant subspaces S_1 and S_2; the action of $G(\tau)$ on S_i ($i = 1, 2$) is equivalent to the action of the operator of rotation by the angle $\omega_i \tau$ ($\omega_{1,2}^2 = \frac{1}{2}(d_1 \pm \sqrt{d_1^2 - 4d_2})$).

Conditions (4.3) are satisfied in a subset D_s of the set D (Figure 3). The subset D_s consists of the square $D_\circ = \{0 < \mu \leqslant 1, 0 < \lambda \leqslant 1\}$ with a deleted point $K_0 = \left(1, \frac{1}{4}\right)$ and an open curvilinear triangle $K_1 K_2 K_3$, $K_1 = (0, 0)$, $K_2 = \left(\frac{1}{3}, -\frac{1}{3}\right)$, $K_3 = (0, \frac{-7 + \sqrt{45}}{2})$.

The behavior of $W(\tau, h)$ at a relatively fast motion of $(p_\alpha(\tau, h), \alpha(\tau, h))$ between the saddle points can be established by considering system (3.1) with $p_\alpha(\tau, h), \alpha(\tau, h)$ substituted by the asymptotic solutions $p_{\alpha_\pm}(\tau), \alpha_\pm(\tau)$. The matrix of fundamental solutions of the system $d\mathbf{w}/d\tau = JH(p_{\alpha_\pm}(\tau), \alpha_\pm(\tau))\mathbf{w}$ will be designated as $W_\pm(\tau)$.

Taking into account the above mentioned properties of $W(\tau, h)$, we can obtain the asymptotic expressions for the monodromy matrix of system (3.1) with parameters $\mu, \lambda \in D_s$. In the case of the planar motion of oscillatory type

$$M(h) = QU_+^{-1}QG(\kappa \ln \zeta)U_- QU_-^{-1}QG(\kappa \ln \zeta)U_+ + O(\zeta \ln \zeta), \tag{4.4}$$

where $\zeta = |h - h_\circ|$,

$$U_\pm = \lim_{\tau \to \infty} G(\tau_\circ - \tau) W_\pm(\tau), \quad \tau_\circ = \frac{\ln(24\mu)}{2\sqrt{3\mu}}, \quad \kappa = -\frac{1}{\sqrt{3\mu}}.$$

Similar formula for rotational long-periodic motions has the form

$$M(h) = \left[Q U_\pm^{-1} Q G(\kappa \ln \zeta) U_\pm \right]^2 + O(\zeta \ln \zeta). \tag{4.5}$$

Here $+(-)$ corresponds to the satellite rotation in the direction coinciding with (opposite to) the direction of rotation of the orbital reference frame.

Concerning the properties of the monodromy matrix for $(\mu, \lambda) \in D \setminus D_s$, we note only that in the general case the matrix $M(h)$ will contain elements that grow unboundedly for $h \to h_\circ$.

5. Stability and Instability Alteration

As it follows from relations (4.4) and (4.5), for $(\mu, \lambda) \in D_s$ the elements of the monodromy matrix and the coefficients of the equation (3.2) will be quasi-periodic functions of $\eta = \ln \frac{1}{|h-h_\circ|}$ (in the general case). The points $(a_1(\eta), a_2(\eta))$ are distributed on the plane $\mathbb{R}^2$ over some region (nonlinear projection of torus $\mathbb{T}^2$ on $\mathbb{R}^2$) and can be located both inside and outside the curvilinear triangle (3.3) (Figure 4). It implies that at continuous variation of η (or, what is the same, of h) the alteration of intervals of (orbital) stability and instability of corresponding long-periodic motions takes place. This phenomenon is the usual property of the family of the periodic solutions contiguous to a loop of a saddle-center separatrix in the Hamiltonian system (Churchill et al., 1980).

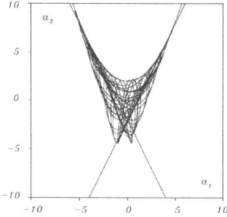

Figure 4. The positions of points determined by the values of the coefficients in equation (3.2) when η takes the values ranging from 10 to 100, $\mu = 0.99$ and $\lambda = 0.154$ (the case of oscillatory motion).

Numeric investigations show that the planar motions of a satellite with parameters $(\mu, \lambda) \notin D_s \bigcup K_0$ are unstable for h close enough to $h_\circ$ (the results of these extensive investigations will be published separately in the future). Nevertheless, if C_1 is any positive number then, under the appropriate choice of μ and λ values, stable motions will satisfy the condition $|h - h_\circ| < C_1^{-1}$, where $h_\circ \sim 1$. It is possible to prove that in the set D the appropriate points (μ, λ) can be found in the vicinity of points $L_1 = (0.34276, -0.34276)$ and $L_2 = (0.60413, -0.60413)$ (Figure 3).

Acknowledgements

This research was supported in part by INTAS grant 97-10771. The third author has been also supported by grants DGICYT PB 94-0215 (Spain) and CIRIT 1998 SGR-00042.

References

Beletsky, V.V.: 1975, *Attitude Satellite Motion in Gravitational Field*, MGU, Moscow (in Russian).

Churchill, R.C., Pecelli, G. and Rod, D.L.: 1980, 'Stability transitions for periodic orbits in hamiltonian systems', *Arch. Rat. Mech. Anal.*, **73**, pp. 313–347.

Kane, T.R. and Shippy, D.J.: 1963, 'Attitude Stability of Spinning Unsymmetrical Satellite in a Circular Orbit', *J.Astonaut. Sci.*, **14**, pp. 114–119.

Markeev, A.P., and Sokolsky, A.G.: 1977, 'Investigation of Stability of Planar Periodic Satellite Motions in a Circular Orbit', *Izv. Akad. Nauk SSSR, Ser. Mekh. Tverd. Tela*, pp. 46–57 (in Russian).

SOLVING A GYLDEN–MEŠČERSKIJ SYSTEM IN DELAUNAY–SCHEIFELE–LIKE VARIABLES

LUIS FLORíA

Grupo de Mecánica Celeste I. Dept. de Matemática Aplicada a la Ingeniería. Escuela Técnica Superior de Ingenieros Industriales. Universidad de Valladolid. E–47 011 Valladolid, Spain

Abstract. For an analytical treatment of Gylden–Meščerskij systems, in order to reduce the Gylden model to a perturbed Keplerian system, a canonical transformation from extended polar nodal variables to universal Delaunay–Similar variables is developed. This TR–like transformation is then followed by a change of independent variable. Our considerations are based on Deprit's approach to Scheifele's reducing TR–mapping and Deprit's treatment of the Gylden problem.

Key words: nonstationary two–body problem, Meščerskij's laws, Delaunay–Scheifele–like canonical variables, uniform treatment of two–body motion.

1. Introduction: On Gylden Systems and TR–Mappings

A *Gylden system* (Deprit 1983) is a two–body Kepler problem whose Keplerian coupling parameter $\mu \equiv \mu(t)$ undergoes variations in time. Inspired by a treatment of time–dependent Delaunay transformations (Deprit 1983, §3), we considered (Floría 1997, 1999) TR–type canonical maps (Deprit 1981; Floría 1994) that generalize the original transformations (Scheifele 1970, 1972) to Delaunay–Scheifele (DS) variables, and applied these mappings to a class of Gylden–type systems.

We study a two–body problem with a time–varying parameter $\mu(t)$ according to the Meščerskij's laws, in which case the problem is integrable. The *first Meščerskij law* for the change of μ yields the integrability case of Gylden systems most often applied in Celestial Mechanics. This model is a first approximation to more complex systems in astronomical, astrophysical and cosmological studies (e.g., isotropic mass loss in binary or multiple star systems), and is a preliminary step prior to the treatment of perturbed motion involving time–dependent gravitational–type systems, given as *perturbed Gylden problems*.

In *extended*, 8–dimensional *phase space* (homogeneous canonical formalism), one considers the canonically conjugate pair (t, p_0): the physical time t as a coordinate, and the negative p_0 of the energy. Scheifele (1970) proposed a canonical set of 8 *Delaunay–Similar* (DS) elliptic Keplerian elements, with the *true anomaly* as the independent variable, giving rise to a TR–theory of orbital motion. In that approach, the *true* anomaly is introduced as a canonical variable of the DS set. Accordingly, powers of $1/r$ are trigonometric polynomials (finite Fourier expansions) in the DS–angles, instead of Fourier series in t, which is more suitable for

Pretka-Ziomek et al./Dynamics of Natural and Artificial Celestial Bodies, 235–240, 2001.
© 2001 *Kluwer Academic Publishers.*

perturbation studies, since certain perturbed Kepler Hamiltonians reduce to Fourier polynomials.

Scheifele's derivation in extended phase–space is based on the Keplerian Hamiltonian in spherical variables, the integration of the Hamilton–Jacobi equation by separation of variables, and the evaluation of intermediate quadratures with the help of auxiliary integration variables (the usual eccentric and true anomalies.) The independent variable (physical time) is changed via a reparametrizing Sundman–type transformation, the new pseudo–time being of the true anomaly type.

The *Hill–Whittaker set* is $(r, \theta, v; p_r, p_\theta, p_v)$. *Deprit's approach* (1981) [also Floría 1994] to the TR–mapping resorts to these (extended) polar nodal variables to define a DS set, the TR–transformation being an extension of Hill's canonical transformation from a 6-dimensional phase space to an enlarged, 8-dimensional one.

We develop, with *circular* and *universal functions*, a canonical TR–like mapping to systematically derive universal DS–like variables for a *uniform* DS–type treatment of a model of time–dependent two–body problem. *Universal* means "independent of the kind of orbit." These variables generalize the sets applied by Scheifele in Artificial Satellite Theory. Our transformations convert the Gylden Hamiltonian into a perturbed Kepler problem, from which a normalizing Lie–type transform process can be started. Our study involves, as our basic *mathematical tools*, universal transcendental *Stumpff c–functions* and *Battin U–functions*, whose definitions and properties are found in, or derived from, Stiefel & Scheifele (1971, §11), Stumpff (1959, Ch. V), Battin (1987, §4.5, §4.6), and universal angle–like parameters (anomalies along a fictitious Keplerian–like motion.)

2. Notations and Development of the Transformation

The homogeneous Hamiltonian of a conventional *Gylden system* is

$$
\begin{aligned}
\mathcal{H}_{0,h} &\equiv \mathcal{H}_0 \left(r, -, -, t; p_r, p_\theta, -, - \right) + p_0 \\
&= (1/2) \left(p_r^2 + p_\theta^2/r^2 \right) - \mu(t)/r + p_0 .
\end{aligned}
\tag{1}
$$

The gravitational coupling parameter $\mu(t)$ is time–dependent. The variation of μ very often follows the so–called *Meščerskij Laws*, which fit into the pattern (where $\mu_0, \alpha, \beta, \delta$ are constants)

$$
\mu(t) = \mu_0/R(t), \quad \mu_0 = \mu(t_0), \quad R(t) = \sqrt{\alpha t^2 + 2\beta t + \delta} :
$$

- First Law: $\beta^2 - \alpha\delta = 0 \Longrightarrow R(t) = At + B$ is a linear function of t.
- Second Law: $\alpha = 0$.
- Unified Law: otherwise (general case).

From the extended polar nodal chart we perform a canonical transformation to DS–type variables. This *TR–like mapping* is defined by

$$S \; : \; (r, \theta, v, t \, ; \; p_r, p_\theta, p_v, p_0) \; \longrightarrow \; (\varphi, l, g, h \, ; \; \Phi, L, G, H) \, ,$$

$$S \equiv S(r, \theta, v, t \, ; \; \Phi, L, G, H) = \theta G + v H + t L + \int_{r_0}^{r} \sqrt{Q} \, dr,$$

$$Q \equiv Q(r, t \, ; \; \Phi, L, G, H) = 2\mu(t)/r - 2L - \gamma^2/r^2 \, ,$$

where $\gamma \equiv \gamma(\Phi, L, G, H)$ and r_0 is a simple zero of the r–equation

$$Q(r, t \, ; \; \Phi, L, G, H) = 0 \, . \tag{2}$$

As in Deprit (1983, §3), "the feature typifying a Gylden system enters the scheme in only one place, namely in the generator S where μ is allowed to be a function of the time." By the theory of canonical transformations, the *generating relations* derived from S are

$$p_r = \partial S/\partial r = \sqrt{Q} \, , \quad p_\theta = \partial S/\partial \theta = G, \quad p_v = \partial S/\partial v = H,$$

$$p_0 = \partial S/\partial t = L + \mu'(t) I_1 \, , \quad \mu'(t) \equiv d\mu(t)/dt \, ,$$

$$\varphi = \partial S/\partial \Phi = -\gamma (\partial \gamma/\partial \Phi) I_2 \, ,$$

$$l = \partial S/\partial L = t - I_0 - \gamma (\partial \gamma/\partial L) I_2 \, ,$$

$$g = \partial S/\partial G = \theta - \gamma (\partial \gamma/\partial G) I_2 \, ,$$

$$h = \partial S/\partial H = v - \gamma (\partial \gamma/\partial H) I_2 \, , \qquad \text{where}$$

$$I_m = \int_{r_0}^{r} \frac{dr}{r^m \sqrt{Q}} \, dr \, , \quad m = 0, 1, 2, \quad \text{and} \; p_r = dr/dt \, .$$

To rewrite Q and calculate the quadratures, introduce some *subsidiary quantities* q, e and p, depending on t and the new momenta:

$$q = \mu(1 - e)/2L \, , \quad \gamma^2 = \mu q (1 + e) \equiv \mu p \, ,$$

$$\implies Q = \mu(r - q)[r(e - 1) + q(e + 1)]/qr^2 \, ,$$

and q is a zero of Eq (2). Quadratures I_m are evaluated in terms of *auxiliary variables* s and f, acting as *universal* parameters: eccentric– and true–like anomaly. Changing $r \to s \equiv s(r, t; \Phi, L, G, H)$ leads to *eccentric–like anomaly developments:*

$$r = q + \mu e s^2 c_2 (2Ls^2) = q + \mu e U_2 (s, 2L) \, ,$$

$$dr = \mu e s c_1 (2Ls^2) \, ds = \mu e U_1 (s, 2L) \, ds \, ,$$

$$r = r_0 = q \implies s_0 = s(r_0) = 0 \; \text{(pericentre)} \, ,$$

$$r \sin f = \gamma U_1 (s, 2L) \, , \quad df = (\gamma/r) \, ds \, ,$$

$$\sqrt{Q} = dr/dt = [\mu e U_1 (s, 2L)] / [q + \mu e U_2 (s, 2L)] \, ,$$

$$I_0 = (1/\gamma) \int_0^f r^2 \, df = q s + \mu e U_3 (s, 2L), \quad I_1 = s \, .$$

The change of integration variable $r \to f \equiv f(r, t; \Phi, L, G, H)$ gives rise to *true–like anomaly developments* summarized in the formulae:

$$r = \frac{q\,(1+e)}{1 + e\cos f}\,, \quad d\left(\frac{1}{r}\right) = -\frac{d\,r}{r^2} = -\frac{e\sin f}{q\,(1+e)}\,d\,f\,,$$

$$r = r_0 = q \implies f_0 = f(r_0) = 0 \ \text{(pericentre)},$$

$$Q = \left[\mu\,e^2\sin^2 f\right]/\left[q\,(1+e)\right] \implies I_2 = f/\gamma\,.$$

Accordingly, we complete the transformation formulae derived from S:

$$p_r = \sqrt{Q}\,, \quad p_\theta = G\,, \quad p_\nu = H\,, \quad p_0 = L + \mu'(t)\,s\,,$$

$$\varphi = -\,(\partial\gamma/\partial\Phi)\,f\,, \quad l = t - [q\,s + \mu\,e\,U_3] - (\partial\gamma/\partial L)\,f\,,$$

$$g = \theta - (\partial\gamma/\partial G)\,f\,, \quad h = \nu - (\partial\gamma/\partial H)\,f\,.$$

Our derivations and results involve *universal functions* of Stumpff and Battin, and give rise to *implicit function and inversion problems*.

3. Transformation of the Gylden Hamiltonian

Under this TR–like transformation, Hamiltonian (1) is converted into

$$\widetilde{\mathcal{H}}_h = \left(G^2 - \gamma^2\right)/\left(2r^2\right) + [d\,\mu(t)/d\,t]\,s\,, \tag{3}$$

which may be simplified via a reparametrizing transformation $t \to \tau$, of the *Sundman type*, introducing a *new independent variable*:

$$d\,t = \widetilde{f}\,d\,\tau \quad \text{with} \quad \widetilde{f} = r^2/\mathcal{G}\,, \quad \mathcal{G} \equiv \mathcal{G}\,(\Phi, L, G, H)\,.$$

The *new Hamiltonian* $\mathcal{K}_0$ corresponding to this pseudo–time τ is

$$\mathcal{K}_0 = \widetilde{\mathcal{H}}_h\,\widetilde{f} = \left(G^2 - \gamma^2\right)/\left(2\mathcal{G}\right) + \left[r^2\,\mu'(t)\,s\right]/\mathcal{G}\,. \tag{4}$$

These Hamiltonians $\widetilde{\mathcal{H}}_h$ and $\mathcal{K}_0$ depend on the canonical momenta (Φ, L, G, H) and on the *non–canonical variables* (r, t, s). By virtue of the theory of TR–transformations and DS variables, $\mathcal{K}_0$ can be viewed as a perturbed Keplerian system in which the Keplerian term, namely $\left(G^2 - \gamma^2\right)/(2\mathcal{G})$, undergoes the perturbations stemming from $r^2\mu'(t)s/\mathcal{G}$. This perturbing part should be developed in terms of the new canonical variables of the DS–type; to solve nested dependences between variables and parameters, *implicit function results, inversion theorems and Fourier analysis of the two–body problem* can be used. In addition to this, as a general rule, variations of μ can be expected to be slow, and the perturbation effects will be small; quantities as $\mu'(t)$ are of the first order in some appropriate small parameter, $\left[\mu'(t)\right]^2$ and $\mu''(t)$ of the second order, etc. Thus, the perturbing terms can be expanded as a series in powers of a small perturbation parameter. Consequently, an approximate analytical solution can be obtained in integrating the Hamilton equations with the help of canonical perturbation techniques (e.g., Prieto & Docobo 1997a,b).

In particular, certain choices of γ and $\mathcal{G}$, as functions of the new momenta, lead to simple forms for the first term (the standard Keplerian–like part) of these Hamiltonians (Floría 1994). For instance, for

- $\gamma(\Phi, G) = G - \Phi$, $\mathcal{G} \equiv 1$ (Scheifele 1970; Scheifele & Stiefel 1972, Part B),
- $\gamma(\Phi, G) = G - \Phi$, $\mathcal{G} = (G + \gamma)/2$ (Sigrist, in Scheifele & Stiefel 1972, Part D):

$$\partial\gamma/\partial\Phi = -1, \ \partial\gamma/\partial G = 1, \ \partial\gamma/\partial L = \partial\gamma/\partial H = 0,$$

and $\varphi = f$, the true-like anomaly is made a canonical DS variable.

4. Some Remarks Concerning Elliptic–Type Motion

These universal–function developments are readily particularized in the case of bounded motion, from which we recover our formulae (Floría 1997, §3, §4), with elliptic eccentric–like anomaly $E = s\sqrt{2L}$, "fictitious" semi–major axis $a = \mu/(2L)$, $\gamma^2 = \mu a(1 - e^2) = \mu p$,

$$p_0 = L + \left[\mu'(t) E/\sqrt{2L}\right],$$

$$l = t - \left[\mu(t)/(2L)^{3/2}\right](E - e \sin E) - (\partial\gamma/\partial L) f,$$

$$\mathcal{K}_0 = (G^2 - \gamma^2)/(2\mathcal{G}) + \left[r^2 \mu'(t) E\right]/\left(\mathcal{G}\sqrt{2L}\right).$$

If $\mu' \equiv 0$, the equations correspond to those of a Kepler problem. Thus, slow variations of μ can be considered as small perturbations, and the Gylden problem can be dealt with and analytically integrated through conventional methods of the theory of perturbations. To this end, $\mathcal{K}_0$ is to be expanded in powers of a small parameter ε such that for $\varepsilon = 0$ the standard Kepler problem is recovered. Postulating, after appropriate choice of units, a law $\mu(t) = 1/(1 + At)$, with $A \ll 1$, this parameter ε should be related to coefficient A. The resulting perturbed Kepler problem can be treated by a normalization process to obtain a Hamiltonian that depends on the canonical momenta only. In our case, the derivatives of μ can be expressed in the form:

$$d^n\mu/dt^n = n!(-A)^n [\mu(t)]^{n+1}.$$

Acknowledgements

Study partially supported by Junta de Castilla y León (Grants VA34/99 and VA68/00), and DGES of Spain (PB98–1576).

References

Battin, R. H.: 1987, *An Introduction to the Mathematics and Methods of Astrodynamics*, American Institute of Aeronautics and Astronautics, New York.

Deprit, A.: 1981, 'A Note Concerning the TR–Transformation', *Celest. Mech.* **23**, pp. 299–305.

Deprit, A: 1983, 'The Secular Accelerations in Gylden's Problem', *Celest. Mech.* **31**, pp. 1–22.

Floría, L.: 1994, 'On the Definition of the Delaunay–Similar Canonical Variables of Scheifele', *Mechanics Research Communications* **21**, pp. 409–414.

Floría, L.: 1997, 'Perturbed Gylden Systems and Time–Dependent Delaunay–Like Transformations', *Celest. Mech. and Dyn. Astron.* **68**, pp. 75–85.

Floría, L.: 1999, 'Generalized Gylden–Type Systems in Universal DS–Like TR–Variables', In: J. Henrard & S. Ferraz–Mello (Eds.), *Impact of Modern Dynamics in Astronomy* (IAU Colloq. 172, Namur, 1998), Kluwer, pp. 461-462.

Prieto, C. and Docobo, J. A.: 1997a, 'Analytical Solution of the Two–Body Problem with Slowly Decreasing Mass', *Astron. and Astroph.* **318**, pp. 657–661.

Prieto, C. and Docobo, J. A.: 1997b, 'On the Two–Body Problem with Slowly Decreasing Mass', *Celest. Mech. and Dyn. Astron.* **68**, pp. 53–62.

Scheifele, G.: 1970, 'Généralisation des éléments de Delaunay en Mécanique Céleste. Application au mouvement d'un satellite artificiel', *Comptes rendues de l'Académie des Sciences de Paris*, série **A, 271**, pp. 729–732.

Scheifele, G. and Stiefel, E.: 1972, *Canonical Satellite Theory Based on Independent Variables Different from Time*, Report to ESRO under ESOC–contract No. 219/70/AR, ETH, Zürich.

Stiefel, E. L. and Scheifele, G.: 1971, *Linear and Regular Celestial Mechanics*, Springer-Verlag, Berlin, Heidelberg, New York.

Stumpff, K.: 1959, *Himmelsmechanik, I*, VEB Deutscher Verlag der Wissenschaften, Berlin.

NEAR-EARTH ASTEROIDS CLOSE TO MEAN MOTION RESONANCES: THE ORBITAL EVOLUTION

LARISA BYKOVA and TATYANA GALUSHINA
Applied Mathematics and Mechanics Institute, Tomsk, Russia

Abstract. The results of study of orbital evolution of some real and fictitious Near-Earth asteroids close to low-order resonances with the Earth or other planets are presented. The investigations were carried out by means of a numerical integration of differential equations, taking into account of the perturbations from major planets (except Pluto) and Moon. For each investigated object an ensemble of 100 test particles with orbital elements nearby those of nominal orbit was constructed and its evolution was retraced on the time span (-3000, $+3000$ years).

Key words: asteroids, resonances, orbital evolution

1. Introduction

The dynamics of Near-Earth asteroids (NEAs) close to low-order resonances with the Earth or other planets is considered. Many papers are devoted to study of population of NEAs (e.g., (Milani et al., 1989; Tancredi, 1998; Wiegert et al., 1998; Chapman and Morrison, 1994)). The aim of this paper is construction of probability domains of the motion of some NEAs near resonances conditioned by commensurabilities of mean motions of the asteroids and major planets, investigation of interaction of close approaches and resonances as well as stability of resonance relations. The paper is the continuation of our previous investigations of the NEAs dynamics (Bykova and Timoshenko, 1998; Bykova and Galushina, 2000).

2. Methods of Investigation

The simulation of the NEAs motion in the long intervals of time by the methods of numerical integration meets certain difficulties connected with peculiarities of NEAs motion: crossing of their orbits with the orbits of major planets, close approaches with planets and large eccentricities. One can point out two problems here. The first problem consists in preserving the acceptable precision of properly numerical integration of the motion equations for long intervals of time. The difficulties arise in simulation of motion of NEAs having many close approaches in connection with strong perturbations and correspondingly with the sharp decrease of the integration step and rapid growth of method and rounding off errors.

The second problem is connected with errors of the initial data. A considerable growth of these errors in time does not enable to investigate a long-term evolution of concrete real object.

Therefore we applied the following technique of the investigation of motion of each real NEA. In order to achieve a more high accuracy of numerical integration, new forms of regularizing and stabilizing equations of motion (Bordovitsyna et al., 1998) obtained by KS-transformation (Stiefel and Scheifele, 1971) were used. Equations of motion of asteroids were integrated numerically by the Everhart method (Everhart, 1985), taking into account perturbations from major planets (except Pluto) and Moon. Then for each investigated object an ensemble of particles with initial parameters close to the nominal orbit was constructed and its evolution was retraced on the time span $(-3000, +3000$ years). The initial set of orbits has been generated for 100 test particles in relation to the selected center (initial epoch) with the help of random number generator on the basis of the normal law of dispersing and a corresponding covariation matrix.

3. Numerical Simulation of Motion of Some NEAs

The described approach was applied to the construction of possible motion domains of the following NEAs: 1996 DH, 1994 RB, 3838 Epona, 2608 Seneca, 1996 AJ1. In the process of numerical integration of motion equations of these objects all close approaches and commensurabilities of low orders with the Earth and other planets were revealed. Commensurabilities were determined by means of an estimation of the resonance band value $\alpha = k_1 n_a - k_2 n_p$, where n_a is the mean motion of the asteroid, n_p is the mean motion of a planet, k_1, k_2 are integers. The initial osculating elements of orbits were taken from the E. Bowell catalogue at the epoch of 22.01. 1999 (ftp://ftp.lowell.edu/pub/elgb/ astorb.dat). The evaluation of probable errors of the initial orbital elements was carried out by the linear least squares method using 120 observations of 1996 DH, 99 observations of 3838 Epona, 53 observations of 2608 Seneca in several oppositions, 21 observations of 1994 RB in the interval of 2 days and 25 observations of 1996 AJ1 in the interval of 16 days.

The results of numerical simulation of orbital evolution for all enumerated NEAs are presented in figures 1–5. Here the close approaches with the Earth and other planets, the evolution of the resonance band α and the osculating orbital elements a, e, q, i are shown for the nominal orbits (black solid lines) and for the sets of 100 test particles (a rarefied background). In the figures d is a planetocentric distance of the NEAs in astronomical units, T is the time in thousands years, $\alpha = k_1 n_a - k_2 n_p$ in sec/day). Initial sets of orbits were generated on the basis of the probable errors: $\Delta r_0 = 6 \cdot 10^{-6}$ AU in coordinates, $\Delta v_0 = 3 \cdot 10^{-8}$ AU/day in velocity components for 2608 Seneca, $\Delta r_0 = 1 \cdot 10^{-6}$ AU, $\Delta v_0 = 1 \cdot 10^{-8}$ AU/day for 3838 Epona, $\Delta r_0 = 1 \cdot 10^{-4}$ AU, $\Delta v_0 = 3 \cdot 10^{-5}$ AU/day for 1994 RB, $\Delta r_0 = 1 \cdot 10^{-6}$ AU, $\Delta v_0 = 5 \cdot 10^{-8}$ AU/day for 1996 DH, and $\Delta r_0 = 2 \cdot 10^{-3}$ AU, $\Delta v_0 = 4 \cdot 10^{-5}$ AU/day for 1996 AJ1.

The investigated asteroids have the following commensurabilities of mean motions: 3838 Epona with Venus 1:3, 2608 Seneca and 1994 RB with Jupiter 3:1,

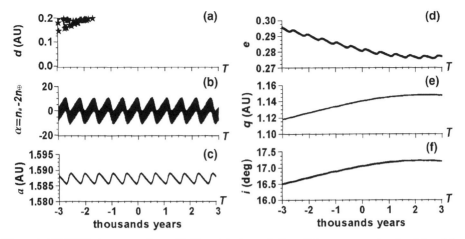

Figure 1. 1996 DH: the approaches with Mars (a), the evolution of the resonance band α (b) and the orbital elements a (c), e (d), q (e), i (f).

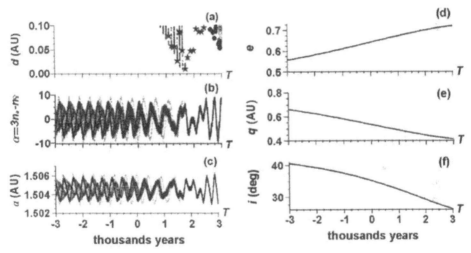

Figure 2. 3838 Epona: the approaches with the Earth("•") and Mars("⋆") (a), the evolution of the resonance band α (b) and the orbital elements a (c), e (d), q (e), i (f).

1996 DH with the Earth 1:2. The evolution of the resonance band α shows that all test particles near 1996 DH, 1994 RB and 3838 Epona regularly pass through the value of precise commensurability $\alpha = 0$, moving away periodically from it at $\pm 11''$ for 1996 DH and 3838 Epona (figures 1b, 2b) and at $\pm 22''$ for 1994 RB (figure 3b). It is well seen (figures 2a, 2b) how a group of close approaches of the asteroid 3838 Epona with Mars in 1390-1780 years ($0.008 < d < 0.03$ AU) influences the regular behaviour of α, however it does not destroy resonance. 2608 Seneca had the very close approach with the Earth in 441 year (figure 4 a1). This approach influences the regular behaviour of α for the nominal orbit, however it

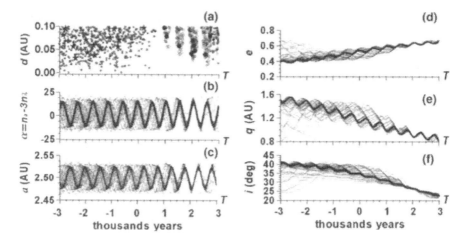

Figure 3. 1994 RB: the approaches with the Earth ("•") and Mars ("⋆"), the evolution of the resonance band α (b) and the orbital elements a (c), e (d), q (e), i (f).

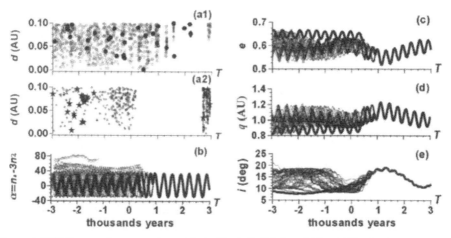

Figure 4. 2608 Seneca: the approaches with the Earth ("•")(a1) and Mars("⋆")(a2), the evolution of the resonance band α (b) and the orbital elements e (c), q (d), i (e).

does not destroy resonance, but some of the test particles near 2608 Seneca move in a rather large neighbourhood of resonance (figure 4b). Quite another picture is observed for asteroid 1996 AJ1 having many close approaches with planets (figure 5). Some of the test particles near 1996 AJ1 reach the closest approaches with the Earth (0.0005 AU) and other planets (e.g., 0.0001 AU with Venus). This object and test particles close to it repeatedly reach precise commensurability with the Earth (2:3) and without delay in it move in a large neighbourhood of the resonance.

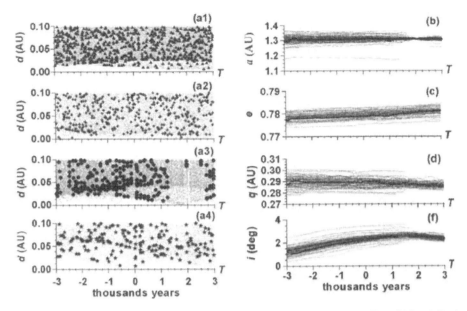

Figure 5. 1996 AJ1: the approaches with Mercury(a1), Venus(a2), the Earth(a3) and Mars(a4), the evolution of the orbital elements a (b), e (c), q (d), i (f).

4. Conclusion

Thus regions of possible motions of the NEAs 2608 Seneca, 3838 Epona, 1996 DH, 1994 RB, 1996 AJ1 constructed on the basis of probable errors of initial data are considered in the interval of time (-3000 years, 3000 years). It is shown that real and fictitious asteroids (generated as described above) close to 1996 DH, 3838 Epona, 1994 RB do not have very close approaches with the Earth and other planets and preserve stable resonance relationships. For the nominal orbit of asteroid 2608 Seneca, having only one very close approach with the Earth in the considered interval, the resonance 3:1 with Jupiter is not destroyed, but some of the test particles near Seneca escape from the resonance region. The NEA 1996 AJ1 and test particles close to it have many very close encounters with the Earth, Venus, Mercury and Mars. Despite the fact that these particles reach the value of precise commensurability 2:3 with the Earth repeatedly they do not enter resonance.

Acknowledgements

We express gratitude to Chernetenko Yu.A.(Institute of Applied Astronomy of RAS) for presenting observations of NEAs at our disposal. We also acknowledge the support coming from the organizers for attending the US-European Celestial Mechanics Workshop. This research was supported by the Russian Foundation for Basic Researches, Grant N 98-02-16491.

References

Bordovitsyna, T. V., Bykova L. E. and Avdyushev V. A.: 1998, 'Problems in applications of regularizing and stabilizing KS-transformations to tasks of dynamics of planets, natural satellites and asteroids', *Astronomy and Geodezy*, Tomsk State University Publishers, **16**, 33–57.

Bykova, L.E. and Galushina, T.Yu.: 2000, 'Numerical simulation of motion of some Near-Earth asteroids close to low-orders resonances', In *Proceedings of the conference 'Near-Earth Astronomy and problems of studying of small bodies of Solar system'*, Cosmosinform, Moscow, 56–64.

Bykova, L.E. and Timoshenko, L.V.: 1998, 'The near-Earth asteroids: encounters with the planets, transformation of the orbital elements', *Astronomy and geodesy*, Tomsk State University Publishers **16**, 183–238.

Chapman, C. R. and Morrison, D.: 1994, 'Impacts on the Earth by asteroids and comets: assessing the hazard', *Nature* **367**, 33–40.

Everhart, E.: 1985, 'An efficient integrator that uses Gauss-Radau spacings', In *Dynamics of comets: their origin and evolution* (A. Carusi and G.B. Valsecchi, Eds.), Reidel, Dortrecht, 185-200.

Milani, A., Carpino, M., Hahn, G. and Nobili, A.: 1989, 'Dynamics of planet-crossing asteroids classes of orbital behaviour. Project SPACEGUARD', *Icarus* **78**, 212–269.

Stiefel, E. L. and Scheifele, G.: 1971, *Linear and regular celestial mechanics*, Springer–Verlag, Berlin, Heidelberg, New York.

Tancredi, G.: 1998, 'An asteroid in a Earth-like orbit', *Celes. Mech. Dyn. Astron.* **69**, 119–132.

Wiegert, P., Innanen, K. A. and Mikkola, S.: 1998, 'The orbital evolution of Near–Earth asteroid 3753', *Astron. J.* **115**, 2604–2613.

LYAPUNOV STABILITY FOR LAGRANGE EQUILIBRIA OF ORBITING DUST

VÍCTOR LANCHARES
Dpto. Matemáticas y Computación. Universidad de La Rioja. 26004 Logroño. Spain

TEODORO LÓPEZ-MORATALLA
Real Instituto y Observatorio de La Armada. 11110 San Fernando. Spain

DAVID FARRELLY
Department of Chemistry and Biochemistry. Utah State University, Logan, Utah 84322-0300

Abstract. We examine the basic dynamics associated with a simple model of a dust particle interacting with a planetary gravitational field, solar radiation pressure and a constant magnetic field. We find the presence of global equilibrium points that are analogous to the Lagrangian equilibria of the circular restricted three-body problem.

Linear stability analysis is carried out for the equilibrium points in order to establish stability conditions in terms of the two free parameters of the problem. However, linear stability does not ensures Lyapunov stability and further analysis is needed. To go further in the analysis, we apply a theorem due to Arnold that ensures Lyapunov stability almost for every pair of the free parameters but for some resonant cases. These resonant cases are analyzed by studying the phase flow on the reduced phase space after normalization.

Key words: Lyapunov stability, Resonances

1. Introduction

The model to account for the effect of radiation pressure on dust particles orbiting about an idealized planet revolving around the Sun on a circle in a fixed plane is characterized (Deprit, 1984) by the Hamiltonian:

$$\mathcal{H} = \frac{1}{2}(X^2 + Y^2 + Z^2) - \frac{\mu}{r} + f\,(x\cos\omega_f t + y\sin\omega_f t).$$

In this context, the origin of coordinates is set at the center of mass in the pair sun-planet; the coordinate plane (x, y) is identified with the plane of the planet's orbit and ω_f stands for the mean motion of the planet around the sun. If, in addition, an homogeneous magnetic field perpendicular to the planet's orbit is accounted, the Hamiltonian characterizing the new model is given by:

$$\mathcal{H} = \frac{1}{2}(X^2 + Y^2 + Z^2) - \frac{\mu}{r} + f\,(x\cos\omega_f t + y\sin\omega_f t) + \frac{\omega_c}{2}(xY - yX) + \frac{\omega_c^2}{8}(x^2 + y^2),$$

where ω_c is the cyclotron frequency. This system was first established in the context of atomic physic by Lee *et al.* (Lee *et al.*, 1995; Lee *et al.*, 1997) to describe the

Pretka-Ziomek et al./Dynamics of Natural and Artificial Celestial Bodies, 247–252, 2001.
© 2001 *Kluwer Academic Publishers.*

dynamics of a Hydrogen atom subjected to a static magnetic field and a circularly polarized microwave field.

We will restrict ourselves to the case of co-planar orbits, that is to say, those orbits for which $z = Z = 0$. Indeed, they are the solutions for the canonical equations derived from the Hamiltonian:

$$
\mathcal{H} = \frac{1}{2}(X^2 + Y^2) - \frac{\mu}{r} + f\,(s\cos\omega_f t + y\sin\omega_f t) +
$$
$$
+ \frac{\omega_c}{2}(xY - yX) + \frac{\omega_c^2}{8}(x^2 + y^2). \tag{1}
$$

The explicit dependence of time in Hamiltonian (1) may be removed by going to a synodic frame of reference that rotates at the constant angular velocity ω_f. The new Hamiltonian becomes:

$$
\mathcal{H} = \frac{1}{2}(X^2 + Y^2) - \frac{\mu}{r} + fx - \left(\omega_f - \frac{\omega_c}{2}\right)(xY - yX) + \frac{\omega_c^2}{8}(x^2 + y^2),
$$

depending on the parameters μ, f, ω_f and ω_c. However, it is possible to reduce the number of parameters by a suitable scaling of coordinates and time arriving to the definitive Hamiltonian:

$$
\mathcal{H} = \frac{1}{2}(X^2 + Y^2) - \frac{1}{r} + \epsilon x - (xY - yX) + \frac{\omega_s^2}{8}(x^2 + y^2), \tag{2}
$$

where ϵ and ω_s are positive dimensionless parameters defined as:

$$
\epsilon = \frac{f}{\mu^{1/3}\omega^{4/3}}, \qquad \omega_s = \frac{\omega_c}{\omega}
$$

being $\omega = \omega_f - \frac{\omega_c}{2}$.

2. Equilibria and Linear Stability Analysis

The equations of the motion are given by the four equations:

$$
\dot{x} = X + y, \qquad \dot{X} = -\frac{x}{r^3} + Y - \epsilon - \frac{\omega_s^2}{4}x,
$$
$$
\dot{y} = Y - x, \qquad \dot{Y} = -\frac{y}{r^3} - X - \frac{\omega_s^2}{4}y. \tag{3}
$$

The presence of equilibrium points, similar to those encountered in the restricted three body problem, has been noticed by several authors (Białynicki-Birula *et al.*, 1994; Lee *et al.*, 1995; Lee *et al.*, 1997; Salas *et al.*, 1999). Indeed, the critical points are obtained by equating to zero the right hand sides of the equations (3) and they must satisfy the following conditions:

$$X = 0, \qquad Y = x, \qquad y = 0,$$

$$\left(1 - \frac{\omega_s^2}{4}\right) x^3 - \epsilon x^2 - 1 = 0, \quad (x > 0),$$

$$\left(1 - \frac{\omega_s^2}{4}\right) x^3 - \epsilon x^2 + 1 = 0, \quad (x < 0). \tag{4}$$

Moreover, these points can be associated to the critical points of an effective potential, also known as a zero velocity surface, given by:

$$V(x, y) = -\frac{1}{r} + \epsilon x - \frac{1}{2}\left(1 - \frac{\omega_s^2}{4}\right)(x^2 + y^2).$$

Depending on the values of the parameters we find two relevant configurations:

a) $(1 - \frac{\omega_s^2}{4}) > 0$, then $V(x, y)$ has a maximum (E_M) and a saddle point (E_s).

b) $(1 - \frac{\omega_s^2}{4}) < 0$ and $\epsilon > \sqrt[3]{\frac{27}{4}\left(1 - \frac{\omega_s}{4}\right)^2}$, then $V(x, y)$ has a minimum (E_m) and a saddle point (E_s).

Linear stability analysis may be performed by a second order expansion of the Hamiltonian (2) around the equilibrium points. In this way, we obtain the quadratic Hamiltonian:

$$\mathcal{H}_2 = \frac{1}{2}(X^2 + Y^2) - (xY - yX) + \frac{1}{2}(\alpha x^2 + \beta y^2),$$

where we have translated the critical point to the origin and dropped the constant term. The two parameters α and β are given by:

$$\alpha = \frac{\omega_s^2}{4} - 3\frac{x_0^2}{r_0^2} + \frac{1}{r_0^3}, \qquad \beta = \frac{\omega_s^2}{4} + \frac{1}{r_0^3}, \tag{5}$$

being x_0 the x coordinate of the equilibrium point and $r_0 = x_0$ for $x_0 > 0$ and $r_0 = -x_0$ for $x_0 < 0$.

We find linear stability if one of the two following conditions is satisfied (see e.g. López-Moratalla, 1997)

i) $\alpha > 1$ and $\beta > 1$. In this case $\mathcal{H}_2$ is positive defined and Lyapunov's stability theorem ensures non-linear stability.

ii) $\alpha < 1$, $\beta < 1$ and $(\alpha - \beta)^2 + 8(\alpha + \beta) > 0$. In this case $\mathcal{H}_2$ is neither positive nor negative defined and linear stability does not imply non-linear stability.

It is not hard to prove that the points labeled as E_s are unstable while E_m is a stable equilibrium satisfying condition i) and, then, stable in the Lyapunov sense. On the other hand, for E_M we find $\alpha < 1$ and $\beta < 1$ and it can be either stable or unstable depending on the sign of $(\alpha-\beta)^2+8(\alpha+\beta)$. Then, $(\alpha-\beta)^2+8(\alpha+\beta) = 0$ defines a stability boundary in the parameter plane given by a curve which consists, in this case, of an upper branch and a lower branch (Lee *et al.*, 1997; Salas *et al.*, 1999):

$$\epsilon = \frac{2 - 3\omega_s^2 \pm (1 - \frac{\omega_s^2}{2})\sqrt{4 - 9\omega_s^2}}{2^{1/3}\omega_s^{2/3}\left(2 \pm \sqrt{4 - 9\omega_s^2}\right)^{2/3}}.$$

The two branches join smoothly at $\omega_s = \frac{2}{3}$, where the problem defined by Hamiltonian (2) becomes integrable (Lee *et al.*, 1997).

3. Non-linear stability analysis

To establish non-linear stability conditions for E_M we use the well known Arnold-Moser theorem (Arnold, 1934). In order to apply the theorem, we have to compute the Birkhoff normal form of the Hamiltonian, in a vicinity of the equilibrium point, up to some order N. Expressed in action-angle variables $(\psi_1, \psi_2, \Psi_1, \Psi_2)$, it is given by:

$$\mathcal{H} = \mathcal{H}_2 + \mathcal{H}_4 + \cdots + \mathcal{H}_{2n} + \bar{\mathcal{H}}, \tag{6}$$

where $\mathcal{H}_{2k} = \mathcal{H}_{2k}(\Psi_1, \Psi_2)$ are homogeneous polynomial of degree k in Ψ_i with real coefficients, $\mathcal{H}_2 = \omega_1\Psi_1 - \omega_2\Psi_2$, $(0 < \omega_2 < \omega_1)$ and $\bar{\mathcal{H}}$ contains terms of order higher than n.

Under these assumptions (Meyer and Schmidt, 1986), the origin of the system of differential equations derived from (6) is stable if for some k, $2 \leqslant k \leqslant n$ $D_{2k} = \mathcal{H}_{2k}(\omega_2, \omega_1) \neq 0$.

Note that some non-resonant assumptions on the frequencies ω_1 and ω_2 are implicit in the conditions of the above theorem. So, in the simplest case, $D_4 \neq 0$, the theorem fails for $q\,\omega_1 = p\,\omega_2$ with $p + q \leqslant 4$, $p, q \in \mathbf{N}$.

The application of the theorem requires a great amount of calculation we have carried out using Mathematica^{TM}. Outside the resonant cases 1:1, 2:1 and 3:1, we obtain $D_4 = \frac{3}{8}\frac{P}{Q}$ with:

$$Q = x_0^5 \left(-8 - 4x_0^3 + x_0^3\,\omega_s^2\right)^2 \left(9 - 8x_0^3 + 4x_0^6\,\omega_s^2\right)$$
$$\left(-864 + 656x_0^3 + 144x_0^6 - 36x_0^3\,\omega_s^2 - 328x_0^6\,\omega_s^2 + 9x_0^6\,\omega_s^4\right)$$

$$\begin{aligned}
P = \; & 995328 - 1142784x_0^3 - 309760x_0^6 - 773888x_0^9 + 1774848x_0^{12} - \\
& 387072x_0^{15} - 165888x_0^{18} - 82944x_0^3\,\omega_s^2 + 1651968x_0^6\,\omega_s^2 - \\
& 1235136x_0^9\,\omega_s^2 - 955648x_0^{12}\,\omega_s^2 + 89088x_0^{15}\,\omega_s^2 + 474624x_0^{18}\,\omega_s^2 + \\
& 7776x_0^6\,\omega_s^4 + 123696x_0^9\,\omega_s^4 + 421344x_0^{12}\,\omega_s^4 - 224768x_0^{15}\,\omega_s^4 - \\
& 248064x_0^{18}\,\omega_s^4 + 25596x_0^9\,\omega_s^6 - 31824x_0^{12}\,\omega_s^6 + 37952x_0^{15}\,\omega_s^6 + \\
& 46272x_0^{18}\,\omega_s^6 - 2187x_0^{12}\,\omega_s^8 + 4680x_0^{15}\,\omega_s^8 - 3048x_0^{18}\,\omega_s^8 + 54x_0^{18}\,\omega_s^{10}
\end{aligned}$$

Thus, we can ensure that E_M is Lyapunov stable in the whole region of linear stability but, may be, in the sets $\{(\omega_s, \epsilon) \mid D_4 = 0\}$, $\{(\omega_s, \epsilon) \mid \omega_1 = 2\omega_2\}$, $\{(\omega_s, \epsilon) \mid \omega_1 = 3\omega_2\}$ and $\{(\omega_s, \epsilon) \mid \omega_1 = \omega_2\}$. The last one coincides with the linear stability boundary and, for this case, E_M is unstable.

3.1. RESONANCE 2:1

We study here the stability at the 2:1 resonance. Now, Arnold's theorem fails and other results must be applied, for example, the construction of an appropriate Lyapunov function (Markeev, 1968). However, we prefer a geometrical approach by using the so called *extended Lissajous variables* (Elipe, 2000), specially useful to handle oscillators in resonance.

Let us assume that there exists two coprime integers p and q such that $q\,\omega_1 = p\,\omega_2$. Then, the transformation:

$$\psi_1 = p(\ell + g), \quad \Psi_1 = \frac{L + G}{2p}, \quad \psi_2 = q(\ell - g), \quad \Psi_2 = \frac{L - G}{2q},$$

puts $\mathcal{H}_2$ in the very simple form:

$$\mathcal{H}_2 = \omega G, \tag{7}$$

being $\omega = \frac{\omega_1}{p} = \frac{\omega_2}{q}$.

The main advantage of this set of canonical coordinates is that the Lie derivative associated with (7) is very simple and, when the perturbation is periodic in g, the normalization procedure results in averaging over g.

In order to get the geometrical approach, attention must be paid on the topological structure of the phase space after normalization. In fact, it is defined by the invariants:

$$I_0 = \tfrac{1}{2}G, \quad I_1 = \tfrac{1}{2}L,$$
$$I_2 = 2^{-(p+q)/2}(L - G)^{p/2}(L + G)^{q/2} \cos 2pq\ell,$$
$$I_3 = 2^{-(p+q)/2}(L - G)^{p/2}(L + G)^{q/2} \sin 2pq\ell,$$

that are not independent, but satisfy $I_2^2 + I_3^2 = (I_1 + I_0)^q (I_1 - I_0)^p$.

Provided that I_0 is an integral of the normalized system, the reduced phase space is a two-dimensional surface of revolution $\mathcal{F}(I_0)$ given by:

$$\mathcal{F}(I_0) \equiv I_2^2 + I_3^2 = (I_1 + I_0)^q (I_1 - I_0)^p.$$

The origin is regarded as the vertex of the surface $\mathcal{F}(0) \equiv I_2^2 + I_3^2 = I_1^{p+q}$ where we are going to study the stability properties of it.

After first order normalization, we find for the resonance 2:1:

$$\mathcal{H} = \frac{1}{2}\omega_2 I_0 + \kappa I_3, \tag{8}$$

where κ is a real number. The orbits on the reduced phase space can be viewed as the intersection of the family of planes given by (8) with $\mathcal{F}(I_0)$. For the case of interest, $I_0 = 0$, we find two asymptotic trajectories to the vertex of the surface (the origin), an ingoing trajectory and an outgoing trajectory. So, the origin must be consider as an unstable critical point provided $\kappa \neq 0$. This result agrees with the conditions of stability established by Markeev (1968) for the 2:1 resonance.

Acknowledgments

This work has been partially supported by Universidad de La Rioja (Project API-99/B18) and by the Spanish Ministry of Education and Science (Project PB98-1576).

References

Arnold, V.I.: 1934, 'The stability of the equilibrium position of a Hamiltonian system of ordinary differential equations in the general elliptic case', *Soviet Math. Dokl.*, **2**, pp. 247–249.

Białynicki-Birula, I., Kalinsky, M. and Eberly, J.H.: 1994, 'Lagrange equilibrium points in celestial mechanics and nonspreding wave packets for strongly driven Rydberg electrons', *Phys. Rev. Lett.*, **73**, pp. 1777–1780.

Deprit, A.: 1984, 'Dynamics of orbiting dust under radiation pressure', In *The Big Bang and George Lemaitre* (A. Berger, Ed.), pp. 151–180.

Elipe, A.: 2000, 'Complete reduction of oscillators in resonance', *Phys. Rev. E*, **61**, pp. 6477–6484.

Lee, E., Brunello, A.F. and Farrelly, D.: 1995, 'Single atom quasi-Penning trap', *Phys. Rev. Lett.*, **75**, pp. 3641–3643.

Lee, E., Brunello, A.F. and Farrelly, D.: 1997, 'Coherent states in a Rydberg atom: Classical mechanics', *Phys. Rev. A*, **55**, pp. 2203–2221.

López-Moratalla, T.: 1997, *Estabilidad orbital de satélites estacionarios*, PhD thesis, Universidad de Zaragoza, Spain.

Markeev, A.P.: 1968, 'Stability of a canonical system with two degrees of freedom in the presence of resonances', *PMM*, **4**, pp. 738–744.

Meyer, K.R. and Schmidt, D.S.: 1986, 'The stability of the Lagrange triangular point and a theorem of Arnold', *J. Diff. Eqns.*, **62**, pp. 222-236.

Salas, J.P., Iñarrea, M., Lanchares, V. and Pascual, A.I.: 1999, 'Electronic traps in a circularly polarized microwave field and a static magnetic field: Stability analysis', *Monografías de la Academia de Ciencias de Zaragoza*, **14**, pp. 113–117.

GLOBAL DYNAMICS IN THE SOLAR SYSTEM

P. ROBUTEL (robutel@bdl.fr) and J. LASKAR (laskar@bdl.fr)
Astronomie et Systèmes Dynamiques, IMC, CNRS UMR 8028, 77 Av. Denfert-Rochereau,
75014 Paris, France

Abstract. Using Laskar's Frequency Map Analysis, we have performed a complete study of the dynamics of massless particles in the Solar System, from Mercury to the outer parts of the Kuiper belt, for all values of the eccentricities. This provides a complete dynamical map of the Solar System which is, in this first step, restricted to mean motion resonances. The same method is also applied to the planetary 3-body system composed of the Sun, Jupiter and Saturn.

1. Introduction

We present here a unified vision of the dynamics of particles in the Solar System, obtained by the construction of a complete cartography of the Solar System dynamics, using Laskar's Frequency Map Analysis (FMA) (Laskar, 1990, 1999). The main advantages of this approach, besides having a rigorous support (Laskar, 1999), is to permit very short integration times of a few million of years and to identify easily the various resonances involved. As we used only a very short integration time, we could include a huge amount of initial conditions, which provides a complete view of the single particle dynamics in the Solar System. In a first stage, which is presented here, we limit ourselves to the consideration of short term dynamics, or more precisely to diffusion and chaotic behavior resulting essentially from short period resonances. Next we apply the same method to planetary 3-body problem.

2. Frequency Map Analysis

For Hamiltonian system on $\mathbf{R}^n \times \mathbf{T}^n$, FMA constructs numerically a map which associates the n-dimensional frequency vector to the action-like variables (see Laskar, 1999 for details). The dynamical behavior of the Hamiltonian system is obtained from the study of the regularity of this frequency map. The construction of the frequency map requires only a very short integration time and allows to get a measure of the diffusion of the trajectories. This diffusion, corresponding to the variation of the fundamental frequencies with respect to time is computed in the following way: the first determination of the frequencies is done on the time interval $[0, T]$ and a second one on $[T, 2T]$. For a quasiperiodic solution, the two frequency vectors are equal, and if it is not the case, theirs difference gives an estimate of the chaotic diffusion of the trajectory over the time T.

Pretka-Ziomek et al./Dynamics of Natural and Artificial Celestial Bodies, 253–258, 2001.
© 2001 *Kluwer Academic Publishers.*

3. Short Time Dynamics of Test Particles in the Solar System

We consider the motion of test particles in the gravitational field generated by the Sun and N planets (see Robutel and Laskar, 2000 for more details). As the motion of the massless body does not affect the planetary motions, the planet frequencies are fixed, and it is sufficient to study the frequency map in a 3-dimensional space of action-like variables. For any particle, these 3 action variables are the usual elliptical elements (a, e, i) (semi-major axis, eccentricity, and inclination), while the 3 angle variables are the associated angles (M, ω, Ω) (mean anomaly, argument of perihelion, and longitude of the node). We fix for $t = 0$ all the initial angles of the particle $(M_0, \omega_0, \Omega_0)$ (see Laskar, 1999), and construct numerically the frequency map which associates the numerically determined (n, g, s) frequencies (associated to mean longitude, perihelion and node) to the initial actions (a_0, e_0, i_0). We are only interested here in the short time dynamics, so only the mean motion frequency n will be determined (the two others are associated to the secular motion). This frequency will be called the "proper mean motion", in the sense that if the motion is quasiperiodic, this quantity is a integral of the motion. Moreover, we will fix the initial value of the inclination i_0, and will thus consider only the map $(a_0, e_0) \longrightarrow n$. If the motion is not quasiperiodic, the two values of $n^{(1)}$ and $n^{(2)}$ obtained over the consecutive intervals $[0, T]$ and $[T, 2T]$ will not be equal in general, and the quantity $\sigma = 1 - n^{(2)}/n^{(1)}$ is computed to provide a measure of the diffusion rate of this trajectory.

The results of this investigation are plotted in Figure 1. The initial conditions for the massless particles are taken in the plane defined by $i_0 = M_0 = \omega_0 = \Omega_0 = 0$, and the initial conditions (a_0, e_0) are chosen on a grid. The test particles in the inner Solar System Fig. 1 a-c have been integrated with the eight planets (Pluto is not considered) during 0.5 Myr, while the outer system Fig. 1 d-h was integrated during 2 Myr taking into account only the four giant planets. For a given particle, three different dynamical situations can occur: First, the particle is ejected or undergoes a close encounter with a planet before the end of the integration; the corresponding dot is plotted in white. Second, the particle survives the integration but is in mean motion resonance with a planet: a black dot is plotted. Third, the trajectory is not resonant: The dot is then colored according to its proper mean motion diffusion rate, from dark grey, for motion close to quasiperiodic $(\sigma \leq 10^{-6})$, to light grey, for strongly irregular motions $(\sigma \geq 0)$. The collision lines with the planets correspond to the black curves. This picture provides a striking global view of the dynamics in the Solar System, and emphasizes the separation of the system in three dynamical structures connected and interpenetrated by the zones of mean motion resonances. Contrarily to the inner Solar System (Figure 1 a-c) and the outer planets region (Figure 1 d-e), these structures appear very clearly in the Kuiper Belt (Figure 1 f–h). Indeed, the outer planet region is too chaotic to contain large regular structures, and the spatial resolution used in the inner part of the system is too coarse to detect resonances with

GLOBAL DYNAMICS IN THE SOLAR SYSTEM 255

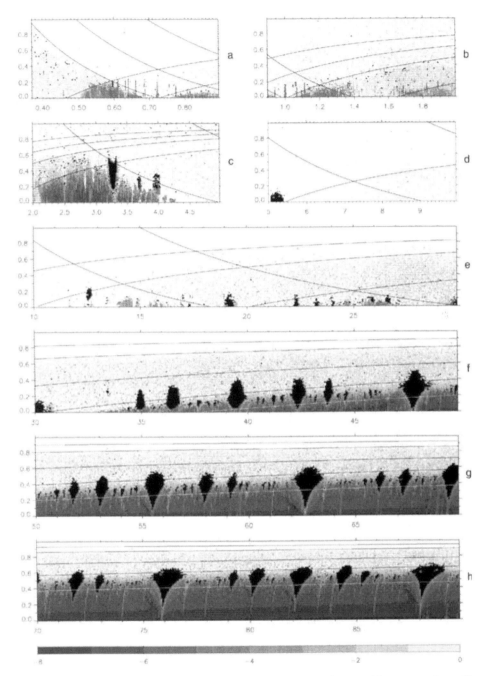

Figure 1. Frequency map of the Solar system in the plane: (a_0, e_0) for $i_0 = M_0 = \omega_0 = \Omega_0 = 0°$. The color code is for $\log \sigma$ (unit: Myr^{-1}).

terrestrial planets. We will only briefly describe here these structures in the Kuiper belt.

The first zone is the collisional region, which lies above the collision line with Neptune. Its main characteristic is strongly irregular dynamics. The particles which have not been ejected over 2 Myr have a very high proper mean motion diffusion rate. This strong chaos is mostly due to close encounters with the planets. The bodies which remain after the 2 Myr integration will probably be removed later from the Solar System after planetary close approach. The only exception to this situation is, as for Pluto, when the particle evolves in the stable area associated to a resonance with Neptune. This situation provides a mechanism which prevent the particle from collision with the considered planet. This is why many resonances penetrate the large chaotic zone above the collision line.

The second region named resonance overlap region is perhaps most interesting. This domain corresponds to a mean motion diffusion rate between $10^{-2.5}$ and $10^{-1.5}$ (between the Neptune's collision curve and the white curve in Figure 1 f–h). Figure 1 clearly shows that the resonant zones intersect in this band. Here the diffusion, which is smaller than in the collisional region, is essentially driven by resonance overlap. The chaotic motions induced by this phenomenon will probably lead to a complete depletion of this area by close encounters. As in the collision domain, this region will certainly be cleared after a sufficiently long time.

The last dynamical domain pointed out in our study is the region of slow diffusion which contains the orbits with $\sigma < 10^{-6}$. For $a < 45$ AU, this region is limited to very small spots with low eccentricity. Due to the possible presence of long time scale chaos (Holman and Wisdom,1993; Levison and Duncan, 1993) generated by secular resonances (not detected in this study), the stability of these orbits should need further studies. For semi major axis greater than 45 AU, the secular frequencies of the massless body are too small (at least outside resonances) to lead to low order secular resonances with the giant planets. For this reason, we can suppose that these stable regions, having a proper mean motion diffusion rate lower than 10^{-6} or 10^{-7}, are stable for a very long time.

4. Frequency Map for the Planetary Three-body Problem

The method presented here being very general, only small modifications were needed to adapt it to the study of the dynamics of a planetary system where all the bodies have non zero mass. In this case, the dimension of the frequency space is much larger than previously and in order to construct a 2-D map, all the initial elliptical elements of the planets will be fixed, except two action-like variables of a selected planet. For the Sun-Jupiter-Saturn case presented in Figure 2, only the initial semi major axis (X-axis) and the initial eccentricity of Saturn (Y-axis) are variable, while the integration time is 2 Myr.

In Figure 2 a, the masses of the two planets are the actual masses of Jupiter and Saturn. This provides a very clear picture of the stability domain of the

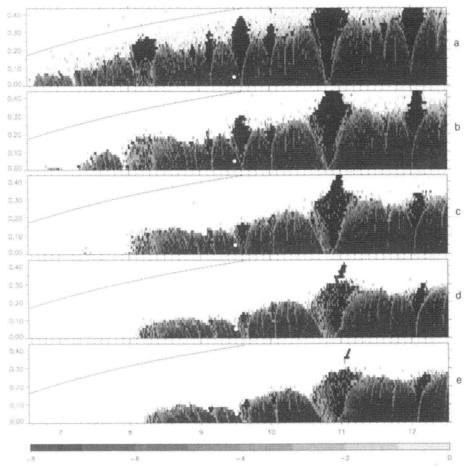

Figure 2. Frequency map of a Sun-Jupiter-Saturn like planetary system, with the present mass values (a), and planetary masses multiplied by a factor 2.5 (b), 5 (c), 7.5 (d) and 10 (e). The white disc represents the actual location of Saturn.

Sun-Jupiter-Saturn system. Close to Jupiter, and up to about 7 AU, there exist no stable regions, while beyond the (1:2) resonance, at 8.4 UA, we have many very large zones of stability. The actual Saturn (marked by a white disc) evolves in one of these regions close to the left edge of the (2:5) resonance. In the case of this 3-body planetary system, these stability results, valid on a short time, can be extended on a very much longer time scales (of the order of billion years). Indeed, outside the mean motion resonances, the secular problem provides a good approximation of the long time dynamics. As in this case, the secular problem possess only two degrees of freedom, the possible chaotic trajectories are confined between invariant tori, and then induced only local and very small destabilizing effect. An extensive study of the dynamics of this secular problem has been done in (Robutel, 1993) for the Jupiter-Saturn like planetary systems, where it was shown that only a few very small chaotic regions exists. A conclusion of this previous study was the following: either a two planet the system is destabilized in a short time scale (due to chaotic behavior resulting from mean motion resonances overlap) or it remains stable over a very long a period of time, eventually comparable to the age of the solar system. Indeed, the results presented in Figure 2 describe precisely the zones of stability where the mean motion resonances have no practical effect. These zones can thus be considered also as zones of very long time stability. This simple study thus provides important information for the long time stability of these systems.

The four next figures (Figure 2 b-e) correspond to the same experiment except for the planetary masses which are magnified by a factor 2.5 (b), 5 (c), 7.5 (d) and 10 (e). This stresses the destabilizing effect of the resonances. Particularly, we observe gaps on the edges of the libration area (in black) of the main resonances ((1:2), (2:5) and (1:3)) which becomes wider as the perturbation increases. This gives also a precise estimate of the shape and of the size of the stability domain with respect to the perturbation size. This aspect will be more developed in some further study.

References

Holman, M. and Wisdom, J.: 1993, 'Dynamical stability in the outer Solar System and the delivery of short period comets'. , *Astron. J.* **105**, 1987–1999.

Laskar, J.: 1990, 'The chaotic motion of the Solar System. A numerical estimate of the size of the chaotic zones'. , *Icarus* **88**, 266–291.

Laskar, J.: 1999, 'Introduction to frequency map analysis', In *NATO ASI Hamiltonian Systems with Three or More Degrees of Freedom* (C. Simò, Ed.), pp. Kluwer, Dordrecht. 134–150.

Levison, H. and Duncan, M.: 1993, 'The gravitational sculpting of the Kuiper Belt'. , *Astrophys. J.* **406**, L035–038.

Robutel, P.: 1993, 'Contribution à l'étude de la stabilité du problème planétaire des trois corps', *Ph.D. Thesis, Observatoire de Paris.*

Robutel, P. and Laskar, J.: 2000, 'Frequency Map and Global Dynamics in the Solar System I : Short Period Dynamics of Massless Particles', *Submitted to Icarus.*

LYAPUNOV STABILITY OF THE LAGRANGIAN LIBRATION POINTS IN THE UNRESTRICTED PROBLEM OF A SYMMETRIC BODY AND A SPHERE IN RESONANCE CASES

KRZYSZTOF GOŹDZIEWSKI

Toruń Centre for Astronomy N. Copernicus University Poland

In papers (Goździewski and Maciejewski, 1998a, 1998b, 1999) inspired by the ideas of Kokoriev and Kirpichnikov (1988, 1988) we investigate unrestricted, planar problem of a symmetric rigid body and a sphere. We call it the KK problem from hereafter. In this paper we study the nonlinear stability of the triangular equilibria existing in this model in the case of low order resonances (up to the order four). Detailed analysis of the existence and bifurcations of the equilibria as well as their linear stability one can find our recently published paper (Goździewski and Maciejewski, 1999).

The KK model is the simplest but highly non-trivial version of the unrestricted two rigid bodies problem. It is described by a very simple Hamiltonian depending on three parameters having clear physical meaning. As it was shown in the cited papers the problem is very reach. It is worth to remind here that the KK problem in its restricted version (i.e., when the point mass vanishes) generalizes two well known classical problems of Celestial Mechanics: the restricted, planar three body problem and the problem of two fixed centers by Euler. There exist certain generalization of the problem. An example of such generalization we propose in our paper mentioned above.

One of goals of this work is to demonstrate that for the triangular equilibria of the KK problem almost all of possible low order resonances can appear, i.e.,

- first order resonance with two null eigenvalues (non-diagonal case),
- first order resonance with one null eigenvalue (diagonal and non-diagonal cases),
- second order resonance (non-diagonal case),
- resonances of the third and fourth order.

The first order resonance may lead to the so called transcendental case. Then all the normal form coefficients which are required for resolving the stability question are vanishing.

Calculations and results presented here are based on the theorems by Sokolsky (1974, 1977, 1980, 1981) and Markeev (1985). They are related to our previous work concerning the limit case of the KK problem (Goździewski and Maciejewski, 1998a). Some of them were derived already by quasi-analytical or numerical means and they were announced (Goździewski and Maciejewski, 1998b), moreover in this work we are focused on new, strictly analytical results.

260 KRZYSZTOF GOŹDZIEWSKI

We found analytical solution to the stability question in all the cases of the first and second order resonances. Detailed description of our calculations we are going to present in (Goździewski, 2000).

Acknowledgements

Some calculations required by this work we performed with help of the MATHEMATICA package. We would like to thank the team of the Regional Multiprocessor Laboratory of N. Copernicus University for the technical support.

References

Goździewski, K.: 2000, 'Lyapunov Stability of the Triangular Libration Points in the Unrestricted Planar Problem of a Symmetric Body and a Point Mass', *Celest. Mech.* submitted.

Goździewski, K. and Maciejewski, A.J.: 1999, 'The Unrestricted Problem a Symmetric Body and and a Point Mass. Triangular Libration Points and Their Stability', *Celest. Mech.* **75**, 251–283.

Goździewski, K. and Maciejewski, A.J.: 1998a, 'The Nonlinear Stability of the Lagrangian Libration Points in the Problem of Chermnykh', *Celest. Mech.* **70**, 41–58.

Goździewski, K. and Maciejewski, A.J.: 1998b, 'The Special Version of the Three Body Problem', In: *NATO ASI Series, Series B: Physics.* New York and London, Plenum Press.

Kirpichnikov, S.N. and Kokoriev, A.A.: 1988, 'On the Stability of Stationary Collinear Lagrangean Motions in the System of Two Attracting Bodies: an Axisymmetrical, Peer-like and Spherically Symmetric', *Vest. Leningrad Univ.* **3**(1), 72–84.

Kokoriev, A.A. and Kirpichnikov, S.N. : 1988, 'On the Stability of Stationary Triangular Lagrangian Motions in the System of Two Attracting bodies: an Axisymmetrical, Peer-like and Spherically Symmetric', *Vest. Leningrad Univ.* **1**(1), 75–84.

Markeev, A.P.: 1985, 'Periodic Motions of a Satellite in a Circular Orbit', **23**(3), 323–330.

Sokolsky, A.G.: 1974, 'On the Stability of an Autonomous Hamiltonian System with Two Degrees of Freedom in the Case of Equal Frequencies', *Prikh. Mat Mech.* **38**, 791–799.

Sokolsky, A.G.: 1977, 'On the Stability of an Autonomous Hamiltonian System with Two Degrees of Freedom at the Resonance of the First Order', *Prikh. Mat. Mech.* **41**, 24–33.

Sokolsky, A.G.: 1980, 'On the Problem of Regular Precessions of a Symmetric Satellite', *Kosm. Issl.* **18**, 698–706.

Sokolsky, A.G.: 1981, 'On the Stability of a Hamiltonian System with Two Degrees of Freedom in the Case of Null Characteristic Exponents', *Prikh. Mat. Mekh.* **45**(3), 441–449.

ON THE BEHAVIOUR OF STATIONARY POINT IN QUAI–CENTRAL CONFIGURATION DYNAMICS

A.E. ROSAEV (rosaev@nedra.yar.ru)
FGUP NPC "NEDRA" Yaroslavl, Russia

We shall consider motion of particle with mass m in field of attraction other $N-1$ particles, placed in vertex of regular polygons and central body with mass M. The stationary solution of N-body problem, at which masses placed in vertex of the regular polygon is well known.

The some authors use different expressions of perturbation function for odd and even number of ring's particles in rectangular coordinates.

On our view, polar coordinates much more suitable for central configuration research. It may be obtained by general way – by differentiation of the potential energy in the system. These equations easy to obtain directly from geometry:

$$(\frac{\partial^2}{\partial t^2} R(t)) - R((\frac{\partial}{\partial t} \phi(t)) + \Omega)^2$$
$$= -\frac{GM}{R^2} - \left(\sum_{j=1}^{N-1} \frac{Gm_j (2 R_0 \sin(a_j)^2 + x)}{((x^2 + 4 R_0^2 \sin(a_j)^2)(1 + \frac{x}{R_0}))^{3/2}} \right),$$

$$R(\frac{\partial^2}{\partial t^2} \phi(t)) + 2((\frac{\partial}{\partial t} \phi(t)) + \Omega)(\frac{\partial}{\partial t} R(t))$$
$$= -\left(\sum_{j=1}^{N-1} \left(2 \frac{Gm_j R_0 \sin(a_j) \cos(a_j)}{((x^2 + 4 R_0^2 \sin(a_j)^2)(1 + \frac{x}{R_0}))^{3/2}} \right) \right),$$

where R_0 — radius of ring (the distance of j—particle from center), $2 * a_j$ — angle between particles, and x distance of test particles from ring, x much smaller then R_0. $R = R_0 + x$, — distance of testing particle from center, Ω — the angular velocity, ϕ—the polar angle of the test particle. It easy to see, that these equations are correct both for odd and even number of particles.

Put masses of all particles equal, $m_j = m$ and $a_j = 6.28318/N + f$, where f is possible angular declination from stationary position. Then require $f = 0$. In this case second equations reduced to conservation of angular moment. In case of the regular rotation of all system, the angular velocity may be determined:

$$\Omega = \frac{GM}{R_0^{3/2}} + \frac{1}{2} \left(\sum_{j=1}^{N-1} \frac{Gm}{\sqrt{R_0^3 \sin(a_j)}} \right).$$

Pretka-Ziomek et al./Dynamics of Natural and Artificial Celestial Bodies, 261–262, 2001.
© 2001 *Kluwer Academic Publishers.*

262   A.E. ROSAEV

So, the accounting of the ring's particles attraction followed to moving resonance position away from planet, but for the small mass of the ring this effect is negligible.

Then the libration points coordinates may be determined. There are two kinds of solution exist. The case of the collinear points of libration are most interesting. It may be reduced to fifth degree polynom:

$$(\Omega^2 + A)x^5 + (3\Omega^2 R_0 + 2AR_0 + B)x^4 + (3\Omega^2 R_0^2 + R_0^2 A + 2BR_0)x^3$$
$$+ (BR_0 + Gm)x^2 + Gm R_0 = 0.$$

In general case, approximately solution of them (at m/M much smaller, then $7.84048/N^3$ and x/R smaller, then $3.1415/N$) is:

$$\frac{m}{M R^3 (3-k)} < x^3, \ k = .3005125000 \ \frac{m N^3}{M \pi^3}.$$

At the limit small m/M, this expression reduced to well-known solution 3-body problem. At the increasing of masses of ring's particles the moving of libration points away from ring's axis take place. The similar behavior observed, when the number of particles decreased. In neighborhood $k = 3$ and at largest k expression for k is unsuitable. Numerical solution at large m/M showed, that the inner libration point coordinate have a limit s, depended from N. It cannot be close to the center of system near then s, even if $M = 0$. Outer point of libration going away from the ring's axis for all time increasing m/M ratio, but at large mass of ring outer libration points are disappeared. Now let us consider quai-central elliptic configurations. Put ring of N particles m with small eccentricity orbits in gravity field of central mass M. The stationary distribution of particles by true anomaly followed from equality zero tangential force on ring's particle. (e, v - eccentricity and true anomaly of particle, δ - angular distance between neighborhood particles). At the suggestion, that most neighbor particles give main contribution in perturbation:

$$\frac{\partial}{\partial v} \delta(v) = -\frac{1}{2} \frac{e \delta \sin(v)}{1 + e \cos(v)}$$

It is simply to obtain the solution of this equation. Based on the Kepler's law for stationary distribution particles on elliptic orbit, we have:

$$\delta = \delta_0 (1 + e \cos(v))^{1/2}$$

The another way to obtain this relation is from the angular momentum conservation law.

A MAPPING MODEL FOR THE COORBITAL PROBLEM

ZSOLT SÁNDOR (sandor@konkoly.hu)[*]
Konkoly Observatory of the Hungarian Academy of Sciences H-1525 Budapest
P.O. Box 67, Hungary

MARIA HELENA MORAIS
Astronomy Unit, Queen Mary and Westfield College, Mile End Road, London E1 4NS,
United Kingdom

In this work we developed a symplectic mapping to study the phase space structure of the coorbital region in the restricted three-body problem (RTBP). This was constructed using the method suggested by Hadjidemetriou (1991), in which the averaged Hamiltonian is the generating function of the mapping. Although this mapping technique has been applied widely for several cases of different mean motion resonances, no application has ever been done for the 1 : 1 resonance, which is the main feature of the coorbital dynamics. However, the 1 : 1 resonance is an outstanding problem (see Érdi (1997) for a recent overview) and it deserves continuous attention.

The Hamiltonian of the RTBP is (see Brown and Shook (1933)):

$$H = 1/[2(x+1)^2] + x + 1 + \mu a' R(x, x_2, \tau, \omega)/r',$$

where R is the perturbing function, μ is the mass parameter, $x = (a/a')^{1/2} - 1$, $x_2 = (x+1)(\sqrt{1-e^2} - 1)$ are action like variables, $\tau = \lambda - \lambda'$ (mean synodic longitude), and ω (argument of the perihelion) are angle variables.

If we keep only the long-period terms and expand R up to $O(e^2)$, then

$$H(x, \tau, x_2) = 1/[2(x+1)^2] + x + 1 + \mu R_0(\tau) + \\ + \mu e^2 R_1(\tau) + \mu e^2 \alpha^2 R_2(\tau)/4,$$

where $\alpha = a'/a$, $R_0 = 1/[(2 - 2\cos\tau)^{1/2}] - \cos\tau - 1/2$, $R_1 = \partial^2 R_0/\partial \tau^2$, and $R_2 = \partial^2 R_0/\partial \alpha^2$. Moreover, noticing that $e^2 = 1 - [x_2/(x+1) + 1]^2$ and $\alpha = (x+1)^{-2}$, we can write the coefficients of R_1 and R_2 as $f_1(x, x_2) = e^2$ and $f_2(x, x_2) = e^2 \alpha^2/4$, respectively. The generating function of the mapping is then:

$$W = x_{n+1}\tau_n + T/[2(x_{n+1}+1)^2] + T(x_{n+1}+1) + \mu T R_0(\tau_n) \\ + \mu T f_1(x_{n+1}, x_2) R_1(\tau_n) + \mu T f_2(x_{n+1}, x_2) R_2(\tau_n),$$

where T is the period in the center of libration, and x_2 is the parameter of the mapping.

The mapping is obtained via eqs. $x_n = \partial W/\partial \tau_n$, $\tau_{n+1} = \partial W/\partial x_{n+1}$:

$$x_{n+1} = x_n - \mu T R_0'(\tau_n) - \mu T f_1(x_{n+1}, x_2) R_1'(\tau_n) - \\ - \mu T f_2(x_{n+1}, x_2) R_2'(\tau_n)$$

[*] Supported by the Hungarian NRF under the grant OTKA F030147.

Pretka-Ziomek et al./Dynamics of Natural and Artificial Celestial Bodies, 263–264, 2001.
© 2001 *Kluwer Academic Publishers.*

$$\tau_{n+1} = \tau_n - T[(x_{n+1} + 1)^{-3} - 1] + \mu T f_1'(x_{n+1}, x_2) R_1(\tau_n) +$$
$$+ \mu T f_2'(x_{n+1}, x_2) R_2(\tau_n),$$

where $R_i'(\tau_n) = dR_i/d\tau_n$, and $f_i'(x_{n+1}, x_2) = \partial f_i/\partial x_{n+1}$. The first equation of the mapping is in implicit form, therefore a numerical procedure should be applied to solve it for x_{n+1}.

A mapping model describes well the original system if both of them have the same fixed points and their stability indices coincide. We studied the stability in the simplified case, when the eccentricity of the test particle is zero. The fixed points of the mapping in that case are $\mathcal{L}_{4,5}(\tau = \pm\pi/3)$ and $\mathcal{L}_3(\tau = \pi)$, and for their stability indices we found: $Tr(\mathcal{L}_{4,5}) = 2 - 27\mu T^2/4$ and $Tr(\mathcal{L}_3) = 2 + 21\mu T^2/8$. It can be shown, that the averaged system has the same fixed points corresponding to the T periodic orbits in the original (non-averaged) system, and the stability indices are the same to $O(\mu^2)$.

The phase space structure of our mapping represents well the features of the original system. The fixed points can be identified as the Lagrangian points $L_{4,5}$ and L_3, and their stability reflects the stability of the corresponding equilibrium solutions ($L_{4,5}$ are stable, L_3 is unstable). Around the stable Lagrangian points there are invariant curves corresponding to the quasi periodic librations in the original system. If μ is small there are horseshoe orbits encircling L_4, L_3, and L_5, and tadpole orbits around L_4 and L_5. If μ is larger, there are only stable tadpole orbits. The phase space structure for $\mu = 0.001$ (Sun-Jupiter case) and $x_2 = -0.002$, obtained by the mapping, is shown in Figure 1.

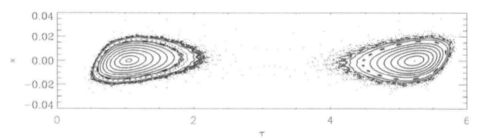

Figure 1. The phase space structure for $\mu = 0.001$

References

Brown, E. W. and Shook, C. A.: 1933, *Planetary Theory*, Cambridge University Press
Érdi, B.: 1997, *Cel. Mech. and Dyn. Astron.* **65**, 149–164.
Hadjidemetriou, J. D.: 1991, in A. E. Roy (ed.) *Predictability, Stability, and Chaos in N-Body Dynamical Systems*, Plenum Press, New York, pp. 157–175.

SELF-SIMILAR STRUCTURE IN THE LINEAR THREE-BODY PROBLEM

HIROAKI UMEHARA (ume@crl.go.jp)

Kashima Space Research Center, Communications Research Laboratory (CRL) Hirai, Kashima, Ibaraki, 314-0012, Japan

(http://www.crl.go.jp/ka/control/people/ume/index-e.html)

What characterizes the self-similarity of a phase structure in the N-body problem? Chaos or collision is considered to be an essential factor of self-similarity. This paper shows that the answer is collision. In the gravitational three-body problem, the self-similar structure of phase trajectories are correlated with triple and binary collisions (Umehara and Tanikawa, 2000). Each set of continuous phase trajectories converges to a triple couision trajectory (Tanikawa and Umehara, 1998). An analysis has shown that collision singularity induces such a distribution (Umehara and Tanikawa, 1999). However, there is also a correlation between self-similarity and triple collision even in the thre body problem with non-singular attractive potential (Nakato and Aizawa, 2000). Here, the system is further simplified. The three-body problem with harmonic potential is analyzed. This is the linear system. Even in this non-chaotic system, the phase structure shows the self-similarity with convergence to a collision orbit.

The Lagrangian of the harmonic three-body system is given by

$$L = \sum_{i=1}^{3} \frac{1}{2} m_i \dot{q}_i^2 - \frac{1}{2} \sum_{i<j}^{3} |\mathbf{q}_i - \mathbf{q}_j|^2, \tag{1}$$

where $\mathbf{q}_i$ is the position vector of a particle i. The free fall problem with three particles is considered ($\dot{\mathbf{q}}_i(0) = 0$ for $i = 1, 2$ and 3). All possible initial configurations are realized if $m_1, m_2,$ and m_3 stand still at

$$\mathbf{q}_1(0) = (\zeta - x_0, \eta - y_0), \quad \mathbf{q}_2(0) = (-0.5 - x_0, -y_0),$$
$$\mathbf{q}_3(0) = (0.5 - x_0, -y_0), \quad \text{where} \quad (\zeta, \eta) \in \{(x, y); y \geqslant 0\}, \tag{2}$$

respectively. Here, (x_0, y_0) is the gravity center of the masses. The solution of the particle i is given by

$$\mathbf{q}_i(t) = \mathbf{a}_i \cos c_- t + \mathbf{b}_i \cos c_+ t, \tag{3}$$

where $\mathbf{a}_i$ and $\mathbf{b}_i$ are the functions of the masses and the initial positions, and c_- and c_+ are dependent on ouly the masses. If c_+/c_- is an irrational number, various relative positions are realized between particles. Let (m_1, m_2, m_3) be $(1/8, 2/8, 5/8)$. Then $c_+/c_- = (1/2)\sqrt{5/3}$.

There is only one triple collision trajectory in the case with non-equal masses. All particles coincide with each other at the initial value:

Pretka-Ziomek et al./Dynamics of Natural and Artificial Celestial Bodies, 265–267, 2001.
© 2001 *Kluwer Academic Publishers.*

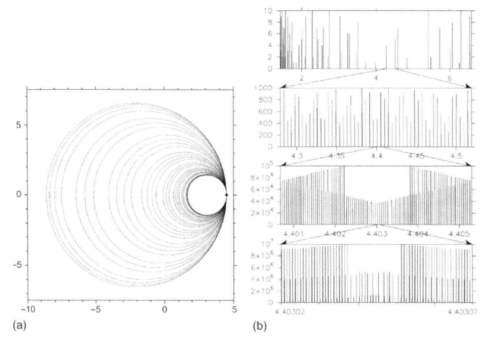

Figure 1. (a) $C(\xi, \eta, t_k)$ on the (ξ, η)-space, (b) Self-similarity of $\sigma(t_k)$.

$$(\xi, \eta) = (\xi_{\text{coll}}, 0), \quad \xi_{\text{coll}} = \frac{m_2 m_3 c_-^2 - 3(m_2 + m_3)/2}{m_3 - m_2}. \tag{4}$$

The initial-value distribution of the following motion is investigated. When m_1 passes between m_2 and m_3, the motions of m_2 and m_3 are parallel. This type of motion is called a neutral slingshot. This type occurs if the systems start from the following initial-value sets:

$$C(\xi, \eta, t_k) = \left\{ (\xi, \eta); \ \left(\xi - \frac{\xi_{\text{coll}} + f(t_k)}{2} \right)^2 + \eta^2 = \sigma^2(t_k) \right\},$$

$$\sigma(t_k) = \frac{\xi_{\text{coll}} - f(t_k)}{2}, \quad t_k = \frac{\pi}{c_-} \left(k - \frac{1}{2} \right), \quad k = 1, 2, \ldots, \tag{5}$$

where $f(t_k)$ is the function of the masses and the time t_k. For every k an initial-value set experiencing the k-th neutral-slingshot is a circle through the initial point experiencing the triple collision. Thus, the trajectory sets converge into the triple collision trajectory. The circles $C(\xi, \eta, t_k)$ for $k = 1, 2, \ldots, 30$ are shown in Figure 1a. One filled circle is the point given by Equation (4). The self-similar distribution of $C(\xi, \eta, t_k)$ is shown in Figure 1b, representing the k-dependent radii. Each abscissa is $\sigma(t_k)$ and each ordinate is the number of triple encounters. A definite region of radius is magnified and the calculation time t is extended. A similar structure can be observed by magnifications.

The author thanks Dr. Yoji Aizawa for giving his original idea and discussing the linear system from the point of view of chaos theory.

References

Nakato, M. and Aizawa, Y.: 2000, *Chaos, Solitons and Fractals* **11**, pp. 171–185.
Tanikawa, K. and Umehara, H.: 1998, *Cel. Mech. Dyn. Astr.* **70**, pp. 167–180.
Umehara, H. and Tanikawa, K.: 1999, *Cel. Mech. Dyn. Astr.* **74**, pp. 69–94.
Umehara, H. and Tanikawa, K.: 2000, *Cel. Mech. Dyn. Astr.*, in press.

MULTIDIMENSIONAL FOURIER ANALYSIS OF THE PLANETARY DISTURBING FUNCTION

SERGEI KLIONER
Lohrmann Observatory, Dresden Technical University, 01062 Dresden, Germany

Abstract. The application of numerical Fourier analysis to compute the coefficients of the Fourier expansions of the planetary disturbing function is discussed. It is argued that this method has recently become feasible for virtually all cases important for Solar system dynamics. A few examples of such calculations are given.

Key words: Fourier transformation, disturbing function, semianalytical theories

It is generally known that the Fourier coefficients of the planetary disturbing function can be computed numerically with the help of numerical Fourier analysis (see, e.g., Brouwer and Clemence, 1961). The idea to use purely numerical methods here goes back to Leonhard Euler (Euler, 1749) who invented numerical Fourier analysis specially for this purpose. What is making this approach feasible nowadays is (1) fast computers with large amount of memory and (2) Fast Fourier Transform (FFT) algorithm. The FFT algorithm has been introduced into modern practice by Cooley and Tukey in 1965. Retrospectively it has become known that the FFT algorithm was invented independently by a dozen of individuals starting from Gauss in 1805 (Heideman *et al.*, 1985). It was exciting also to learn that Gauss has invented the algorithm to compute Fourier expansions of the planetary disturbing function.

The most general 5-dimensional Fourier expansion of planetary disturbing function in standard notations reads

$$
R = \frac{G\,\mathcal{M}'}{a'} \sum_{i=-\infty}^{\infty} \sum_{j=-\infty}^{\infty} \sum_{k=-\infty}^{\infty} \sum_{l=-\infty}^{\infty} \sum_{m=0}^{\infty} B_{ijklm}(\alpha, e, e', i, i') \times
$$
$$
\times \cos(i\,M + j\,M' + k\,\omega + l\,\omega' + m(\Omega - \Omega')), \tag{1}
$$

$\alpha = a/a'$ being the ratio of the two semi-major axes.

It is well known that coefficients B_{ijklm} can be represented analytically in various forms: (1) as a series in powers of eccentricities e and e' and inclinations i and i' with coefficients depending on Laplace coefficients and their derivatives as functions of α (Le Verrier-type expansion; see also Murray and Harper (1993)) and (2) as a series in powers of α with coefficients containing Hansen coefficients of e and e' and Kaula inclination functions of i and i' (Kaula-type expansion; see, e.g., Brumberg (1995)). In many cases the problem of convergency (both in the mathematical sense and in the sense of practical rapidity of convergency) makes these expansions difficult to use.

It is the aim of this work to show that the direct approach of numerical Fourier analysis of R has become practical for calculations of B_{ijklm} and their partial de-

Pretka-Ziomek et al./Dynamics of Natural and Artificial Celestial Bodies, 269–272, 2001.
© 2001 *Kluwer Academic Publishers.*

rivatives with respect to α, e, e', i and i'. Numerical Fourier transformation of R allows one to compute coefficients B_{ijklm} and their derivatives for any admissible numerical values of α, e, e', i, i'. Note that the series (1) converges for any α, e, e', i, i' which correspond to the orbits not intersecting in space for arbitrary M, M', ω, ω', Ω and Ω'. That is, it converges for any i and i' and $\alpha < (1 - e')/(1 + e)$.

Numerical computation of 5-dimensional series (1) has become feasible on a typical server-class computer quite recently, the main problem being the large amount of the required memory. It is clear that due to intrinsic aliasing errors of the FFT for any given amount of memory there exists a region in the 5-dimensional parameter space, for which B_{ijklm} can be computed with a given accuracy. In order to compute B_{ijklm} with a given accuracy for the values of parameters which lie out of that region more memory is required. The region of allowed values of parameters is getting larger with increasing the amount of available memory. However, the amount of memory available on modern computers nowadays allows one to compute the series with sufficient accuracy for many cases which are important for practice. The computer used here for test calculations was an SGI Origin2000 with 48 R10000 processors running at 195MHz. On that computer it was possible to use up to 8Gb RAM for a single calculation. A special research code which computes coefficients B_{ijklm} for 5 given numerical arguments and estimates the resulting accuracy of calculations by accounting for all kinds of numerical errors, has been written. Further details and the code itself are available at http://rcswww.urz.tu-dresden.de/~klioner/fftpert.html (see also Klioner, 2000).

Let us give two examples. In order to compute series (1) for the perturbations of Jupiter on Veritas ($\alpha \approx 0.609$, $e \approx 0.1009$, $e' \approx 0.0485$, $i \approx 9°16'$, $i' \approx 1°18'$) with an accuracy of 10^{-14} the FFT size should be taken as $56 \times 52 \times 48 \times 24 \times 208$ which requires about 5.4 Gb RAM and 30 minutes of computing time (single processor mode). In this example, the number of coefficients B_{ijklm}, absolute value of which is greater than 10^{-14}, is 1 083 126. The computation of the perturbations of Jupiter due to Saturn ($\alpha = 0.54$, $e = 0.048$, $e' = 0.054$, $i = 1.305°$, $i' = 2.485°$) requires to take the FFT size $42 \times 54 \times 28 \times 28 \times 180$ (which takes 2.2 Gb RAM and 10 minutes CPU time) and ends up with 280 735 coefficients larger than the 10^{-15} threshold. For both examples the number of coefficients larger than ε for ε from 0.1 to 10^{-14} is shown on Fig.1. More examples and further details of the algorithm can be found in Klioner (2000). Parallel code for SMP computers is also available and can be used to speed up the calculations by a factor which in some cases almost equal to the number of used processors (the actual gain depends on the dimensionality of the FFT to be performed and, therefore, on the parameters).

The current version of the code uses a real-to-complex FFT algorithm which automatically accounts for the fact that R is a real function. This reduces both the required memory and the required CPU time by a factor of 2 as compared to the general complex-to-complex FFT. Taking into account further symmetries of B_{ijklm}

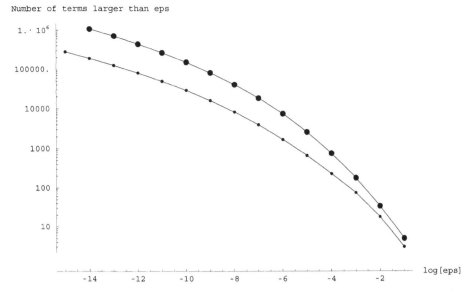

Figure 1. The number of coefficients, absolute value of which is larger than ε for ε from 10^{-1} to 10^{-15}. Upper curve corresponds to the perturbations of Veritas due to Jupiter while the lower one shows the perturbations of Jupiter due to Saturn (see text).

1. B_{ijklm} are real (real-to-real (cosine) FFT is sufficient) and
2. $B_{ijklm} = 0$ if k and l are of different parity (one can set $k = l + 2s$),

the algorithms can be improved to require up to 4 times less memory and up to ~ 3 times less computing time as compared to the current version (2 and 1.5 times less, respectively, if only the first property is accounted for). The work in this direction is underway.

To make the results of calculations of B_{ijklm} useful it is very important to estimate the accuracy of calculations and to keep the aliasing errors under control. As it is shown in Klioner (2000) in order to estimate the accuracy of calculations and to avoid the aliasing errors one can use additional symmetry properties of B_{ijklm} and/or the requirement that the spectrum of R must be below the noise level near the borders of the grid.

This approach to computing the Fourier expansions can be considered as an alternative to standard algorithms for numerical calculations of special functions of celestial mechanics (like Hansen coefficients). It allows one also to avoid the problem of [slow] convergency of the classical power expansions and to concentrate on the operations with Fourier series (1). In this sense the suggestion is to consider B_{ijklm} as a family of special functions of their five arguments, to retain them in symbolic form in analytical calculations (as one often does for, e.g., Hansen coefficients) and, if required, to compute them and/or their partial derivatives numerically with the aid of the numerical Fourier analysis.

Our experience shows that calculation of even the most general 5-dimensional Fourier expansion of the planetary disturbing function by a direct 5-dimensional FFT is quite feasible. Of course, if one is interested in 4- or 2-dimensional expansions of R or similar functions the amount of required computations reduces drastically and such computations represent no difficulty even for a modern PC-class computer. Therefore, numerical Fourier analysis of the disturbing function can become a useful tool of celestial mechanics in the near future.

References

Brouwer, D. and Clemence, G.M.: 1961, 'Methods of Celestial Mechanics', Academic Press, New York.

Brumberg, V.A.: 1995, 'Analytical Techniques of Celestial Mechanics', Springer, Berlin.

Euler, L.: 1749, 'Sur les inégalités du mouvement de Saturn et de Jupiter', Paris, **Section 29**, pp. 30.

Heideman, M.T., Johnson, D.N. and Burrus, C.S.: 1985, 'Gauss and the History of the Fast Fourier Transform', *Archive for History of Exact Sciences*, **34**, pp. 265–277.

Klioner, S.A.: 2000, 'Numerical Fourier Expansions of Planetary Disturbing Function', *Celestial Mechanics and Dynamical Astronomy*, accepted.

Murray, C.D. and Harper, D.: 1993, 'Expansion of the Planetary Disturbing Function to Eighth Order in the Individual Elements', *QMW Maths Notes*, **15**, Queen Mary and Westfield College, London, pp. 436.

ASTEROID ORBITS IN SOME THREE–BODY RESONANCES

AKMAL VAKHIDOV

Lohrmann Observatory, Technical University Dresden, Germany

Abstract. With the help of numerical experiments we investigate the dynamics of orbits of minor planets in the vicinity of some selected mixed resonances "Jupiter–Saturn–Asteroid". We analyze the behaviour of Lyapunov–Exponent as a function of eccentricity and inclination of asteroid orbits in domains of mixed resonances considered and investigate, under what conditions the mixed resonances can be responsible for the origin of chaos in motion of minor planets. Some preliminary results of our numerical experiments are presented.

Key words: minor planets, mixed resonances, Lyapunov–Exponent.

Chaotic structure of the main asteroid belt was investigated in detail in many papers (see, for example, Murray et al., 1998; Nesvorný and Morbidelli, 1998; Šidlichovský, 1999), and we know now, that the different mean–motion resonances and also secular resonances are responsible for the chaos in motion of minor planets and for the formation of chaotic regions in the main asteroid belt. The recent investigations presented in (Nesvorný and Morbidelli, 1998; Morbidelli and Nesvorný, 1999) show also, that in many cases the so called mixed–resonant perturbations can also lead to the origin of chaos in motion of minor planets. In particular, it was shown, that the mixed resonances can be responsible for the phenomenon of "stable chaos" in motion of minor planets. We notice, that this phenomenon was investigated in detail in many papers (see, for example, Milani and Nobili, 1992; Milani et al., 1997) and was explained earlier as a result of high–order resonances with Jupiter, where chaoticity arises from overlapping the subresonances forming the resonance multiplet structure.

As mixed resonances we call commensurabilities between mean motions of at least three celestial bodies. In particular, the mixed resonances "Jupiter–Saturn–Asteroid" are defined by the condition $k_1 n_J + k_2 n_S + k_3 n_A = 0$, where n_J, n_S, n_A are the mean motions of Jupiter, Saturn and an asteroid, respectively, k_1, k_2, k_3 are some integers. According to (Morbidelli and Nesvorný, 1999) we call the parameter $\Phi = |k_1 + k_2 + k_3|$ the order of mixed resonance.

The aim of our research is to investigate in detail the evolution of asteroid orbits near the different mixed–resonances "Jupiter–Saturn–Asteroid", to analyze the influence of joint perturbations from Jupiter and Saturn on the evolution of orbits of minor planets and on the value of Lyapunov–Exponent as a function of eccentricity and inclination, and to determine finally, under what conditions the joint Jovian–Saturnian perturbations (in particular, mixed–resonant perturbations) can be responsible for the chaos in motion of minor planets.

For solving this problem we integrate numerically several thousands of fictitious asteroid orbits injected into domains of different mixed resonances "Jupiter–

Pretka-Ziomek et al./Dynamics of Natural and Artificial Celestial Bodies, 273–280, 2001.
© 2001 *Kluwer Academic Publishers.*

Saturn–Asteroid" with the simultaneous computation of Lyapunov–Exponent. The integration time was from 1 Myrs till 10 Myrs. For each asteroid orbit we studied three dynamical models: 1) Jupiter and Saturn as disturbing bodies; 2) only Jupiter as disturbing body; 3) only Saturn as disturbing body. The third case is important only near the secular resonances ν_6 and ν_{16} for separating the effects caused by secular resonances and joint Jovian–Saturnian perturbations. In the vicinity of strong Jovian resonances the model with only Saturn was not considered. In our numerical experiments we investigated all mixed resonances "Jupiter–Saturn–Asteroid" of order zero and also several resonances of the first order on the interval of semi–major axes between 1.7 AU and 3.3 AU with order numbers $|k_x| < 8$ ($x = A, J, S$). Other resonances of the first order (as for example, the resonance $k_J:k_S:k_A = -4:2:1$) are subject of future investigations. Initial conditions for eccentricity e and inclination i were chosen distributed on the following intervals: $e_0 = 0.01 - 0.20$; $i_0 = 0.5^o - 25.0^o$. The initial value of semi–major axis was put directly into the center of resonant domain. Only orbits remaining near the mixed resonance during the integration time were analyzed, all other orbits were excluded from consideration.

Let us present now some preliminary results of our numerical experiments. At first we consider the resonances of order zero, after this – some selected resonances of the first order.

Mixed resonance $k_J:k_S:k_A=-7:6:1$ (a=1.883 AU). This resonance is quite stable and characterized by Lyapunov times between 150 000 and 500 000 years. The joint Jovian–Saturnian perturbations cause no influence on the value of Lyapunov time, but in the model with two disturbing bodies we observe large variations of orbit elements (eccentricity and inclination), which do not appear in the models with one disturbing body. The amplitudes of these variations increase with increasing the eccentricity and decreasing the inclination. These variations practically disappear at inclinations larger than 20°–22°. The variations of eccentricity can be so large that they are able to cause the close encounters of asteroids with Mars, and it can make this region extremely unstable, if we take into consideration the terrestrial planets of the Solar system. It can explain the absence of real asteroids in this region on orbits with small inclinations. In the domain of large inclinations the situation is opposite and no large variations of orbit elements can be observed. It explains the possibility of existing a stable family of asteroids in this region on orbits with high inclinations ("Hungarias"). An example of an asteroid orbit in this resonance is presented in Fig. 1 for models with only Jupiter (+) and Jupiter–Saturn (x).

Mixed resonance $k_J:k_S:k_A=-6:5:1$ (a=2.067 AU). This region is very complicated for investigation because of presence here of a strong Jovian resonance 4:1 and secular resonances ν_6 and ν_{16}. The Lyapunov times are between 6 000 and 500 000 years. The dynamics of orbits is determined here mainly by secular resonances and Jovian resonance 4:1, the orbit elements undergo extremely large variations. The influence of mixed–resonant perturbations is negligible small in comparison with the effects caused by 4:1 Jovian resonance and by secular resonances.

Mixed resonance $k_J:k_S:k_A=-5:4:1$ (a=2.303 AU). In the case of Jupiter as the only disturbing body the motion of asteroids here is quite stable with Lyapunov times larger than 200 000 years. In the case of two disturbing bodies (Jupiter and Saturn) we can see three dynamical regions: 1) a thin, extremely chaotic layer at inclinations about $14°-15°$, corresponding to the secular resonance ν_6 with Lyapunov times about 30 000 – 50 000 years, here we can see large variations of orbit elements; 2) many stable regions with Lyapunov times larger than 200 000 years without large variations of orbit elements; 3) several thin relative chaotic layers at eccentricities $e > 0.12$ at different values of inclination with Lyapunov times between 50 000 and 100 000 years without large variations of orbit elements. These chaotic layers appear only, if we consider the joint influence of two disturbing bodies, and do not appear in the models with one disturbing body.

Mixed resonance $k_J:k_S:k_A =-4:3:1$ (a=2.621 AU). This resonance is similar to the previous one, but all effects here are amplified because of smaller distance to giant planets. In the model with only Jupiter the motion is stable and characterized by Lyapunov times larger than 200 000 years. In the model with two disturbing bodies we can see: 1) a very chaotic layer near the secular resonance ν_6 with Lyapunov times between 8 000 and 30 000 years and large variations of orbit elements; 2) stable regions with Lyapunov times larger than 150 000 years without large variations of orbit elements; 3) several thin chaotic layers at different values of inclination with Lyapunov times from 10 000 till 50 000 years, sometimes we can observe here large–periodic variations of eccentricity with moderately large amplitudes. An example of such an orbit is presented in Fig. 2.

Mixed resonance $k_J:k_S:k_A =-7:5:2$ (a=2.826 AU). This region is extremely complicated because of presence here of a very strong Jovian resonance 5:2. The Lyapunov times here are between 2 000 and 150 000 years both in the case of one disturbing body (Jupiter) and two disturbing bodies (Jupiter and Saturn). The influence of mixed–resonant perturbations is negligible small in comparison with large perturbations caused by resonance 5:2.

Mixed resonance $k_J:k_S:k_A =-3:2:1$ (a=3.078 AU). This resonance is located near the Jovian resonance 11:5 (at $a=3.075$ AU), and some asteroids can be captured into this Jovian resonance. It leads to large variations of eccentricity of captured orbits and to far encounters of asteroids with Jupiter. The Lyapunov times of such asteroid orbits are between 4 000 and 20 000 years. Without capture into 11:5 resonance the Lyapunov times in the model with only Jupiter are larger than 200 000 years, in the model with two disturbing bodies – between 2 500 and 120 000 years. So high chaoticity of asteroid orbits can be caused both by mixed resonance $k_J:k_S:k_A=-3:2:1$ and by influence of Saturn introducing into Jovian 11:5 resonance many asteroids from neighbouring regions. For separating these effects we should use semianalytical methods, where we can see explicitly, what terms of disturbing function are responsible for chaos in asteroid motion in this region.

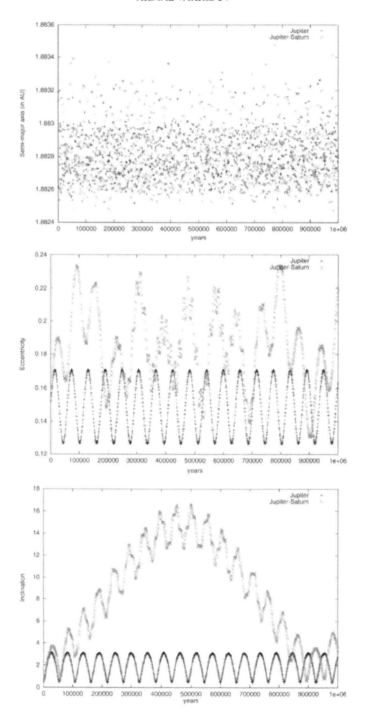

Figure 1. Semi–major axis (a) eccentricity (b) and inclination (c) of an asteroid orbit in resonance J:S:A -7:6:1.

ASTEROID ORBITS IN SOME THREE–BODY RESONANCES 277

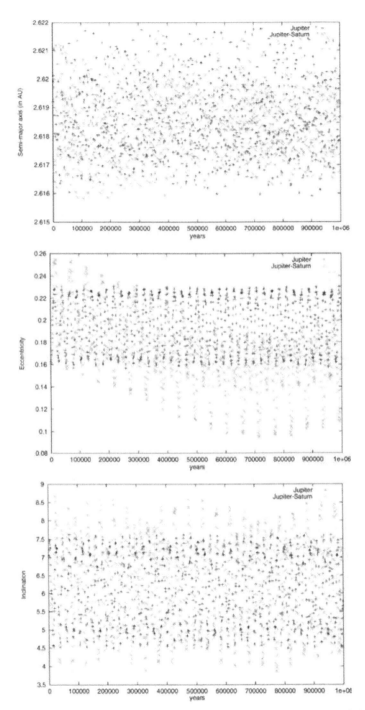

Figure 2. Semi–major axis (a) eccentricity (b) and inclination (c) of an asteroid orbit in resonance J:S:A -4:3:1.

Mixed resonance $k_J:k_S:k_A = -7:5:1$ (a=1.781 AU). This resonance is situated near the strong Jovian resonance 5:1 (at $a=1.779$ AU), and sometimes we observe the capture of asteroids into this Jovian resonance. The Lyapunov times of captured asteroids are between 10 000 and 100 000 years, in the cases without capture into 5:1 resonance the Lyapunov times are always larger than 200 000 years. Also in this resonance (as in the resonance –7:6:1) we observe the large variations of asteroid orbit elements. Amplitudes of these variations increase with increasing the eccentricity and decreasing the inclination. The variations become much smaller at inclinations about 18°–20° and practically disappear at inclinations larger than 23°–25°.

Mixed resonances $k_J:k_S:k_A = -6:4:1$ (a=1.939 AU) and $k_J: k_S: k_A = -7:7:1$ (a=2.001 AU). In both these resonances the behaviour of orbits is similar to the resonances –7:6:1 and –7:5:1. The motion of asteroids is quite stable and the Lyapunov times are always larger than 150 000 years, but there are large variations of orbit elements, the amplitudes increase with increasing the eccentricity and decreasing the inclination. The variations become much smaller at inclinations $i > 20^o - 22^o$ and practically disappear at inclinations $i > 25^o - 27^o$. These large variations are observed only in the model with two disturbing bodies (Jupiter and Saturn) and do not appear, if we consider the sum of perturbations from Jupiter and Saturn independently.

Mixed resonance $k_J:k_S:k_A = -3:1:1$ (a=2.752 AU). This resonance is situated near the direct Jovian resonance 13:5. Many asteroids undergo here the capture into this Jovian resonance; it can lead to increasing the eccentricity, and, as a result, to far encounters of asteroids with Jupiter. The Lyapunov times of such orbits are from 3 500 till 20 000 years. For orbits without capture into 13:5 Jovian resonance the Lyapunov times in the model with one disturbing body (Jupiter) are always larger than 200 000 years, in the model with two disturbing bodies (Jupiter and Saturn) – between 20 000 and 50 000 years. Near the secular resonance ν_6 we can see large variations of eccentricity. In other regions the variations of eccentricity are very small, and even the far encounters of asteroids with Jupiter are impossible. Nevertheless, the Lyapunov times of asteroid orbits here are small, and we observe the phenomenon of so called "stable chaos" (Šidlichovský, 1999), which often appears in domains of mixed resonances.

Mixed resonance $k_J:k_S:k_A=-5:2:2$ (a=3.174 AU). In the model with only Jupiter this region is quite stable with Lyapunov times larger than 200 000 years. In the case of large eccentricities the far encounters of asteroids with Jupiter become possible, then the Lyapunov time can be smaller. In the model with two disturbing bodies the motion of asteroids in this region is always chaotic with Lyapunov times between 5 000 and 70 000 years. The high chaoticity of asteroid orbits here is caused by joint perturbations from Jupiter and Saturn. Here we also observe the phenomenon of "stable chaos", and the motion of asteroids is chaotic even without encounters with major planets.

Analyzing the results of our numerical experiments we can make the following preliminary conclusions.

For orbits with semi–major axes smaller than 2.1 AU the joint action of perturbations from Jupiter and Saturn leads to large variations of orbit elements (eccentricity and inclination). The amplitudes of these variations slowly increase with increasing the eccentricity and drastically increase with decreasing the inclination. These variations become much smaller at inclinations $i \approx 18^o - 20^o$ and practically disappear at inclinations $i > 25^o$. These variations are observed already at semi–major axes $a \approx 1.78$ AU so that they cannot be caused by secular resonances ν_6 or ν_{16}. There is no correlation between oscillations in eccentricity and inclination, so that this effect is not caused by Kozai resonance. Because these large variations appear only in the model with two disturbing bodies (Jupiter and Saturn), we can conclude that these effects are caused by mixed–resonant perturbations. Of course, not only mixed mean–motion resonances, but also mixed secular resonances can be responsible for the large variations of orbit elements here. Such large variations in the eccentricity are able to cause the close encounters of asteroids with Mars, that makes this region extremely unstable, if we take into consideration the influence of terrestrial planets of the Solar system. It can explain the absence of real asteroids with semi–major axis $1.7 < a < 2.1$ AU on orbits with small inclinations. The motion of asteroids with large inclinations is stable, and orbit elements do not undergo any large variations. It can explain the presence of many real asteroids in the region $1.7 < a < 2.1$ AU on orbits with high inclinations (Hungarias). More detailed analysis should be made using semianalytical methods in order to analyze, what terms of disturbing function are responsible for these large variations of orbit elements.

In the main asteroid belt ($2.1 < a < 3.3$ AU) the joint action of perturbations from Jupiter and Saturn leads to formation of chaotic regions with Lyapunov times from 10 000 till 70 000 years at different values of inclination and at moderately large eccentricities. The number of these layers and their width increase with increasing the eccentricity and with increasing the semi–major axis. In difference from secular resonance ν_6 with the similar Lyapunov time, the other chaotic layers are not characterized by large variations in eccentricity and therefore cannot be connected with far or close encounters with Jupiter. The observed here phenomenon of "stable chaos" (Šidlichovsky, 1999) is caused quite probably by mixed–resonant perturbations, which provide the high chaoticity of asteroid orbits in this region.

On the basis of our numerical experiments we are able to explain now the possibility of existing the family of asteroids "Hungarias" on orbits with large inclinations.

The author is very grateful to M.Soffel and I.Tupikova for their help in this research and many stimulating discussions. The author would like to thank also D.Nesvorný for his attention to this work and useful discussions. The author thanks also A.Morbidelli for useful remarks and suggestions. The author thanks DFG (Deutsche Forschungsgemeinschaft) for the financial support of this research.

References

Milani, A. and Nobili, A.: 1992, 'An example of stable chaos in the Solar System', *Nature* **357(6379)**, pp. 569–571.

Milani, A., Nobili, A. and Knezevic, Z.: 1997, 'Stable chaos in the asteroid belt', *Icarus* **125**, pp. 13–31.

Morbidelli, A. and Nesvorný, D.: 1999, 'Numerous weak resonances drive asteroids toward terrestrial planets orbits', *Icarus* **139**, pp. 295–308.

Murray, N., Holman, M. and Potter, M.: 1998, 'On the origin of chaos in the asteroid belt', *Astron. Journal* **116**, pp. 2583–2589.

Nesvorný, D. and Morbidelli, A.: 1998, 'Three–body mean–motion resonances and chaotic structure of the asteroid belt', *Astron. Journal* **116**, pp. 3029–3037.

Nesvorný, D. and Morbidelli, A.: 1998, 'An analytic model of three–body mean–motion resonances', *Celest. Mech. and Dynam. Astron.* **71**, pp. 243–271.

Šidlichovský, M.: 1999, 'On stable chaos in the asteroid belt', *Celest. Mech. and Dynam. Astron.* **73**, pp. 77–86.

DETERMINATION OF MASSES OF SIX ASTEROIDS FROM CLOSE ASTEROID–ASTEROID ENCOUNTERS

GRZEGORZ MICHALAK

Wrocław University Observatory, Kopernika 11, 51-622 Wrocław, Poland.
E-mail: michalak@astro.uni.wroc.pl

New masses of six asteroids: (6) Hebe, (10) Hygiea, (15) Eunomia, (52) Europa, (511) Davida and (704) Interamnia were determined within the ongoing Asteroid Mass Determination Program, started in 1998 in the Wrocław University Observatory. The masses were calculated by means of the least-squares method as weighted means of the values found separately from the perturbations on several single asteroids. Special attention was paid to the selection of the observations of the asteroids. For this purpose, a criterion based on the requirement that the post-selection distribution of the $(O - C)$ residuals should be Gaussian, was implemented. Asteroid encounters enabling mass determinations were found as a result of an extensive search for large asteroidal perturbations. In the search, gravitational perturbations exerted by 250 most massive asteroids on 4500 numbered minor planets (called hereafter test asteroids) were determined. The perturbations were calculated as differences in right ascension between perturbed and not perturbed orbits of the test asteroids when integrated backward. As candidates for the mass determinations were used the test asteroids, for which the maximum perturbation in whole time interval covered by its observation was large (more than about 1 arcsecond). Most of them were never used before for this purpose. Details of the mass determinations are given in Table I. As an outcome of the search for possible perturbers among 912 asteroids with diameters larger than 50 km, for all test asteroids under consideration, correct dynamical models including important perturbers, were proposed. So far, the program gave also reliable mass estimates for the three largest asteroids: (1) Ceres, (2) Pallas and (4) Vesta. The mass of Ceres, $(4.70 \pm 0.04) \times 10^{-10}$ M$_\odot$, was obtained from 25 individual solution, that of Pallas, $(1.21 \pm 0.26) \times 10^{-10}$ M$_\odot$, was determined from 2 new test asteroids, and for Vesta, $(1.36 \pm 0.05) \times 10^{-10}$ M$_\odot$, was obtained as a weighted mean of 26 individual solutions (Michalak 2000, Astronomy and Astrophysics, in press). This work was supported by the Wrocław University grant No. 2041/W/IA/2000.

TABLE I

The results of the mass determinations for six minor planets from perturbations on individual test asteroids. M_p is the largest perturbation effect in right ascension multiplied by $\cos \delta$. m_0 are initial messes for which M_p were calculated

Test asteroid	Date of the closest approach	Min. dist. [AU]	$|M_p|$ ['']	Time interval covered by observations	Mass $[10^{-11}\ M_\odot]$
(6) Hebe ($m_0 = 0.72 \times 10^{-11}\ M_\odot$)					
(1150) Achaia	1943.10.22	0.002	3.2	1929–1997	0.76 ± 0.28
(234) Barbara	1946.09.07	0.023	2.9	1883–1997	0.88 ± 0.40
				Weighted mean mass of (6) Hebe	**0.80 ± 0.23**
(10) Hygiea ($m_0 = 4.7 \times 10^{-11}\ M_\odot$)					
(7) Iris	1928.01.18	0.072	1.7	1848–1997	4.6 ± 0.7
(20) Massalia	1933.11.05	0.150	2.9	1853–1997	5.3 ± 1.4
(829) Academia	1927.05.19	0.006	3.3	1914–1998	5.9 ± 1.8
(111) Ate	1878.02.11	0.094	4.9	1870–1998	8.6 ± 1.9
(69) Hesperia	1951.09.05	0.086	1.9	1861–1997	4.7 ± 2.6
(209) Dido	1958.01.14	0.179	1.3	1879–1994	3.5 ± 2.6
(116) Sirona	1939.05.21	0.050	3.2	1872–1998	4.2 ± 2.6
(60) Echo	1867.05.07	0.211	3.0	1861–1997	6.3 ± 3.0
				Weighted mean mass of (10) Hygiea:	**5.1 ± 0.5**
(15) Eunomia ($m_0 = 1.6 \times 10^{-11}\ M_\odot$)					
(1284) Latvia	1960.04.06	0.009	3.5	1925–1998	1.4 ± 0.3
(7) Iris	1873.09.06	0.315	0.8	1848–1997	1.3 ± 0.6
(1313) Berna	1955.05.11	0.031	4.2	1911–1998	1.0 ± 0.7
				Weighted mean mass of (15) Eunomia:	**1.3 ± 0.3**
(52) Europa ($m_0 = 2.4 \times 10^{-11}\ M_\odot$)					
(993) Moultona	1966.08.31	0.022	2.0	1928–1998	2.8 ± 0.9
(1023) Thomana	1971.05.31	0.007	4.2	1924–1998	2.1 ± 1.4
(84) Klio	1994.03.09	0.150	3.5	1865–1998	7.7 ± 1.6
(306) Unitas	1945.01.14	0.098	2.8	1892–1997	3.6 ± 1.6
				Weighted mean mass of (52) Europa:	3.6 ± 0.6
				Weighted mean without solution obtained from (84) Klio:	**2.8 ± 0.7**
(511) Davida ($m_0 = 3.0 \times 10^{-11}\ M_\odot$)					
(532) Herculina	1963.04.04	0.031	3.2	1904–1997	3.5 ± 0.3
(89) Julia	1957.10.27	0.039	1.0	1866–1998	3.7 ± 1.4
(1847) Stobbe	1974.09.20	0.049	1.3	1902–1998	4.0 ± 3.2
				Weighted mean mass of (511) Davida:	**3.5 ± 0.3**
(704) Interamnia ($m_0 = 3.5 \times 10^{-11}\ M_\odot$)					
(95) Arethusa	1985.03.13	0.144	5.9	1872–1998	3.5 ± 1.2
(37) Fides	1957.08.19	0.057	2.3	1856–1998	4.0 ± 1.5
(993) Moultona	1973.11.23	0.014	1.5	1928–1998	4.8 ± 1.9
				Weighted mean mass of (704) Interamnia:	**3.9 ± 0.8**

SCHWARZSCHILD NONEQUATORIAL PERIODIC MOTION ABOUT AN ASTEROID MODELED AS A TRIAXIAL ROTATING ELLIPSOID

OLGA O. VASILKOVA

Institute of Applied Astronomy RAS, 10, Kutuzov quay, St.Petersburg, 191187, Russia,
e-mail: 1021@ita.spb.su

Abstract. A homogeneous uniformly rotating triaxial ellipsoid is considered. A closed form of its gravitational potential is taken to compute the 3 – dimensional periodic motions of Schwarzschild's type in the vicinities of two ellipsoid models with semi-axes 70, 60, 50 km and 140, 60, 25 km using the Poincare's periodical solutions theory which was adapted by Batrakov (1957) for the case of motion in the vicinity of triaxial ellipsoid. All found periodic solutions are of Schwarzschild's type as their periods are not equal to those of generating periodic solutions. The numerical method for finding and adjusting such the solutions is developed. It is in a way alike to the method of successive approximations. The method is in fact a method of numerical continuation with a certain parameter or alternatively with a number of parameters (eccentricity, argument of pericenter, longitude of ascending node and inclination of the generating orbit). The table lists the initial elements of the found generating orbits for ellipsoid model with semi-axes 70, 60, 50 km and commensurabilities 1/1 and 2/1.

Key words: Poincare's periodical solutions theory, Schwarzschild's periodical solution, generating periodical solutions, motion about the triaxial ellipsoid

1. Introduction

In studying particle and satellite periodic motion about any body there are many perturbations which must be accounted for. The external perturbations are the solar tide, solar radiation pressure, perturbations from the planets and others. The internal perturbations are those due to nonspherical shape of a body. When fairly far from the Sun and close to the body, perturbations from nonspherical shape dominate the orbital dynamics. Given the three major dimensions of a body it is convenient to present it as a triaxial ellipsoid model which is significant as it includes the effect of the major shape variations and can be specified based on optical observations alone. The first to advance the problem of periodical motion about triaxial rotating ellipsoid was Batrakov who in 1957 proved the existence of three-dimensional periodic motion in the vicinity of rotating triaxial ellipsoid of near-spherical shape at certain commensurabilities. Since that time many studies touched upon the problem of dynamics about triaxial ellipsoid. These are works by Abalakin (1957), Demin and Aksenov (1960), Aksenov (1960), Duboshin (1961), Antonov (1961), Danby (1965), Kammeyer (1978), Dobrovolskis and Burns (1980), Kosenko (1981), Zhuravlev and Zlenko (1983, 1983a), Mulder and Hooimeyer (1984), German and Friedlander (1991), Chauvineau, Farinella, Mignard (1993), Scheeres (1993, 1994), Petit, Greenberg, Geissler (1994), Vasilkova (1999,

Pretka-Ziomek et al./Dynamics of Natural and Artificial Celestial Bodies, 283–288, 2001.
© 2001 *Kluwer Academic Publishers.*

1999a) and others. Most of the works have focused on the planetary case where the body oblateness is small and effects of nonspherical shape are relatively negligible compared to the attraction of the central body. The planetary case model allows to express the ellipsoid gravitational potential as a truncated harmonic expansion including usually up to the second harmonic. But when orbiting such irregularly shaped and elongated small bodies as asteroids these analyses may no longer apply due to the relatively large perturbations seen by orbiters. It is known that truncated harmonic expansion diverges when inside the sphere which is circumscribing for an asteroid. Even when out of the sphere but close to it, this series converges extremely slowly. That is why for the asteroid case it is reasonable to leave harmonic formulation aside and concentrate on the closed form of the ellipsoid gravitational potential. The closed form potential may be viewed as an idealization lying between simple spherical models and actual gravity field potential. Some studies have taken an advantage of using a closed form potential to search particle and satellite dynamics about an ellipsoid. Among them are works by Duboshin (1961), Antonov (1961), Danby (1965), Dobrovolskis and Burns (1980), Kosenko (1981), Mulder and Hooimeyer (1984), German and Friedlander (1991), Chavineau, Farinella, Mignard (1993), Petit, Greenberg, Geissler (1994), Vasilkova (1999, 1999a). A detailed study of the problem can be found in the papers by Scheeres (1993, 1994).

The current study also uses the closed form of triaxial ellipsoid potential. It is based on Poincare's theory of periodic solutions which was adapted by Batrakov (1957). for the case of triaxial ellipsoid model. The intent of this paper is to search actual three-dimensional periodic motions in the vicinity of certain models of triaxial ellipsoid. These periodic motions are of Schwarzschild's type as their period is not necessarily equal to the period of initial generating Keplerian orbit. The method for computing such periodic motions developed in the current study is generally applicable to any principal-axis rotating triaxial ellipsoids.

2. General Equations

Define a body-fixed coordinate system in the ellipsoid. The x-axis lies along the largest dimension, the y-axis lies along its intermediate dimension, and z-axis lies along its smallest dimension. This analysis assumes that the ellipsoid rotates uniformly about its largest moment of inertia, thus the ellipsoid rotates uniformly about the z-axis. Given a constant density for the asteroid and its three major dimensions there is the classical formula for the gravity field potential of an ellipsoid:

$$V = k^2 \pi \rho abc \int_{\lambda}^{+\infty} \left(1 - \frac{x^2}{a^2 + s} - \frac{y^2}{b^2 + s} - \frac{z^2}{c^2 + s} \right) \frac{ds}{R(s)}, \tag{1}$$

where k is the Gauss constant, $R(s) = \sqrt{(a^2 + s)(b^2 + s)(c^2 + s)}$, $a > b > c$ are the semi-axes of the ellipsoid. Parameter λ is a function of x, y, z and is solved from equation:

$$\frac{x^2}{a^2 + \lambda} + \frac{y^2}{b^2 + \lambda} + \frac{z^2}{c^2 + \lambda} = 1.$$

It defines the ellipsoid passing through the point x, y, z which is confocal to the body's ellipsoid. This equation is a cubic one and has a unique positive root when outside the ellipsoid and root equal to 0 when the particle lies on the ellipsoid surface. For searching the periodical solutions in the vicinity of the ellipsoid in a rotating body-fixed coordinate frame it is convenient to express equations of motion of a particle in terms of Delaunay's canonical elements. The appropriate Hamiltonian system is the following:

$$
\begin{aligned}
\dot{L} &= F_l & \dot{l} &= -F_L \\
\dot{G} &= F_g & \dot{g} &= -F_G \\
\dot{H} &= F_h & \dot{h} &= -F_H,
\end{aligned}
\tag{2}
$$

where left sides are the time derivatives and right sides are partial derivatives with respect to the Delaunay's elements. In accordance with Poincare's theory of periodic solutions the function F should be represented as a power series with respect to a small parameter θ. Note that the series may consist of a finite number of terms. In the current work this expansion consist of two terms:

$$F = F_0 + \theta F_1.$$

The first term F_0, according to the Poincare's theory, should be independent of the angular Delaunay's elements, l, g, h, where l is the mean anomaly, g is the argument of pericenter, and h is the longitude of ascending node. The second term must be periodical with respect to the angular variables l, g, h. Here it is presented as:

$$\theta F_1 = V - k^2 M / r = V_1,$$

where V is the potential of the ellipsoid, k is the Gauss constant, M is the mass of the ellipsoid, and r is the particle position radius.

The periodic solution to the system of equations above corresponding to $\theta = 0$ is usually called the generating periodical solution. The Poincare's conditions for periodic solutions existence adapted by Batrakov for the case of motion in the vicinity of triaxial ellipsoid are the following:

$$\frac{\partial [V_1]}{\partial l_0} = 0, \quad \frac{\partial [V_1]}{\partial g_0} = 0, \quad \frac{\partial [V_1]}{\partial h_0} = 0, \quad \frac{\partial [V_1]}{\partial G_0} = 0,$$

$$[V_1] = \frac{1}{T_0} \int_0^{T_0} V_1 dt,
\tag{3}$$

where l_0, g_0, h_0, G_0 are the elements of the orbit which generates the periodic motion and T_0 is the period of the generating orbit which is not necessarily the same as the period of the final orbit, T.

The method for finding the three-dimensional periodic solutions developed here is suitable for computing the periodic solutions described above. Briefly, the method consists in the following. For initial elements g_0 and h_0 given at 5 degree intervals, we compute partial derivatives shown in the left sides of equations (3). Those values of g_0 and h_0 are then chosen for which the Poincare's conditions are optimally fulfilled. Then these values are numerically adjusted by use of the 19th order Everhart integration method. The results of this adjustment are considered satisfactory if the square root of the sum of squares of differences between the initial and final coordinates do not exceed 1–2 mm. The adjustment method used in this work is in a way alike to the method of successive approximations. The method is in fact a method of numerical continuation with a certain parameter or alternatively with a number of parameters (eccentricity, argument of pericenter, longitude of ascending node and inclination of the generating orbit).

3. Results

The main results obtained in the study are the following. We managed to find a family of non-circular symmetrical polar orbits at commensurability 1/1 with the ellipsoid rotation rate which persists also at commensurability 2/1. Initial conditions for two examples of generating orbits for these families are presented in the Table I. The 3th and 4th columns of the Table I demonstrate the initial conditions for retrograde orbits. The sixth asymmetrical non-equatorial solution is obtained from the fifth one by the method of numerical continuation described above. In the same way the fourth solution is obtained from the third. So given one periodic orbit, other members of the family may be found via numerical continuation of the orbit with respect to some parameter. The periods of all the solutions generated by unperturbed orbits demonstrated in the table are not equal to the periods of the generating Keplerian motions.

TABLE I

Initial elements of generating orbits in the vicinity of an ellipsoid of density $3g/cm^3$ with semi-axes 70, 60, 50 km

n_1/n	$h_0[°]$	$g_0[°]$	e_0	$i_0[°]$	$\Delta\,()$
1/1	0	0	0.02244	90	0.000003
2/1	0	0	0.00600	90	0.000001
1/1	5	49.3	0.15610	180	0.000001
1/1	24.19	65.69	0.15623	179.9	0.0000006
1/1	5.03	174.86	0.00000	89.2543	0.000979
1/1	5.03	174.86	0.00010	89.2541	0.000979

It is necessarily to notice that the computation of these orbits requires a precision integration routine and a set of software tools to force the end-points of an orbit to coincide. That is why the number of the periodic orbits actually computed with satisfactory precision is relatively small, despite the fact that the number of found orbits suspected to be periodic was rather great.

The list of references is not complete owing to lack of space.

References

Abalakin, V.K.: 1957, 'On stability of equilibrium points of a triaxial rotating ellipsoid', *Bull. I.*, **6(8)**, pp. 543–549 (in Russian).

Aksenov, E.P.: 1960, 'On particle periodical motions in gravity field of rotating body', *Vestnik Moskovskogo Universiteta. Phisika, astronomiya* **4**, pp. 86–95 (in Russian).

Antonov, V.A.: 1961, 'Hydrodynamic models with singular points', *Vestnik Leningradskogo Universiteta* **13**, pp. 157–160 (in Russian).

Batrakov, Yu.V.: 1957, 'Periodic motions of particle in gravity field of rotating triaxial ellipsoid', *Bull. I.* **6(8)**, pp. 524–542 (in Russian).

Chauvineau, B., Farinella, P. and Mignard, F.: 1993, 'Planar orbits about a triaxial body: application to asteroidal satellites', *Icarus* **105(2)**, pp. 370–384.

Danby, J.M.A.: 1965, 'The formation of arms in barred spirals', *Astron. J.* **70(7)**, pp. 501–512.

Demin, V.G. and Aksenov, E.P.: 1960, 'On particle periodical motions in gravity field of slowly rotating body', *Vestnik Moskovskogo Universiteta. Phisika, astronomiya* **6**, pp. 87–96 (in Russian).

Dobrovolskis, A.R. and Burns, J.A.: 1980, 'Life near the Roche Limit: Behavior of ejecta from satellites close to planets', *Icarus* **42**, pp. 422–441.

Duboshin, G.N.: 1961, 'Theory of gravitation attraction', *Gos. izdanie Phys.-math. literature* (in Russian).

German, D. and Friedlander, A.A.: 1991, 'A simulation of orbits around asteroids using potential field modeling', *Proc. AAS/AIAA Spaceflight Mechanics Meeting, Houston, TX, February 11-13*.

Kammeyer, P.C.: 1978, 'Periodic orbits around a rotating ellipsoid', *Celest. Mech.* **17(1)**, pp. 37–48.

Kosenko, I.I.: 1981, 'The libration points in the problem of the triaxial graviting ellipsoid. Geometry of the stability area', *Cosmic Research* **19(2)**, pp. 200–209 (in Russian).

Mulder, W.A. and Hooimeyer, J.R.A.: 1984, 'Periodic orbits in a rotating triaxial potential', *Astron. and Astrophys.* **134(1)**, pp. 158–170.

Petit, J.M., Greenberg, R. and Geissler, P.: 1994, 'Orbits around a small, highly elongated asteroid: Constraints on Ida's Moon', *Bulletin of the American Astronomical Society* **26(3)**, pp. 1157–1158.

Poincaré, H.: 1892, 'Méthodes nouvelles de la méchanique céleste', Paris, 1.

Scheeres, D.J.: 1993, 'Satellite Dynamics About Tri-Axial Ellipsoids', *Advances in Non-Linear Astrodynamics, University of Minnesota, Nov 8–10* (personal communication).

Scheeres, D.J.: 1994, 'Dynamics about uniformly rotating triaxial ellipsoids: applications to asteroids', *Icarus* **110(2)**, pp. 225–238.

Subbotin, M.F.: 1949, 'Course of celestial mechanics', *GTTI*, 3 (in Russian).

Vasilkova, O.O.: 1999, 'Equilibrium points of a rotating triaxial ellipsoid', *Proceedings of Institute of Applied Astronomy* **4**, pp. 246–259 (in Russian).

Vasilkova, O.O.: 1999, 'Equilibrium points of an asteroid modeled as a rotating triaxial ellipsoid', *Asteroids, Comets, Meteors Cornell University, July 26-30*, Abstract 24.09. http://scorpio.tn.cornell.edu/ACM.

Zhuravlev, S.G. and Zlenko, A.A.: 1983, 'On stationary solutions in a problem of a translation-rotational motion of a triaxial satellite of a triaxial planet', *Astron. J.(Russian)* **60**(2), pp. 367–374 (in Russian).

Zhuravlev, S.G. and Zlenko, A.A.: 1983, 'Conditional periodic translational-rotational motions of a satellite around a triaxial planet', *Astron. J.(Russian)*, **60**(6), pp. 1217–1222 (in Russian).

A QUANTITATIVE APPROACH OF THE ORBITAL UNCERTAINTY PROPAGATION THROUGH CLOSE ENCOUNTERS

ŞTEFAN BERINDE (sberinde@math.ubbcluj.ro)
Babeş-Bolyai University, Cluj-Napoca, Romania

Abstract. The greatest impediment against collisional predictions of Near-Earth Asteroids with the Earth is their orbital uncertainty and its divergence in time. Repeated close encounters with terrestrial planets generate chaotic motions which make unpredictable the future orbits of these objects. This paper deals with a quantitative analysis of the orbital uncertainty propagation through close encounters. We approach this problem analytically (using the Opik's formalism of close encounters) and numerically (through Monte Carlo simulations). We emphasize the progressive degradation of the orbital uncertainty from one encounter to another, identifying the way in which such an encounter acts to increase this uncertainty. Also, an oscillatory behaviour of the uncertainty is discovered and explained.

Key words: NEA, close encounter, orbital uncertainty

1. Introduction

The orbital uncertainty of a Near-Earth Asteroid (NEA) arises from the fact that, initially, its orbit is determined with finite precision. Later this uncertainty diverges because of the chaotic feature of motion, orbital parameters being slightly changed in an unpredictable way - as long as the orbit is not continuously improved (usually, only when the object approaches the Earth). The aim of this paper is to explore the propagation process of the orbital uncertainty after the last orbital improvement. This is a major task in evaluating future collisional threats with the Earth.

2. Time-growth of Orbital Uncertainty

The orbital uncertainty propagation can be thought as a dispersion process in the space of orbital elements of a swarm of *virtual asteroids* all compatible with the observations of the real one (Milani et al., 2000). An accessible method to carrying out such a process is through a Monte Carlo numerical simulation, using a well designed integrator for a rigorous treatment of close encounters (Everhart, 1985). Doing that, it can be shown that the dispersion of the orbital elements increases suddenly (one or more orders of magnitude) after each close encounter with terrestrial planets (figure 1a).

Between encounters the dispersion of the orbital elements is kept almost constant. There is an exception for the dispersion in mean anomaly, which grows up

Pretka-Ziomek et al./Dynamics of Natural and Artificial Celestial Bodies, 289–294, 2001.
© 2001 *Kluwer Academic Publishers.*

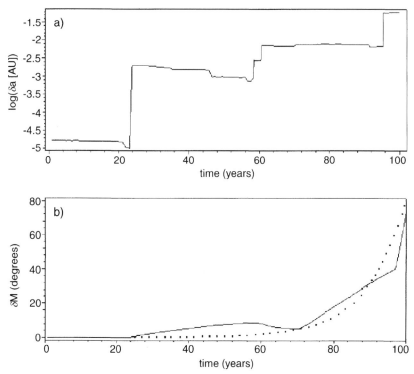

Figure 1. Time-growth of uncertainty a) in semimajor axis (logarithmic scale) and b) in mean anomaly, for the asteroid 1998 HH$_{49}$.

continuously as $\delta M(t) = \delta M_0 + \delta n \cdot (t - t_0)$, where δM_0 is the initial dispersion at t_0 and δn is the dispersion in mean motion. Per one revolution the dispersion increases as $\delta M_{rev} = 3\pi \delta a/a$, where δa denotes the dispersion of semimajor axis a. In short time the orbital uncertainty is mainly due to the dispersion in mean anomaly. This looks like a "string" formed by the virtual asteroids along the average orbit. They are ordered by their own semimajor axis, i.e. the first one has the smallest semimajor axis.

After several close encounters occurring at the moments $t_1, t_2, \ldots$ with the following dispersions in semimajor axis $\delta a_1 < \delta a_2 < \ldots$, the expression of the dispersion in mean anomaly becomes

$$\delta M(t) = \delta M_0 + \frac{3}{2}\sqrt{\mu} a^{-5/2} \sum_{i>0} (t_i - t_{i-1}) \delta a_i, \qquad (1)$$

(μ being the heliocentric gravitational parameter) which is nothing more than a first order spline approximation of the exponential curve (figure 1b)

$$\delta M(t) = \delta M_0 \, e^{\chi(t-t_0)}. \qquad (2)$$

Examining Eq. (1), we see that higher dispersions in semimajor axis produce steeper linear variations and, consequently, a higher value for the Lyapunov exponent χ. It seems that these sporadic dispersions play an important role in the propagation process of the orbital uncertainty. So, it is worthily to understand, at least geometrically, how a close encounter produces such a dispersion in semimajor axis.

3. Orbital Uncertainty Propagation Through a Close Encounter

We proceed with this analysis in the framework of the piecewise two-body model of close encounters - often entitled the *Opik's formalism* of close encounters (Carusi et al., 1990). This model has the major advantage to be approachable through analytic formulas with the price of some simplificative assumptions. We use the following set of assumptions: (**i**) the motion is divided in two regions, heliocentric and planetocentric ones, depending where the object is situated in respect to the sphere of action of the perturbing planet; (**ii**) the planet is moving on a circular orbit around the Sun (a_p is the radius of the orbit); (**iii**) the close encounter takes place near one of the orbital nodes of the perturbed body (let be d the distance from the orbital node to the planet's trajectory and d_p the distance from the planet to the asteroid's line of nodes when it reaches its node - figure 2a); (**iv**) during the close encounter the heliocentric unperturbed trajectories of encountering bodies are supposed to be linear and their heliocentric velocities v and v_p are constant; (**v**) finally, the close encounter acts as an instantaneous impulse and displacement given to the object when it reaches the minimum planetocentric distance b (the *impact parameter*), in such a way that it is moved from one asymptote of its planetocentric hyperbolic orbit to the other one. Let be (θ, ϕ) and (θ', ϕ') the angles quantifying the position of the unperturbed planetocentric velocity u before and, respectively, after the encounter (figure 2b).

If a notes the pre-encounter semimajor axis of the asteroid, its post-encounter value is

$$a' = a_p / \left[1 - 2\left(\frac{u}{v_p}\right)\cos\theta' - \left(\frac{u}{v_p}\right)^2 \right], \qquad (3)$$

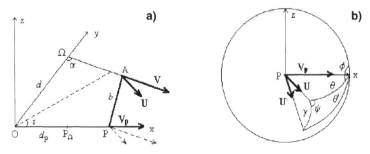

Figure 2. The geometry of a close encounter. a) heliocentric branch of motion, b) planetocentric branch of motion.

where $\cos \theta' = \cos \theta \cos \gamma + \sin \theta \sin \gamma \cos \psi$. γ is the *deflexion angle* and ψ is a measure of the *planetocentric orbital inclination*, having the following values

$$\tan \frac{\gamma}{2} = \frac{r_p}{2b} \left(\frac{v_{\text{par}}}{u} \right)^2, \tag{4}$$

$$\cos \psi = \frac{d \cos \theta \sin \phi + d_p \sin \theta}{b}, \tag{5}$$

where r_p is the radius of the planet and v_{par} is its parabolic velocity at the surface.

The impact parameter b is obtained from $b^2 = b_{\text{moid}}^2 + (d \sin \phi \cos \theta + d_p \sin \theta)^2$, where b_{moid} is the such called *minimal orbital intersection distance* (Carusi and Dotto, 1996).

We suppose that the initial orbital uncertainty is exclusively due to the dispersion in mean anomaly δM. Thus, in the geometry of the encounter the quantity d_p is the only one which is not constant. Its dispersion is

$$\delta d_p = a \left(\frac{a}{a_p} \right)^{3/2} \delta M. \tag{6}$$

Differentiating all previous written equations we obtain the final dispersion in semimajor axis as

$$\delta a' = C_1 \left[C_2 \left(\frac{r_p}{b} \right) + C_3 \left(\frac{b_{\text{moid}}}{b} \right)^2 \right] \left(\frac{a'}{b} \right) \delta M, \tag{7}$$

where the quantities

$$\begin{cases} C_1 = 2a' \left(\dfrac{a}{a_p} \right)^{3/2} \left(\dfrac{u}{v_p} \right) \sin \theta \\[2ex] C_2 = \left(\dfrac{v_{\text{par}}}{u} \right)^2 (\cos \theta \sin \gamma - \sin \theta \cos \gamma \cos \psi) \cos \psi \\[2ex] C_3 = \sin \theta \sin \gamma \end{cases} \tag{8}$$

are, in general, of the order of unity.

After the last orbital improvement, typical values for δM are between $10^{-7} - 10^{-5}$ radians. For the ratio (a'/b) typical values are between $10^2 - 10^4$. For evaluating the order of magnitude of the quantity in the square brackets of Eq. (7) we distinguish two cases: firstly - when the encounter is almost closest ($b \approx b_{\text{moid}}$) the expected order of magnitude is near 1, secondly - when the encounter is far of being closest ($b \gg b_{\text{moid}}$) then the quantity $(b/b_{\text{moid}})^2$ is negligible and the expected order of magnitude is given by the order of (r_p/b), with typical values of $10^{-3} - 10^{-1}$. Thus, the dispersion in semimajor axis could reach a value around 10^{-2} AU after just one close encounter. As time goes by the dispersion δM becomes higher, allowing for the next close encounter to produce a dispersion in semimajor axis several orders of magnitude higher than the previous one (figure 1a).

4. Oscillatory Behaviour of the Uncertainty in Mean Anomaly

During the close encounter, it is possible that the virtual asteroids forming the string along the average orbit receive different degrees of perturbations in such a way that their arrangement against semimajor axis is altered. In this case, the rearrangement process begins and the string collapses for a period of time (figure 3). Then, the entire order throughout the string is reversed. Generally, this is happening in the case of closest approach, when $\delta a' > 0$ as it results from Eq. (7) and Eq. (8). If the encounter is far of being closest, $\delta a'$ has the sign of C_2 which, in this case, is given by the sign of the quantity $-\sin(\theta \pm \gamma)$, generally being negative. The arrangement against semimajor axis is then preserved.

This oscillatory behaviour is common for many asteroids, at least for several tens of years after the last orbital improvement (figure 4). The major consequence is that the dispersion in mean anomaly is kept at smaller values for some time, and this is a satisfactory behaviour for the future attempts of recovering the asteroid (Milani, 1999).

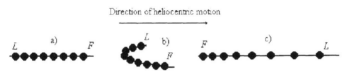

Figure 3. The rearrangement process throughout the string of virtual asteroids (F denotes the first asteroid in the string and L the last one).

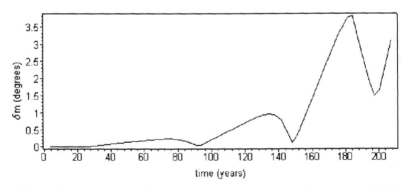

Figure 4. The oscillatory behaviour of the uncertainty in mean anomaly for the asteroid 1999 AN$_{10}$.

5. Conclusions

The orbital uncertainty propagation is mainly governed by the sporadic dispersions of the semimajor axis during the close encounters. The encounters for which $b \approx b_{\text{moid}}$ are more effective in producing higher dispersions but, on the other hand, they reverse the order of the virtual asteroids filling the confidence region and, consequently, an oscillatory behaviour of the uncertainty in mean anomaly appears afterwards.

Acknowledgements

I acknowledge here the partial support come from the organizers in order to attend the US-European Celestial Mechanics Workshop. I also acknowledge a very helpful discussion with Prof. Andrea Milani and Dr. Giovanni Valsecchi.

References

Carusi, A., Valsecchi, G.B. and Greenberg, R.: 1990, 'Planetary Close Encounters: Geometry of Approach and post-Encounter orbital parameters', *Celestial Mechanics and Dynamical Astronomy* **49**, pp. 111–131.

Carusi, A. and Dotto, E: 1996, 'Close Encounters of Minor Bodies with the Earth', *Icarus* **124**, pp. 392–398.

Everhart E.: 1985, 'An efficient integrator that uses Gauss-Radau spacings', In *Dynamics of Comets: Their Origin and Evolution* (A. Carusi and G.B. Valsecchi, Eds.), IAU Colloquium 83, Reidel, Dordrecht, pp. 185–202.

Milani, A.: 1999, 'The Asteroid Identification Problem I: Recovery of Lost Asteroids', *Icarus* **137**, pp. 269–292.

Milani, A., Chesley, S.R., Boattini, A. and Valsecchi, G.B.: 2000, 'Virtual Impactors: Search and Destroy', *Icarus* **145**, pp. 12–24.

PREDICTION, OBSERVATION AND ANALYSIS OF ASTEROID'S CLOSE ENCOUNTERS

J.A. MORAÑO FERNANDEZ (jose.a.morano@uv.es)
Astronomical Observatory of Valencia University Valencia, Spain

A. LOPEZ GARCIA (alvaro.lopez@uv.es)
Astronomical Observatory of Valencia University, Astronomy and Astrophysics Department.
Valencia University Valencia, Spain

J. PASTOR ERADE
Astronomical Observatory of Valencia University Valencia, Spain

YU.D. MEDVEDEV
Institute of Applied Astronomy. IAA, RAS St.Petersburg, Russia

Abstract. Masses of large asteroids are needed for improving modern planetary ephemerides. 239 minor planets with diameter bigger than 100 km have been selected for mass determination through close encounters with all numbered and unnumbered asteroids. To this aim, the orbits of 11000 numbered and 35000 unnumbered minor planets in the time interval 1985-2010 have been integrated. The influence of the epoch of observations and the relation between the accuracy in the mass determination and data distribution are investigated.

1. Introduction

Up to now the masses of several asteroids have been determined: (1) Ceres, (2) Pallas, (10) Hygiea, (11) Parthenope, (15) Eunomia, (704) Interamnia and (24) Themis (see references in (Lopez Garcia et al., 1997)) appear more frequently in the literature.
The usual practice of getting periodically 'new orbits', smoothing residuals and laying aside the faint effect of nearby asteroids, masks its influence and avoids any possibility of mass determination (Lopez et al., 1996).

Recent investigations, including other papers presented in this meeting (Michalak, 2000), improve masses of several asteroids, but accurate mass values of a larger number of asteroid should be of great interest.

2. Close Encounters Prediction

In this presentation we update several encounters predictions (Hilton et al., 1995), (Lopez Garcia et al., 1997), for the period 1985 - 2010, interval with a wide collection of past observations and good possibilities to arrange future observing campaigns.

Pretka-Ziomek et al./Dynamics of Natural and Artificial Celestial Bodies, 295–300, 2001.
© 2001 *Kluwer Academic Publishers.*

We have chosen 239 'selected minor planets' with diameter bigger than 100 km as candidates for getting its masses from close encounters. To this list will be added some other asteroids of big mass although their predicted diameter is rather uncertain. High precision osculating elements of these 'selected asteroids' are calculated for the full time interval and stored in computer files.

Rectangular co-ordinates of 13472 numbered and 35944 unnumbered minor planets orbits have been integrated every ten days by Bulirsh-Stoer method and have been also stored. Perturbation from the nine major planet are taken into account from DE403 theory.

Close encounters are detected when distance between both asteroids is below 0.03 AU. Several parameters are calculated for each encounter: epoch (EP), minimum distance, duration (in intervals of 10 days) and deflection angle (θ) at the encounter's epoch. It is also obtained the geocentric ephemerides for the small asteroid, considering in one case only the perturbation of the big planets (O) and adding in the other case the effect of the perturbing asteroid (M). Accurate co-ordinates of this one are obtained from 'asteroid elements file'.

The maximum value of (M-O) difference for the interval EP-500 to EP+1500 and its epoch are included in encounters records as an estimate of the 'real' effect of close encounter on small asteroid geocentric position.

The angle of deflection, θ, is obtained by the formula:

$$\tan \frac{\theta}{2} = 1.84 \times 10^{-21} \times \frac{r^3}{b \cdot v^2} \tag{1}$$

where:

- r is the approximate radius for large asteroid, in km;
- b is the minimum distance between asteroids in au;
- v is the relative velocity of asteroids in au/day;

We assume a density of asteroids of 3 g/cm^3.

3. List of Close Encounters

Most relevant encounters with numbered (N) and unnumbered (U) asteroids for the period 1985-2010 are presented in Table I and II, ordered by (M-O)max and θ. From our full lists, 24 'selected asteroids' are candidates to mass determination: 1, 4, 7, 10, 13, 15, 16, 20, 29, 45, 46, 48, 52, 65, 104, 141, 150, 175, 221, 268, 423, 451, 532, 1389, in the considered time interval.

Big asteroids with several different encounters (1, 4, 10, 15, 16, 20, 29, 268) hold a good opportunity to improve the mass, although its small predicted value in some cases.

TABLE I

Close encounters with $(M - O)$ max > 2"

A1	A2	EPOCH	DIST	DUR	θ	M-O EPOCH	(M-O)max
1	5303 N	2450337	0.0057	3	3.66	2451607	21.63
1	8231 N	2453687	0.0163	3	2.13	2454887	4.82
1	2933 N	2450097	0.0196	3	1.46	2451207	4.28
4	8331 N	2447177	0.0083	7	5.07	2448227	3.79
1	11319 N	2449527	0.0178	3	1.28	2450757	3.71
4	17 N	2450247	0.0195	7	1.67	2451697	3.45
1	7738 N	2451277	0.0101	3	2.05	2452187	2.76
1	7264 N	2454537	0.0229	2	0.67	2455447	2.66
1	8363 N	2454527	0.0179	3	0.51	2455337	2.30
1	8047 N	2450197	0.0239	2	0.53	2451267	2.15
1	11506 N	2446217	0.0221	2	0.11	2447487	2.02
1	27548 U	2454207	0.0062	11	18.52	2455327	24.43
1	2092 U	2451677	0.0118	4	3.22	2453077	15.64
1	30591 U	2447197	0.0084	2	0.22	2448697	14.56
1	26546 U	2448687	0.0177	4	2.5	2449797	9.5
1	12870 U	2450837	0.0293	2	3.86	2452097	6.58
1	32675 U	2446787	0.0066	3	2.61	2447517	6.53
1	4942 U	2454627	0.0138	2	0.93	2455377	6.44
1	8766 U	2454607	0.0133	3	1.17	2455437	5.82
1	12620 U	2447947	0.024	2	0.26	2449257	5.63
1	7064 U	2447167	0.0082	3	0.22	2448307	5.22
4	15516 U	2447707	0.0058	4	0.52	2448757	4.48
1	35379 U	2454777	0.0146	3	1.14	2455397	4.11
1	25992 U	2446977	0.0238	2	0.42	2448377	4.08
1	2430 U	2452927	0.0177	1	0.19	2454007	4.02
1	10967 U	2447787	0.0277	1	1.46	2449247	4.02
1	3456 U	2451287	0.0176	2	0.62	2452167	3.94
1	3791 U	2450357	0.0209	2	0.52	2451217	3.63
1	7478 U	2447637	0.0234	1	0.17	2448797	3.3
4	10474 U	2447997	0.0082	3	0.12	2449497	3.29
1	12765 U	2450227	0.0253	2	0.98	2451547	2.81
1	25839 U	2452057	0.0226	2	0.23	2453507	2.69
1	24974 U	2452907	0.0259	1	0.12	2454407	2.5
1	12164 U	2454517	0.0139	2	0.37	2455327	2.49
1	5661 U	2453747	0.0278	2	0.79	2454927	2.46
1	22642 U	2449497	0.0264	1	0.13	2450997	2.26
1	1933 U	2451917	0.0053	2	1.51	2453417	2.25
1	24843 U	2451197	0.0254	2	0.32	2452567	2.25

TABLE I
continued

A1	A2	EPOCH	DIST	DUR	θ	M-O EPOCH	(M-O)max
4	845 U	2446377	0.0137	3	0.68	2447687	2.22
1	4936 U	2446767	0.0204	2	0.55	2447917	2.21
1	1954 U	2450257	0.021	2	0.53	2451757	2.18
1	21951 U	2454957	0.0044	3	1.53	2455547	2.16
1	13230 U	2451997	0.0197	2	0.97	2453057	2.14
4	15380 U	2453077	0.0259	4	0.24	2454577	2.04

TABLE II
Close encounters with $\theta > 1"$

A1	A2	EPOCH	DIST	DUR	θ	M-O EPOCH	(M-O)max
4	8331 N	2447177	0.0083	7	5.07	2448227	3.79
1	5303 N	2450337	0.0057	3	3.66	2451607	21.63
1	1847 N	2454977	0.0256	3	2.22	2455387	0.78
1	8231 N	2453687	0.0163	3	2.13	2454887	4.82
1	7738 N	2451277	0.0101	3	2.05	2452187	2.76
10	3946 N	2450907	0.0145	10	1.84	2452157	1.13
4	17 N	2450247	0.0195	7	1.67	2451697	3.45
1	2933 N	2450097	0.0196	3	1.46	2451207	4.28
1	11319 N	2449527	0.0178	3	1.28	2450757	3.71
1	27548 U	2454207	0.0062	11	18.52	2455327	24.43
1	25273 U	2455367	0.0229	4	5.19	2455547	0.2
1	12870 U	2450837	0.0293	2	3.86	2452097	6.58
1	2092 U	2451677	0.0118	4	3.22	2453077	15.64
1	24032 U	2455307	0.0142	3	2.66	2455467	0.17
1	32675 U	2446787	0.0066	3	2.61	2447517	6.53
1	26546 U	2448687	0.0177	4	2.5	2449797	9.5
1	19528 U	2455087	0.0072	4	1.94	2455547	0.89
29	23547 U	2448737	0.0295	6	1.68	2450107	0.36
1	21951 U	2454957	0.0044	3	1.53	2455547	2.16
1	1933 U	2451917	0.0053	2	1.51	2453417	2.25
1	10967 U	2447787	0.0277	1	1.46	2449247	4.02
4	33743 U	2446887	0.027	5	1.19	2447657	0.34
1	8766 U	2454607	0.0133	3	1.17	2455437	5.82
1	35379 U	2454777	0.0146	3	1.14	2455397	4.11
4	11599 U	2449617	0.0087	4	1.13	2451087	1.39
15	13362 U	2449957	0.0259	11	1.01	2451117	0.47

4. Analysis of Close Encounter

In our algorithm 7 parameters (6 motion parameters for the smaller asteroid and the mass of the larger) are determined. A set of 2N (N: number of observations) equations produce a final set of 7 linear equations giving the 7 corrections to motion parameters and perturbing mass. Observations with large (O-C) values are previously eliminated.

A new algorithm, including observations of several close encounters with one big asteroid, is under progress.

The coefficients of conditional equations for improving the motion parameters (dc/dE) and the mass (dc/dm) are calculated by numerical integration (Sitarski, 1995).

The derivatives of spherical co-ordinates with respect to mass (ds/dm) are given by the formula:

$$ds/dm = ds/dc.dc/dm$$

where c is the heliocentric vector of position for smaller asteroid and m is the mass of the larger asteroid.

5. Observation of Close Encounters

As an example of the results that are expected to be reached, we summarize our investigation on (24) Themis (Lopez Garcia et al., 1997) through its encounter with (2296) Kugultinov in 1975.

The distribution of observations for (2296) was rather uniform and 8 of the 62 observations were close to the moment of encounter. Corrections to the 6 components of the initial orbital vector of (2296) Kugultinov and the mass of (24) Themis were determined. The results of these calculations give a mean square error of 50 % in (24) Themis mass value and the set of obtained parameters represents the observations of (2296) Kugultinov with m.s.e. of 1.5".

In other mass determination from close encounters of (10) Hygiea with several asteroids, the dispersion of results was very big, due to the non uniform distribution of the observations, the lack of observations near the moment of some encounters and the short observation arc for other asteroids after the encounter. With the addition of three artificial observations, the mass of (10) Hygiea improved to a rather good value, with a m.s.e. of 50%.

6. Conclusions

A list of asteroid encounters between a 'selected group' of 239 'big' minor planets and all numbered and unnumbered asteroids has been obtained for the time interval 1 Jan 1985 to 31 Dec 2010. To this aim, the orbits of 13472 numbered and 35944 unnumbered asteroids have been integrated for that period.

The mass of a set of 24 'selected asteroids' can be improved from their encounters with numbered and unnumbered asteroids.

The examples included show that it is very important the longitude of the observational arc, the uniform distribution of observations and the presence of observations close to the encounter's epoch.

To get the best possible results, it is necessary to consider all previous observations as well as to prepare observing campaigns for detected future close encounters. Due to the small effect of close encounters on asteroids co-ordinates, the most accurate astrometric techniques should be applied.

References

Hilton, J.L, Middour, J. and Seidelmann, P.K.: 1995, 'Prospects for determining masses for asteroids.', *Abstracts, Symposium 172 de L'UAI, Paris*, pp. 61.

Lopez Garcia, A., Medvedev, Yu.D. and Morano, J.A.: 1996, 'Influence of observations on the minor planets mass determination', *Fourth Intern. Workshop on Positional Astronomy and Celestial Mechanics* pp. 245–248.

Lopez Garcia, A., Medvedev, Yu.D. and Morano Fernandez, J.A.: 1997, 'Using close encounters of minor planets for the improvement of their masses', In *Dynamics and Astrometry of Natural and Artificial Celestial Bodies. Proceedings of IAU Coll. 165* (Wytrzyszczak, I.M., Lieske, J.H., Feldman, R.A., Eds.), Kluwer Acad. Publ., pp. 199–204.

Michalak, G.: 2000, 'Determination of Masses of Six Asteroids from Close Asteroid-asteroid Encounters', *Oral communication at this meeting*.

Sitarski, G.: 1995, 'Determination of Masses of Mercury and Venus from Observations of Five Minor Planets', *ACTA ASTRONOMICA*, **Vol 45**, pp. 665–672.

THE RECONSTRUCTION OF GENETIC RELATIONS BETWEEN MINOR PLANETS, BASED ON THEIR ORBITAL CHARACTERISTICS

A.E. ROSAEV (rosaev@nedra.yar.ru)
FGUP NPC "NEDRA" Yaroslavl, Russia

Abstract. The new numeric method of the fast determination minimal distance between of elliptic orbits is presented. The very anisotropy distribution of the three orbits crossing points is found. It may be explained by supposing one or more catastrophic events close to the Earth orbit during last few thousands years.

Key words: Orbits intersections, Near–Earth asteroids, Minimal distances, Collisions

1. Introduction Main Equations

The problem of the origin of Near–Earth Asteroids (NEA) has a great application for the Earth's protection against asteroid's danger. In spite of the close relation with main belt and comets, the orbits of Aten's association are considerably differ from orbits of another minor bodies. Each theory of origin NEA must take into accounting this fact. On our opinion, the mutual collisions of minor planets put a very remarkable contribution in the forming of recent orbits of NEA. So, few of them appeared as independent objects directly in neighborhood of the Earth. We suppose the new method of numeric solution of the problem of determination of the minimal distance between elliptic orbits, based on a suggestion, that minimal distance reach close to point, where heliocentric distances rm and longitudes l of each objects are approximate equal. Finally, it leads to conditions:

$$d^2 = r^2 \left(\sum_{i=1}^{3} (a_i \cos(l_{min}) + b_i \sin(l_{min})) \right)$$

$$l_{min} = \frac{1}{2} \text{atg} \left(2 \frac{\sum_{i=1}^{3} a_i b_i}{\sum_{i=1}^{3} (a_i^2 - b_i^2)} \right)$$

$a_1 = \cos(\Omega_1)^2 + \sin(\Omega_1)^2 \cos(i_1) - \cos(\Omega_2)^2 - \sin(\Omega_2)^2 \cos(i_2)$
$b_2 = \sin(\Omega_1)^2 + \cos(\Omega_1)^2 \cos(i_1) - \sin(\Omega_2)^2 - \cos(\Omega_2)^2 \cos(i_2)$
$a_2 = \sin(\Omega_1) \cos(\Omega_1) (1 - \cos(i_1)) - \sin(\Omega_2) \cos(\Omega_2) (1 - \cos(i_2))$
$b_1 = \sin(\Omega_1) \cos(\Omega_1) (1 - \cos(i_1)) - \sin(\Omega_2) \cos(\Omega_2) (1 - \cos(i_2))$

Pretka-Ziomek et al./Dynamics of Natural and Artificial Celestial Bodies, 301–303, 2001.
© 2001 *Kluwer Academic Publishers.*

$$a_3 = -\sin(\Omega_1)\sin(i_1) - \sin(\Omega_2)\sin(i_2)$$
$$b_3 = \cos(\Omega_1)\sin(i_1) - \cos(\Omega_2)\sin(i_2)$$

where Ω_1 and Ω_2 – longitude of ascending node first and second objects; i_1 and i_2 – inclinations of orbit first and second objects.

Here we need to verify 8 cases due to arctg function properties. Then, for more accurate determination of the minimal distance value, we apply classical numeric scheme with step 0.03 degree by true anomaly in small region (1.2 degree) of l_{min}, obtained by our method. Evidently, the such scheme make possible to significantly reduce time of calculation (more, than in 10 times). This method is applicable in most cases, except orbits with high mutual inclinations.

2. The Results and Discussion

First of all, we calculate minimal distances between all NEA, including non–numbered. Then the calculation of distance d between 82 numbered asteroids was carried out. Only 154 cases with d smaller then 0.035 a.u. took into account. In results, the very non-casual distribution of crossing points was obtained. The remarkable concentration of them observed near heliocentric longitude $320 - -360$ degrees. In addition, few another associations of asteroid's orbits may be detected at different directions (different l). Of course, there are many casual events, not related with collisions, between 154 crossing orbits. For exception of them, the cases, when 3 or more orbits are crossed in small space ($D_l = 5$ degrees, $D_r = 0.03$ a.u., $D_z = 0.03$ a.u.) are obtained. The results are shown in Table 1.

It is evident, that position of points of triple crossing is not casual, and closed to parts of two orbits, which crossed very near (smaller then 0.01 a.u.) from Earth's orbit. So, it may be suggested, that Earth, by it's own gravitational field, assist to

TABLE I

N	Name of minor planets	Heliocentric distance, a.u.	Longitude, degree
1	Geographos Florence 1986 PA	1.059	336.95
2	Geographos Khufu Florence	1.056	334.80
3	Pan Florence Wilson-H'ton	1.033	342.5
4	Magellan Nefertiti Pele	1.234	341.96
5	Don Quixot Mithra Ubasty	1.321	346.21
6	Midas Cuyo 1980 PA	1.092	1.1
7	Toutatis Verenia Midas	1.066	356.45

the destruction of unknown encountered celestial body. Then, fragments of this body collided one another.

3. The Electronic Databank

The created electronic databank include: 1) database of NEA's orbits intersections (include non–numbered) – 3179 cases; 2) database of NEA's and fireballs intersections; 3) database of fireballs's orbits mutual intersections; 4) database of three NEA's orbits intersections. Our bases may be updated daily. It gives us ability to immediately take into account all changes in near Earth space and in main belt. The electronic databank placed in Internet on address: http://www.chat.ru/˜ringsbro.

The updates may be useful, for example, for investigation of the mutual perturbations of minor planets at encounters and related problem of mass determinations.

4. Conclusions

The new numeric method of the fast determination minimal distance between of elliptic orbits is presented. The significant anisotropy in distribution of the three orbits crossing points is found. It may be explained by supposing one or more catastrophic events close to the Earth orbit during last few thousands years.

The currently updated electronic database of crossing orbits is created and presented in Internet. The updates give the ability to detect strongly perturbed orbits and study the minor planets mutual encounters.

DYNAMICALLY NEW COMETS IN THE SOLAR SYSTEM

PIOTR A. DYBCZYŃSKI and HALINA PRĘTKA–ZIOMEK
Astronomical Observatory of A. Mickiewicz University ul. Słoneczna 36, 60-286 Poznań, Poland
e-mail: dybol@amu.edu.pl, pretka@amu.edu.pl

Traditional definition of a dynamically 'new' comet, which is the comet for the first time visiting our planetary system, is that it should have its $1/a < 1 \times 10^{-4}$ AU^{-1} (eg. Oort and Schmidt, 1951). This definition is based on the value of semimajor axis of the comet at the moment it is observed, but it does not take into regard its dynamical history. Such a long–period comet has semimajor axis larger than 10 000 AU so, while travelling through the space, is influenced by gravitational attraction from all the visible and invisible matter in the Galaxy. Constant galactic perturbations as well as the stochastic influence of individual stars passing close to the Sun can change the cometary orbit, in particular its perihelion distance, which is the critical parameter, determining whether the comet will or will not enter into the planetary region.

We have investigated the dynamical history of the observed long–period comets going backwards to its previous return to the Sun. We studied past evolution of 307 long–period comets from the Catalogue of Cometary Orbits (Marsden and Williams, 1997) and 20 comets that have been newly discovered after publishing the Catalogue. For each comet we calculated at first its barycentric, original orbit, that is the orbit the comet had before it entered the planetary system. We traced backwards the motion of comets, taking into account gravitational influence of all planets and the Moon as well as relativistic effects according to the dynamical model used by (Yau et al., 1994). We stopped the integration when the comet reached the distance of 250 AU from the Sun. From further investigation we rejected 44 comets that did not reach the limit of 250 AU and 28 comets having original hyperbolic orbits. As a result we obtained 255 comets with original elliptical orbits, among them 85 comets had $a > 10\,000$ AU.

Then, starting from their original orbits, we traced the past evolution of all comets with elliptical orbits, including the gravitational attraction of the Galaxy. As the dynamical model we used the model proposed by (Heisler and Tremaine, 1986), restricted only to the galactic disk which is considered to be the most dominant part of the tidal galactic potential in the vicinity of the Sun. We integrated the motion of comets backwards to their previous return to the planetary region, namely to their previous perihelion passage.

For 170 comets that had $a < 10\,000$ AU galactic perturbations were very weak and did not change the cometary orbital elements significantly. These comets are 'old' since during the previous return to the Sun they had their perihelion distances small enough to be perturbed by planets. For the rest of the comets, called 'new' in

---

Pretka-Ziomek et al./Dynamics of Natural and Artificial Celestial Bodies, 305–306, 2001.
© 2001 *Kluwer Academic Publishers.*

the Oort sense, the situation is more complicated. In Table I we present results of our calculation for 85 elliptical comets having their original $a > 10\,000$ AU. Additional, we show here the subset of comets with the observed perihelion distance greater than 1.5 AU and orbits of class 1. These comets have their orbits very well determined and, thanks to having their perihelion distances large enough, they do not pass close to the Sun, so they do not suffer from nongravitational effects.

TABLE I

The distribution of the previous perihelion distance for long–period comets, called 'new' in the Oort sense.

Previous q [AU]	$q_{obs} > 1.5$ AU class 1	all comets
$0 - 2$	3	10
$2 - 5$	24	31
$15 - 40$	6	12
> 40	20	32
Total number	53	85

One can see that from 85 comets called 'new' in the Oort sense as many as 41 comets passed closer to the Sun than 15 AU during their previous return. These comets should be considered as dynamically 'old' ones since they had passed the planetary region deep enough to have their previous orbits perturbed by the planets. Only 44 comets from the whole population can be called 'new' in the dynamical sense. A detailed description of this investigation including also a discussion of the stellar perturbations and physical properties of long–period comets can be found in (Dybczyński, 2001).

References

Dybczyński, P.A.: 2001, submitted to *Astron. and Astrophys.*
Heisler, J. and Tremaine, S.D.: 1986, *Icarus*, **65**, 13.
Oort, J.H. and Schmidt, M.: 1951, *Bull. Astron. Inst. Nether.*, **11**, 259.
Yau, K., Yeomans, D. and Weissman, P.: 1994, *MNRAS*, **266**, 305.

THE PECULIAR ORBIT OF VYSHESLAVIA: FURTHER HINTS FOR ITS YARKOVSKY DRIVEN ORIGIN?

MIROSLAV BROŽ (mira@sirrah.troja.mff.cuni.cz) and DAVID VOKROUHLICKÝ
(vokrouhl@mbox.cesnet.cz)
Institute of Astronomy, Charles University V Holešovičkách 2, CZ-180 00 Prague 8,
Czech Republic

Abstract. The orbit of asteroid 2953 Vysheslavia is presently locked in a tiny chaotic zone very close to the 5/2 mean motion Jovian resonance. Its dynamical lifetime is estimated to be of the order of only about 10 Myr. Such a conclusion poses a problem, since Vysheslavia is a member of the Koronis family which is likely more than 1 Gyr old. Three main hypotheses were developed to solve this apparent contradiction: (i) Vysheslavia might be an outcome of a recent secondary fragmentation event in the family, (ii) Vysheslavia might have been placed on its peculiar orbit by close encounters with nearby massive asteroids, or (iii) the asteroid might have been transported by a slow inward–drift of the semimajor axis due to the Yarkovsky effect. Though we cannot disprove the first two possibilities, here we bring evidence for the third scenario.

1. Introduction and Motivations

The most obvious evolution processes that take place in the main belt asteroid families are the collisional grinding (i. e. secondary collisions between family members, see Marzari et al., 1995, 1999) and the chaotic diffusion, which works in chaotic regions associated with resonances and which affects mainly eccentricities and inclinations of asteroid orbits (see Milani and Farinella, 1994; Milani et al., 1997; Nesvorný and Morbidelli, 1998). In addition, an interesting mechanism has been recently proposed by Farinella and Vokrouhlický (1999): the size-dependent semimajor axis diffusion due to the Yarkovsky effect. Here we test the Yarkovsky diffusion hypothesis in the case of the Koronis family.

The fundamental argument of our work derives from a peculiar finding of Milani and Farinella (1995) who noticed that asteroid 2953 Vysheslavia is located in a tiny chaotic zone (about 10^{-3} AU wide) very close to the outer border of the strong 5/2 mean motion resonance with Jupiter. The expected dynamical lifetime of such an orbit before falling into the resonance is of order of only 10 Myr, what is in an apparent contradiction with the estimated age of the whole Koronis family $1-2$ Gyr (Greenberg et al., 1996). Vysheslavia is presumably its member: the spectroscopic analysis indicates that Vysheslavia is an ordinary S–type asteroid (Bus, 1999) and the statistical analysis of the Koronis family predicts very few interlopers of the Vysheslavia size (Migliorini et al., 1995). Knežević et al. (1997) tentatively identified two other Koronis members very close to the mentioned chaotic zone.

Milani and Farinella (1995) proposed two hypotheses: (i) Vysheslavia might be an outcome of a recent secondary fragmentation event in the family, or (ii) Vysheslavia might have been placed on its present orbit by close encounters with

Pretka-Ziomek et al./Dynamics of Natural and Artificial Celestial Bodies, 307–312, 2001.
© 2001 *Kluwer Academic Publishers.*

massive asteroids. However, none of these two possibilities were found very likely: the probability of the disruption of a large (25–50 km) parent asteroid during the last 100 Myr is less than 5% and the gravitational fluctuations in the semimajor axis due to Ceres and Pallas may statistically accumulate to $\approx 10^{-3}$ AU over the age of the Solar System.

In Vokrouhlický et al. (2001) we proposed the Yarkovsky-driven origin of the Vysheslavia's metastable orbit. We have checked that the Yarkovsky force is able to change the semimajor axis of multikilometer size asteroids by 0.01 – 0.02 AU within its expected collisional lifetime. Objects originally located on stable orbits further from the 5/2 resonance, may be thus brought to its vicinity. In course of this evolution they may be temporarily captured in chaotic regions related to weaker resonances and thus explain origin of the orbit of Vysheslavia. The probability of this process appears much higher than in the previously mentioned possibilities. Here we report 9 – 14 more asteroids (likely members of the Koronis family) that have the same metastable orbit as Vysheslavia, and we preliminarily argue that this brings more evidence for the Yarkovsky-driven origin of these orbits, including Vysheslavia's.

2. Numerical Simulations and Discussion

We have implemented linearized versions of both variants of the Yarkovsky effect (diurnal and seasonal) in different numerical integrators; most importantly, we created a quasi-symplectic integrator swift_rmvsy (based originally on swift_rmvs3 code by Levison and Duncan, 1994) that allows fast simulations. Though the properties of symplecticity are violated due to the weak dissipation caused by the Yarkovsky effect, we have extensively tested our code (both by comparing its results with more precise integrators and by reproducing the analytic results when available). These tests, as well as the Yarkovsky effect implementation, are listed in Brož (1999).

Due to its peculiar location, the Lyapounov time of the Vysheslavia's orbit is ≈ 27 kyr only. This is much shorter than the time span of most of our simulations, and thus the simulations have a statistical meaning only. This is, however, not an obstacle for our work, since our fundamental conclusions are of statistical nature. To make a statistical sense of our work we have used a technique of integrating neighbour orbits. We used two levels of "zooming" in this respect: (i) "fictitious neighbours" (FNs) cover a larger area around a given orbit (typically up to displacements $\Delta a = 10^{-3}$ AU in the semimajor axis and $\Delta e = 10^{-3}$ in eccentricity), and (ii) "close clones" (CCs) that span basically the 3σ uncertainty ellipsoid of the given orbit initial data (notably up to displacements $\Delta a = 10^{-7}$ AU and $\Delta e = 10^{-6}$).

The orbits of FNs and CCs were integrated both with and without the influence of the Yarkovsky effect (we typically used 4 outer planets in our simulations, but

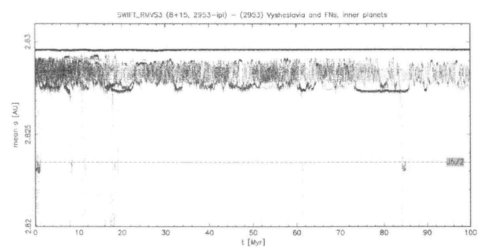

Figure 1. Mean semimajor axis (in AU) vs. time (in Myr) for Vysheslavia and its 14 FNs. The Yarkovsky effect is not included in this simulation, but perturbations due to all planets (except Pluto) are considered. The median lifetime against the fall to the 5/2 resonance for this integration is 10.9 Myr, while lifetime of the objects temporarily residing inside the 5/2 resonance is less than 1 Myr.

checks with 8 planets except Pluto and even with two massive asteroids Ceres and Pallas were systematically performed). Since we are dealing with multikilometer asteroids, we assumed low surface thermal conductivity ($K = 0.0015$ W/m/K), as it is indicated by many observations, but again checks with higher value of K were performed. The major unknown factor that influences the strength of the Yarkovsky effect is then the orientation of the asteroid spin axis (and its possible temporal evolution). For that reason, we have performed tests with different orientations of the spin axis.

Vysheslavia's chaotic zone. Figure 1 shows evolution of the mean semimajor axis of Vysheslavia and its 14 FNs without the Yarkovsky effect. One recognizes the chaotic zone, where the semimajor axis randomly fluctuates. Notice also sporadic instants, when some of the integrated objects fall into the 5/2 resonance. The uppermost FN has been placed on a stable orbit, to contrast its behavior with the others.

Transport towards the 5/2 resonance. Figure 2 indicates the importance of the Yarkovsky effect for overall mixing of small, multikilometer, asteroids in families (here the Koronis asteroid 7340 is integrated with its 14 FNs of different spin orientations; low surface conductivity is assumed). Several facts may be concluded from this integration: (i) the Yarkovsky effect may spread semimajor axes of small asteroids, initially of the same orbit, by as much as ≈ 0.05 AU within their estimated collisional lifetime, (ii) the asteroid 7340 may slowly evolve toward the Vysheslavia's dynamical state (and thus Vysheslavia might have been originally on a similar orbit as 7340), and (iii) even if the asteroids drift outward from the 5/2 resonance, their origin in past is constrained by the presence of the resonance.

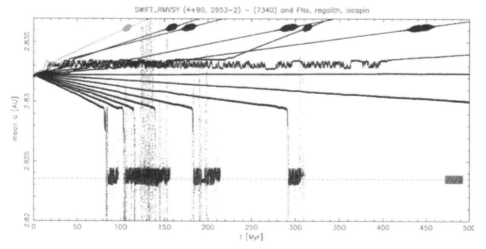

Figure 2. Mean semimajor axis (in AU) vs. time (in Myr) for asteroid 7340 (1991 UA$_2$) and its 14 FNs. The low value $K = 0.0015$ W/m/K of the surface conductivity is assumed, the orientations of the spin axes are distributed isotropically in space (FNs here mean that the initial conditions were the same for all objects, but the orientation of their spin axes was different). Apart from the Vysheslavia's chaotic zone, one may notice another tiny zone at about 2.8333 AU. Given the orientation of the spin axis, the asteroids may drift outwards or inwards. While being originally on a stable orbits, about a half of the FNs reaches the Vysheslavia's chaotic zone (and then the 5/2 resonance) in 70–500 Myr.

Other asteroids on metastable orbits. We have selected ≈ 400 objects in the outer vicinity of the 5/2 resonance from the `astorb.dat` catalogue (vers. Jul 6, 2000; Bowell et al., 1994) and integrated them for about 10 Myr (Yarkovsky effect was not included in these integrations). After eliminating objects initially inside the 5/2 resonance and those on stable orbits, we have identified 9 objects with "metastable" orbits similar to that of Vysheslavia (5 more candidates with poorly known orbits were discarded from our considerations). Moreover, we found 3 objects on unstable orbits in between the Vysheslavia's chaotic zone and the 5/2 resonance and 1 asteroid just at the upper edge of the Vysheslavia's zone (having an unstable orbit on a very long time span; 5 more candidates of this type have been also found but not integrated for sufficiently long time interval). We preliminarily selected some of these objects and integrated them (with limited number of CCs) up to 500 Myr. Figure 3 shows characteristic behaviour of the three classes of orbits and in Table I we give some information about the objects whose orbits have been integrated.

None of these new objects on metastable orbits is of Vysheslavia size, but there are about 7 of them with half of its size (i. e. radii between 3.3 – 4 km). Others are smaller, with typical radii between 1 – 2.7 km. Their entire estimated mass is by about 20 % larger than that of Vysheslavia. These facts seem to weaken the secondary collisional hypothesis mentioned above. Assumption, that they have been created in a single collisional event, inevitably leads to many more fragments

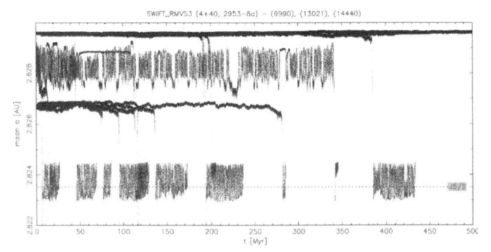

Figure 3. Mean semimajor axis (in AU) vs. time (in Myr) of asteroids 9990, 13021, 14440, each with 4 CCs. Yarkovsky effect is not included here; 9990 (upper body) is on a nearly stable orbit but some of the clones drop time from time to the 5/2 resonance, 13021 (middle body) indicates behaviour very similar to that of Vysheslavia (see Figure 1), and 14440 (lower body) is on a peculiar orbit very close to the 5/2 resonance. None of these objects could be located on its orbit for the entire lifetime of the Koronis family ($\approx 1 - 2$ Gyr).

TABLE I

The sample of asteroids located on metastable orbits close to the 5/2 resonance (most likely Koronis family members); radii estimated from absolute magnitudes and geometric albedo 0.2, dynamical lifetimes against the fall into the 5/2 resonance is given as a median over 5 integrated CCs, proper elements for numbered asteroids from AstDyS (http://newton.dm.unipi.it/asteroid/).

asteroid	radius [km]	lifetime [Myr]	proper a [AU]	proper e
2953	7.2	11	2.82767	0.0531
6814	3.6	20	2.82856	0.0417
9631	2.7	268	2.82736	0.0393
9990	3.3	395	2.82967	0.0606
13021	3.6	147	2.82786	0.0550
14440	2.7	116	2.82692	0.0286
1996 TE	2.1	14	2.82816	0.0503
2000 GQ$_6$	3.4	166	2.82796	0.0344

of the Vysheslavia size (at least 5). Note that the area of the metastable orbits covers less than $\approx 10\%$ of the area to which the collision would spread the ejecta and one should not assume too steep size distribution of the fragments (here we assumed power -3 of the cumulative distribution). The primordial body should have been larger and its disruption less likely (by about a factor 2). Even if the new objects were assumed to be created in different collisional events, we would lack the number of expected parent bodies (according to the latest debiased statistical analyses). On the other hand, since these smaller objects are more mobile by the Yarkovsky effect, we can expect more objects driven to metastable orbits. A preliminary estimate indicates about twice more objects on metastable orbits when Yarkovsky effect supports their repopulation and this may explain their relatively large number.

References

Bowell, E.K., Muinonen, K. and Wasserman, L.H.: 1994, in: *Asteroids, Comets and Meteors 1993*, (A. Milani, M. Di Martino, and A. Cellino, Eds.), Kluwer Acad. Publ., Dordrecht, pp. 477–481.

Brož, M.: 1999, *Orbital Evolution of the Asteroid Fragments due to Planetary Perturbations and Yarkovsky Effects* (Diploma Thesis, Charles Univ., Prague).

Bus, S.J.: 1999, *Compositional Structure in the Asteroid Belt: Results of a Spectroscopic Survey* (PhD Thesis, Massachusetts Institute of Technology).

Farinella, P. and Vokrouhlický, D.: 1999, *Science* **283**, pp. 1507–1511.

Greenberg, R., Bottke, W.F., Nolan, M., et al.: 1996, *Icarus* **120**, pp. 106–118.

Knežević, Z., Milani, A. and Farinella, P.: 1997, *Planet. Space Sci.* **45**, pp. 1581–1585.

Levison, H. and Duncan, M.: 1994, *Icarus* **108**, pp. 18–36.

Marzari, F., Davis, D.R. and Vanzani, V.: 1995, *Icarus* **113**, pp. 168–187.

Marzari, F., Farinella, P. and Davis, D.R.: 1999, *Icarus* **142**, pp. 63–77.

Migliorini, F., Zappalà, V., Vio, R. and Cellino, A.: 1995, *Icarus* **118**, pp. 271–291.

Milani, A. and Farinella, P.: 1994, *Nature* **370**, pp. 40–42.

Milani, A. and Farinella, P.: 1995, *Icarus* **115**, pp. 209–212.

Milani, A., Nobili, A.M. and Knežević, Z.: 1997, *Icarus* **125**, pp. 13–31.

Nesvorný, D. and Morbidelli, A.: 1998, *Astron. J.* **116**, pp. 3029–3037.

Vokrouhlický, D., Brož, M., Farinella, P. and Knežević, Z.: 2001, *Icarus*, in press.

KUIPER-BELT OBJECTS: DISTRIBUTION OF ORBITAL ELEMENTS AND OBSERVATIONAL SELECTION EFFECTS

JOZEF KLAČKA (klacka@fmph.uniba.sk) and ŠTEFAN GAJDOŠ
Astronomical Institute, Faculty of Mathematics and Physics, Comenius University, Mlynská dolina, 842 48 Bratislava, Slovak Republic

Abstract. The influence of the observational selection effects in the set of bodies known also as the Edgeworth-Kuiper belt objects (EKOs) is investigated. The most important observational selection effect is closely connected with the fact that precise orbits are known mainly for objects with smaller perihelion distance, discoveries of the objects were done i) near their orbital nodes, or, ii) objects with small inclinations are known.

Distribution of EKOs in perihelion distance is not comparable with any known set of objects in the solar system. Mass index exhibits different values for the inner and outer zones of the known Edgeworth-Kuiper belt.

Key words: Kuiper-Belt Objects, orbital elements, selection effects

1. Introduction

It is very important to know the influence of observational selection effects on physical interpretation of the observational data. Our experience shows that observational selection effects may dramatically change physical conclusions done from the observational data. This is motivation for investigation of possible observational selection effects in the set of known Edgeworth-Kuiper belt objects (EKOs).

We used the set of the EKOs data adopted from NEO web-page of Minor Planets Center, as reported to the date June 7, 2000. It consists of 279 EKOs.

The most important observational selection effects are connected with the fact that discoveries of the solar system minor bodies are done near their perihelia at favourite observational conditions. This is the reason why we concentrate on distribution of perihelia in EKOs.

2. EKOs and their Distribution in Perihelion Distance

The current data yields for the number of objects within perihelion distances:

$q \in (26.2, 27.0)$ AU   5 objects
$q \in (28.1, 36.3)$ AU   73 objects
$q \in (36.9, 37.2)$ AU   4 objects
$q \in (37.7, 46.7)$ AU   195 objects
$q \in (47.1, 47.4)$ AU   2 objects
Total set: $q \in (26.2, 47.4)$ AU   279 objects

Pretka-Ziomek et al./Dynamics of Natural and Artificial Celestial Bodies, 313–320, 2001.
© 2001 *Kluwer Academic Publishers.*

314   JOZEF KLAČKA

Total number of currently known EKOs is 4-times higher than the number known 1.5 years ago (Klačka and Gajdoš, 1998). As for the group with $q < 36.3$ AU, this number is 2.5-times higher; as for the group with $q > 37.7$ AU, this number is 5-times higher.

THE LEAST-SQUARE METHOD FIT

Definition is following: $N(< q; q > q_{min}) = C (q - q_{min})^{\alpha}$, where $N(< q; q > q_{min})$ is number of objects with perihelia distances less than q and greater than q_{min}.

1. $q \in (28.1, 36.3)$ AU: $q_{min} = 28.1$ AU, $\ln C = 1.992 \pm 0.036$, $\alpha = 1.190 \pm 0.024$

2. $q \in (37.7, 46.7)$ AU: $q_{min} = 37.7$ AU, $\ln C = 2.742 \pm 0.031$, $\alpha = 1.188 \pm 0.020$

The results can be summarized as follows:

- Distribution of EKOs in perihelion distance corresponds to the same value $\alpha = 1.19$ for both large groups.
- Comparison with situation at the end of 1998: 68 known objects yielded $\alpha = 1$ for both large groups (Klačka and Gajdoš 1998). Larger number of known objects gives greater value of α.
- The value $\alpha = 1.19$ for EKOs is very small in comparison with other objects in the solar system: long-period comets ... $\alpha = 1.31$, near-Earth asteroids ... $\alpha = 1.44$, numbered periodic comets $\alpha > 1.5$ (Klačka 1999).

As for comparison with new comets, one obtains on the basis of Marsden and Williams's (1997) catalogue:

$q \in (0.0, 1.8)$ AU: $q_{min} = 0.0$ AU, $\ln C = 3.134 \pm 0.022$,
                                    $\alpha = 1.323 \pm 0.027$;

$q \in (0.0, 2.5)$ AU: $q_{min} = 0.0$ AU, $\ln C = 3.078 \pm 0.022$,
                                    $\alpha = 1.255 \pm 0.027$;

the value 1.32 corresponds to the value for long-period comets; smaller value $\alpha = 1.25$ is caused by observational selection effects – small number of known comets with larger values of perihelion distance. (Perihelion distances distribution of new and long-period comets can be found also in Fernández and Gallardo (1999)).

What is the physics of the fact that the obtained value of α for EKOs is smaller than for the other known groups of objects in the solar system? Probably orbital evolution. As for near-Earth asteroids and numbered periodic comets this may be acceptable – planets (except Neptune) do not play such a significant role in the orbital evolution of EKOs. However, the reason for smaller value of α for EKOs than for new comets and long-period comets, is unknown – some observational selection effect, or any physics?

3. Perihelia Distances and Eccentricities

Total set of known EKOs yields that 188 objects exhibit eccentricities $e \in (0, 0.41)$; the rest 91 objects have $e = 0$, which corresponds to not precisely determined orbits.

$e \in (0.0, 0.1)$ AU   82 objects

$e \in (0.0, 0.2)$ AU   141 objects

$e \in (0.0, 0.3)$ AU   175 objects

The largest concentration in eccentricity is for:

$e \in (0.02, 0.14)$ AU   110 objects.

Up to the limit of $q = 40.6$ AU only one object exhibits:

$e = 0$ ($q = 35.25$ AU).

The distribution $N(< q; q > q_{min}) = C (q - q_{min})^{\alpha}$ yields the following results:

I. group with $e = 0$ (less precise orbits):

$q \in (40.6, 46.7)$ AU: $q_{min} = 40.6$ AU, $\ln C = 2.076 \pm 0.042$, $\alpha = 1.399 \pm 0.034$ ... 88 objects;

II. group with $e > 0$ (only 5 objects with $q > 43.6$ AU):

$q \in (37.7, 43.6)$ AU: $q_{min} = 37.7$ AU, $\ln C = 2.725 \pm 0.032$, $\alpha = 1.091 \pm 0.026$ ... 102 objects.

Observational selection effects in these two groups are evident. The result of the first group – not precise orbits – corresponds to the fact that large number of not precise orbits is awaited for large distances. The result of the second group – more precise orbits – corresponds to the fact that large number of precise orbits is awaited for small distances, i. e., greater concentration of objects with smaller perihelion distances should occur in the data.

Figure 1a depicts two-dimensional distribution in perihelion distance and eccentricity. We do not know EKOs with $e > 0.5$; this corresponds to the condition of the long-term stability of the system (as presented by various authors at this workshop). General trend is: the larger perihelion distance the more circular orbit. This situation is of physical origin (Marsden 1999): i) the lower line bordering the belt in the figure corresponds to the relation $q = a(1 - e)$, where $a \approx 39.4$ AU is the resonance $2 : 3$ – Plutinos, and, ii) the upper line bordering the belt in the figure corresponds to the relation $q = a(1 - e)$, where $a \approx 47.7$ AU is the resonance $1 : 2$; there are also more distant population of nonlibrating objects with low orbital eccentricities (cubewanos, $e < 0.14$). The object lying outside the belt in the figure correspond to Centaurs or to scattered disk objects.

4. Perihelia Distances and Inclinations

The inner zone of EKOs defined by $q \in (28.1, 36.3)$ AU yields:

i) the total set consisting of 72 known objects ($e > 0$) fulfills $i < 32°$,

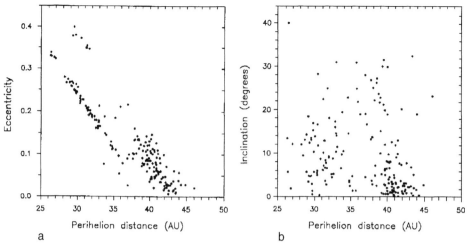

Figure 1. Perihelion distance versus eccentricity (a) and inclination (b).

ii) 12 objects have $i < 4°$,
iii) 37 objects have $i < 10°$,
iv) 64 objects have $i < 20°$.

The outer zone of EKOs defined by $q \in (37.7, 46.7)$ AU yields:

i) the total set consisting of 107 known objects ($e > 0$) fulfills $i < 33°$,
ii) 57 objects have $i < 4°$,
iii) 80 objects have $i < 10°$,
iv) 91 objects have $i < 20°$.

Also Figure 1b shows strong correlation between perihelion distance and inclination for known objects. The most evident result is that objects with the largest observed perihelion distances are concentrated towards the plane of ecliptic. This is closely connected with searches of new objects (Klačka and Gajdoš 1998 have presented that more than 70% of the known objects at the end of 1998 were discovered near their orbital nodes and the other 15 % exhibited inclinations less than 5°).

5. Physics

5.1. SEMIMAJOR AXIS

Figure 2 shows histogram for semimajor axis for known EKOs (only objects with sufficiently known orbits are considered, i. e., with nonzero eccentricities $e > 0$). Fig.3 shows two graphs ($e > 0$): one of them depicts eccentricity versus semimajor axis, the other depicts inclination versus semimajor axis.

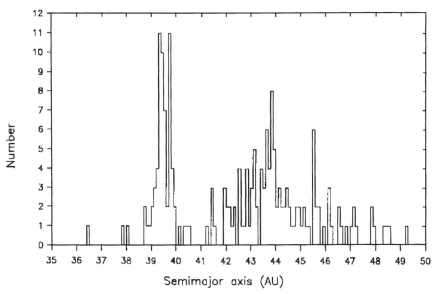

Figure 2. Distribution of known EKOs in semimajor axis.

5.2. ABSOLUTE MAGNITUDE

Absolute Magnitude and Perihelion Distance

The dependence of the cumulative number of EKOs with sufficiently determined orbits ($e > 0$) is presented for both zones: the inner zone – $q \in (28.1, 36.3)$ AU, and, the outer zone – $q \in (37.7, 46.7)$ AU. Fig.4a shows the dependence 'logarithm of the cumulative number on absolute magnitude' for the both zones. Moreover, Fig.4b shows also diagram for absolute magnitude versus perihelion distance.

The approximation $N(> D) = C\, D^{-n}$, $k = 1 + n/3$ is used. $N(> D)$ is the cumulative number of objects of diameter greater than D, n is the population index and k is the (differential) mass index. Supposing that all EKOs in a given zone exhibit similar albedos, we use absolute magnitudes for determining the values of k. The results are:

$$k = 1.92 \pm 0.06\,,\quad H \in\ <6.5, 7.8>\,,\quad \textit{inner zone}$$
$$k = 2.67 \pm 0.05\,,\quad H \in\ <6.4, 7.0>\,,\quad \textit{outer zone.} \qquad (1)$$

The value for the inner zone is comparable with the value $k = 11/6$ of Dohnanyi (1969) for collisionally stable system of objects. However, the value for the outer zone is even greater than the values for large asteroid families, where $k \in (2.15, 2.45)$ (Klačka 1992, 1995).

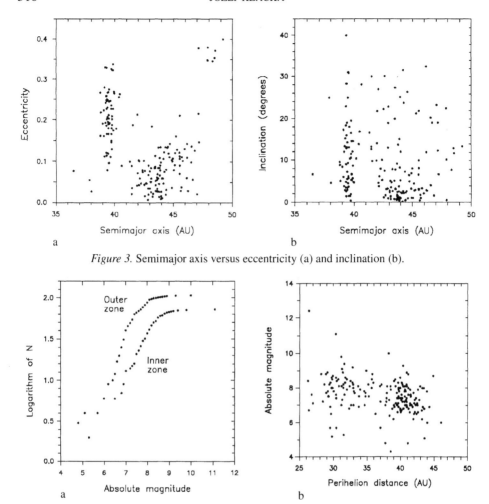

Figure 3. Semimajor axis versus eccentricity (a) and inclination (b).

Figure 4. Distribution for absolute magnitude and perihelion.

Absolute Magnitude and Semimajor Axis

The dependence of the cumulative number of EKOs with sufficiently determined orbits ($e > 0$) is presented for the following two zones: the inner zone – $a <$ 39.9 AU, and, the outer zone – $a \in$ (41.4, 49.3) AU. The inner zone contains 61 EKOs, the outer zone contains 120 EKOs. Only 7 EKOs is known in the zone $a \in$ (39.9, 41.4) AU. Fig.5a shows the dependence 'logarithm of the cumulative number on absolute magnitude' for the both zones. Moreover, Fig.5b shows also diagram for absolute magnitude versus semimajor axis.

Supposing that all EKOs in a given zone exhibit similar albedos, we use absolute magnitudes for determining the values of k. The results are:

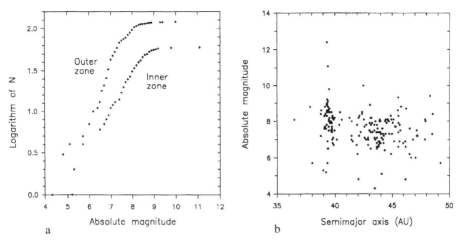

Figure 5. Distribution for absolute magnitude and semimajor axis.

$$k = 2.20 \pm 0.06, \quad H \in \; <7.4, 7.7>, \quad \textit{inner zone}$$
$$k = 2.63 \pm 0.05, \quad H \in \; <6.4, 7.0>, \quad \textit{outer zone}. \qquad (2)$$

The value for the inner zone is not reliable since very small interval of absolute magnitudes is used. These results must be taken as a crude information, since the assumption on the same albedos for the used set of objects is not fully correct due to diversities in surface compositions of EKOs (Luu and Jewitt 1996).

6. Conclusion

We have presented some statistical analyses of the currently known data for the EKOs. The set of EKOs is influenced by observational selection effects, mainly as for the fact that discoveries of objects were done near their orbital nodes, and, objects with small inclinations are known. Less precise orbits are known for objects with greater perihelion distance. Distribution of EKOs in perihelion distance is not comparable with any known set of objects in the solar system. Mass index exhibits different values for the inner and outer zones of the known Edgeworth-Kuiper belt.

Acknowledgements

The authors want to thank to the Organizing Committees for the financial support which enabled our participation at the Conference. The paper was supported by the Scientific Grant Agency VEGA (grant No. 1/7067/20).

References

Dohnanyi, J. S.: 1969, 'Collisional Model of Asteroids and Their Debris', *J. Geophys. Res.* **74** pp. 2531–2554.

Fernández, J.A. and Gallardo, T.: 1999, 'From the Oort cloud to Halley-type comets', In: *Evolution and Source Regions of Asteroids and Comets, Proc. IAU Coll. 173. J. Svoreň*, (E. M. Pittich and H. Rickman, Eds.), Astron. Inst. Slovak Acad. Sci., Tatranská Lomnica, pp. 327–338.

Klačka, J.: 1992, 'Mass distribution in the asteroid belt', *Earth, Moon, and Planets* **56**, pp. 47–52.

Klačka, J.: 1995, 'Mass distribution and structure of the asteroid belt', *Acta Astron. et Geophys. Univ. Comenianae* **XVII**, pp. 62–76.

Klačka, J.: 1999, 'Distribution of long-period comets in perihelion distance', In: *Evolution and Source Regions of Asteroids and Comets, Proc. IAU Coll. 173. J. Svoreň*, (E. M. Pittich and H. Rickman, Eds.), Astron. Inst. Slovak Acad. Sci., Tatranská Lomnica, pp. 277–278.

Klačka, J. and Gajdoš, Š.: 'Kuiper-Belt Objects: Distribution of Orbital Elements and Observational Selection Effects', In: *Proceedings of ESO Workshop on Minor Bodies in the Outer Solar System* (in press).

Luu, J. and Jewitt, D. : 1996, 'Color diversity among Centaurs and Kuiper belt objects', *Astron. J.* **112**, pp. 2310–2318.

Marsden B. G.: 1999, 'Small bodies in the outer solar system', *Cel. Mech. and Dyn. Astron.* **73**, pp. 51–54.

Marsden, B. G. and Williams, G. W.: 1997, *Catalogue of cometary orbits*, Minor Planet Center, Cambridge, MA.

SHORT ARC CCD OBSERVATIONS OF CELESTIAL BODIES: NEW APPROACH

OLEG BYKOV
Pulkovo Astronomical Observatory, Russia

Abstract. We propose to use the Apparent Motion Parameters Method and the Pulkovo Ephemeris Program for Objects of Solar system for the fast orbit determination of any moving celestial object. The modern CCD positional observations may be used for the purpose. The short description of the method and an example of its application are given.

Key words: Asteroids, CCD positional observations, Short arc, Preliminary orbit determination

At present, the positional CCD observations allow us to get the sets of the asteroid coordinates as dense as we want and to process them in the interactive mode. Such observational data contain important information concerning space motion of an observed object, namely the Parameters of the Apparent Motion: the topocentric angular velocity μ, acceleration $\dot{\mu}$, positional angle ψ and curvature C of its trajectory. These values are obtained from each α, δ set by the time polynomial approximation technique.

At the Pulkovo observatory, the Apparent Motion Parameters Method (henceforth the APM-method) was developed by Dr. A.Kiselev and his colleagues to derive the initial elliptical orbit of an observed celestial body from the values of $(\alpha, \delta, \mu, \dot{\mu}, \psi$ and $C)$. The AMP method is a further development of the known Laplacian orbit determination method, and it was successfully applied to various tasks of astrometry – the Artificial Earth Satellites, Space Debris, Small Solar System Bodies and Double Stars [1,2].

The supremacy of the CCD technique, the existence of rich and accurate star catalogues and the Pulkovo investigations of the super short arc information allow us to elaborate a new approach for processing the observations of any celestial body moving at the background of stars. This approach includes:

a) fast analysis of the CCD observations of moving object,
b) preliminary orbit determination,
c) identification of the object if it is known,
d) further CCD observations if the observed object is not known.

The necessary ephemerides may be provided by the AMP method in the interactive mode.

The algorithms of the Laplacian and AMP methods were investigated by the author. The Ephemeris Program for Objects of Solar system (henceforth the EPOS Software) aiming at the fast analysis of any CCD frame where the moving objects could be detected was developed by Dr. Victor L'vov and his colleagues at Pulkovo

Pretka-Ziomek et al./Dynamics of Natural and Artificial Celestial Bodies, 321–326, 2001.
© 2001 *Kluwer Academic Publishers.*

OLEG BYKOV

TABLE I
CCD observations of the *2000 NM* (real geocentric positions)

Obs.	Date (DT)	α	$(O-C)''$	δ	$(O-C)''$	Z
823	2000 07 03.07957	18 10 46.102	0.13	−7 48 53.36	1.25	59
823	2000 07 03.08313	18 10 45.670	0.32	−7 48 16.99	1.04	58
823	2000 07 03.08790	18 10 45.095	0.64	−7 47 28.31	0.71	57
823	2000 07 03.09664	18 11 44.018	0.88	−7 45 58.07	1.09	56
854	2000 07 03.22192	18 10 28.246	0.06	−7 24 24.34	0.85	44
854	2000 07 03.22340	18 10 28.061	0.09	−7 24 08.62	1.21	44
854	2000 07 03.22450	18 10 27.931	0.20	−7 23 57.11	1.32	44
854	2000 07 03.23523	18 10 26.590	0.36	−7 22 05.99	1.08	42
854	2000 07 03.23806	18 10 26.231	0.31	−7 21 37.16	0.52	42
854	2000 07 03.23960	18 10 26.059	0.64	−7 21 21.05	0.64	42
854	2000 07 03.28525	18 10 20.276	0.21	−7 13 26.68	0.20	39
854	2000 07 03.28683	18 10 20.056	−0.10	−7 13 10.18	0.24	39
682	2000 07 03.23330	18 10 26.805	−0.06	−7 22 25.13	1.97	47
682	2000 07 03.26535	18 10 22.806	0.51	−7 16 54.09	−0.05	44
682	2000 07 03.29471	18 10 19.146	1.16	−7 11 49.77	−1.47	44
670	2000 07 03.26536	18 10 22.820	0.73	−7 16 53.72	0.21	43
670	2000 07 03.27313	18 10 21.849	0.86	−7 15 32.97	0.12	42
670	2000 07 03.32789	18 10 14.940	0.96	−7 06 01.95	0.10	41
670	2000 07 03.40050	18 10 05.748	1.03	−6 53 21.56	0.06	52
900	2000 07 03.61157	18 09 38.838	1.04	−6 16 10.38	−0.34	41
900	2000 07 03.61725	18 09 38.101	0.91	−6 15 09.91	−0.35	41
900	2000 07 03.62469	18 09 37.137	0.75	−6 13 50.45	−0.16	42
900	2000 07 03.63036	18 09 36.419	0.89	−6 12 49.88	−0.03	42

observatory [3]. In addition, the exact ephemerides of the Asteroids, Comets, Planets, Sun and Moon may be calculated. The EPOS software is a powerful tool for testing modern CCD observations and estimating their accuracy immediately after they were carried out.

It can be expected that many CCD frames obtained in the course of various astrophysical programs with large or small telescopes contain short visible trails produced by moving objects. It's very important that our approach if applied to such frames is able to extract an additional information from these observations [4].

An example of what the EPOS software can do is given in the Tables I–VII. The CCD observations of the fast Near Earth Asteroid with *2000 NM* nomination

TABLE II

CCD observations of the *2000 NM*: normal places and their errors

Obs.	j	Date (DT)	α	$(O-C)''$	δ	$(O-C)''$	$\sigma''_{1\alpha}$	$\sigma''_{1\delta}$
823	4	2000 07 03.08681	18 10 45.221	0.49	− 7 47 39.18	1.02	± 0.33	± 0.23
				± 0.17		± 0.11		
854	8	2000 07 03.24434	18 10 25.431	0.22	− 7 20 31.39	0.76	± 0.22	± 0.43
				± 0.08		± 0.15		
682	3	2000 07 03.26445	18 10 22.919	0.54	− 7 17 03.00	0.15	± 0.61	± 1.73
				± 0.35		± 1.00		
670	4	2000 07 03.31672	18 10 16.339	0.90	− 7 07 57.55	0.12	± 0.13	± 0.06
				± 0.06		± 0.03		
900	4	2000 07 03.62096	18 09 37.624	0.90	− 6 14 30.15	0.22	± 0.12	± 0.15
				± 0.06		± 0.08		

TABLE III

CCD observations of the *2000 NM*: the first derivatives and their errors calculated for several observatories

Obs.	j	2000 July (DT)	$\dot{\alpha}^s$	$(O-C)''$	$\dot{\delta}''$	$(O-C)''$
823	4	3.08681	-122.073	45.07	10270.03	-8.30
854	8	3.24434	-126.170	-2.56	10382.80	-6.29
682	3	3.26445	-124.727	20.85	10345.97	-57.26
670	4	3.31672	-126.348	1.15	10449.66	9.64
900	4	3.62096	-128.843	-9.31	10671.66	17.25

were taken from the Minor Planet Circular. The *2000 NM* object was discovered by an amateur (MPC code 823, USA) at night 2/3 July 2000. In the Table I the first CCD observations are presented. Only 23 positions distributed on a super short arc are used by us. For each position the residuals $(O - C)$ were calculated with the Pulkovo EPOS Software. The system of orbital elements was taken from Bowell's Orbital Catalogue where the orbit for the *2000 NM* was based on the 646 positions obtained by the observatories through the world in July 2000. The accuracies of the used CCD observations are presented in Tables I, II, III in the columns $(O - C)$.

TABLE IV

A comparison of positions, the first and the second derivatives of the *2000 NM* with exact EPOS calculations and observational values (Epoch 2000 07 03.30000 DT)

	α	$\dot{\alpha}$	$\ddot{\alpha}$
Observ.	$18^h 10^m 18^s.434$	$-2^m 06^s.234$	$-5^s.762$
	$\pm 0^s.012$	$\pm 0^s.066$	$\pm 0^s.286$
EPOS	$18^h 10^m 18^s.399$	$-2^m 06^s.326$	$-5^s.876$
	δ	$\dot{\delta}$	$\ddot{\delta}$
Observ.	$-7^o 10'52''.32$	$2^o 53'45''.99$	$11'35''.23$
	$\pm 0''.26$	$\pm 1''.37$	$\pm 5''.97$
EPOS	$-7^o 10'53''.11$	$2^o 53'48''.26$	$11'44''.03$

One can see that the observations made with the MPC codes 823, 854, 670 and 900 are of a good accuracy. The only observatory with MPC code 682 has a large error for δ in the Table I and also for normal place and $\dot{\delta}$ in the Tables II and III.

TABLE V

The Apparent Motion Parameters calculated by EPOS and statistical processing of the real one night CCD observations (Epoch 2000 07 03.30000 DT)

	EPOS	AMP	Approx.
j	646	23	23
μ, $''/day$	10596.37	10597.84	10593.90
ψ, deg.	349.7804	349.8531	349.7855
$\dot{\mu}''$	710.5027	712.1872	701.5419
C	1.000374	1.001900	1.000378
d	0.167929	0.163537	—
$\dot{d}$	−0.005931	−0.005655	—

TABLE VI

Orbital elements of the *2000 NM* calculated by EPOS Software with the use of 646 positions and by AMP method with the use of 23 positions per night (Epoch 2000 07 03.30000 DT)

	EPOS	AMP
j	646	23
M, deg.	348.249387	346.65793
ω, deg.	69.921580	69.81610
Ω, deg.	274.899485	274.92582
i, deg.	22.482097	22.01773
e	0.66018447	0.63052668
a, a.e.	2.69195834	2.48814190

The mean normal places, the first and the second derivatives were calculated with the use of data given in the Table II. These values are shown in the Table IV. They are compared with the same ones given by the EPOS Software. The Apparent Motion Parameters are presented in the Table V. They coincide with the exact AMP calculated by EPOS Software. And at last, our preliminary AMP orbit of the *2000 NM* is given in the Table VI. It was derived from only 23 positions (5 normal places) obtained during nearly 12 hours (see Table I, II). We can see a good coincidence of the exact and our orbits. With our AMP elements the positions and two derivatives valid for a week or for a month were calculated for the *2000 NM*. These values are presented in the Table VII, and their quality enables us to say that we can find this object in the 1 square degree sky field due to the derivatives' coincidence – one month after its short arc CCD observations.

TABLE VII

Prediction of the positions and derivatives of the *2000 NM* with the use of AMP orbit (Table VI). Dates: 2000 07 10.00000 and 2000 07 28.86000 DT

	α	$\dot{\alpha}$	$\ddot{\alpha}$
AMP	$17^h 53^m 56^s.74$	$-2^m 49^s.3$	$-7^s.9$
EPOS	$17^h 53^m 46^s.18$	$-2^m 52^s.8$	$-8^s.5$
AMP	$16^h 11^m 19^s.81$	$-10^m 52^s.6$	$-67^s.2$
EPOS	$16^h 04^m 13^s.67$	$-12^m 02^s.0$	$-77^s.9$
	δ	$\dot{\delta}$	$\ddot{\delta}$
AMP	$16^o 06' 51''.1$	$3^o 55' 16''.0$	$+4' 06''.9$
EPOS	$16^o 14' 44''.0$	$3^o 57' 48''.2$	$+4' 30''.4$
AMP	$72^o 02' 15''.7$	$1^o 30' 36''.5$	$-7' 57''.1$
EPOS	$72^o 45' 47''.5$	$1^o 29' 36''.6$	$-8' 22''.4$

The majority of our articles dealing with the investigation of the Laplacian and AMP methods for initial orbit determination was published in Russian. Some of them may be found in the "Astronomical and Astrophysical Abstracts" for 1974–2000 yrs.

References

Bykov, O.P. *et al.*: 1998, 'Observations of Kuiper Belt objects with Russian BTA', *Letters to the Russian Astron. J.*, **vol. 24, No. 3**, pp. 220–225.

Kiselev, A.A. and Bykov, O.P.: 1973, 'The determination of the satellite orbit by a single photograph with many satellites trails', *Sov. Astron. J.* **vol. 50**, pp. 1298–1308 (in Russian).

Kiselev, A.A. and Bykov, O.P.: 1976, 'The determination of a satellite elliptical orbit with the use of Parameters of the Satellite's Apparent Motion', *Sov. Astron. J.* **vol. 53**, pp. 879–888 (in Russian).

L'vov, V.N., Smekhacheva, R.I. and Tsekmejster, S.D.: 2000, 'EPOS — the Ephemeris Software Package for Solar System Bodies Research', *User's Guide*, Pulkovo, St.-Petersburg, pp. 1–24.

NEO OBSERVATION PROGRAM AT THE ASTRONOMICAL OBSERVATORY IN MODRA

ŠTEFAN GAJDOŠ (gajdos@fmph.uniba.sk)

Astronomical Institute, Faculty of Mathematics and Physics, Comenius University, Mlynská dolina, 842 48 Bratislava, Slovak Republic

Key words: astrometry, NEO observation

1. Introduction

The Astronomical and Geophysical Observatory (AGO) of the Faculty of Mathematics and Physics (FMPh) is a complex of many scientific facilities and devices devoted for research in the field of space and Earth sciences. The Astronomical Institute of the Faculty operates personally and technically its main part - Astronomical Observatory (AO). The research work is aimed on both the solar physics and interplanetary matter areas.

All the observations are performed at the 0.6-m (in diameter) Carl-Zeiss reflector. Its rebuilted electronic control system is supplemented by CCD-camera (in disposal are SBIG cameras, ST-6 and ST-8) in the primary focus (f/5.5). For a description and system parameters see Kalmančok et al. (1994, 1995), Kornoš and Zigo (1997). The instrumentation is devoted for the minor bodies astrometry, as well as for a physical study of comets and asteroids. The AO equipment is supplied with two all-sky cameras (one guided) incorporated in the European meteoritic network. Meteor trail observations using the forward-scatter radar system on the baseline Lecce-Bologna-Modra are performed in cooperation with FISBAT/CNR, Bologna, Italy. For a view through the activity in the Solar system minor bodies research check the URL: http://www.fmph.uniba.sk/ AGO.html.

2. Astrometry

Astrometry program began in the mid of 1994, and serious observations under Minor Planet Center (MPC) observatory code **118** were conducted since end of 1995. The primary observation program covered main-belt and "ITA" asteroids, as well as those ones from the *Minor Planet Circulars*, objects from *Obsplanner Service*, and comets. Later, objects from *M.P.E.C.* were included, mainly those published in both monthly critical, and unusual list of observable objects, respectively. After recognizing of the important role of minor bodies follow-up and confirmation observations, we oriented on the most needed among them, on Near-Earth Objects (NEOs). The gradual increasing rate of observed NEOs among all observed objects, in course of six years of minor bodies astrometry activity at AO in Modra, is

Pretka-Ziomek et al./Dynamics of Natural and Artificial Celestial Bodies, 327–330, 2001.
© 2001 *Kluwer Academic Publishers.*

TABLE I

Summary of the astrometry results gained at the AO in Modra during last years. The second row denotes new minor planets discoveries, the term *Other* covers the another categories of unusual asteroids, values in the last two rows are adopted by Nakano's annual reports on the activity in minor planets astrometry (1996, 1997, 1998, 1999, 2000)

		Positions of minor planets				NEO's from	Rank in	
Year	Disc.	MB	NEOs	Other	Total	total sum	Europe	world
1995	10	449	78	47	574	13.59%	9	25
1996	18	673	474	125	1272	37.26%	2	13
1997	36	1237	597	250	2084	28.65%	4	12
1998	48	1409	1077	197	2683	40.14%	4	12
1999	24	1279	1497	294	3070	48.76%	7	14
2000	6	472	1096	56	1624	67.49%	?	?

declared in Table 1. Also resulted is a "by-product" of the astrometry program, a discovery above hundred new minor planets. Recently, raised-up results ranged us to a few of leading observatories as to amount of minor bodies precise positions in Europe, see Table I. Such a progress inspired us to apply for The Planetary Society Gene Shoemaker NEO Grant, in the category *NEOs follow-up and confirmation observations*.

3. NEOs Follow-up and Confirmation Program

In such circumstances we were awarded by The Planetary Society receiving The Gene Shoemaker NEO Grant, to enhancement support of our NEO activity. Its objectives are:

- follow-up astrometry of Near-Earth Objects, with the aim of getting of these minor bodies accurate positions, to contribute to international activity in the field, to support and extend knowledges about orbit characteristics of the known NEO population,
- confirmation and follow-up astrometry of newly-discovered NEOs, as well as poorly observed NEOs; as a by-product, discovery of new NEOs,
- preparation of the our NEO-program's second stage, the NEOs photometry.

A budged of The Planetary Society NEO Grant helped us to upgrade recently operated astrometric program, by improvement of our technical and logistical conditions. Additional technical steps follow. The wither technical catch-up we understand as a preparation for minor planets CCD photometry.

4. Results

As an research-educational observatory, the AO in Modra has no fixed observational schedule. Therefore, we can fully employ advantages of a "small observatory" equipped by the low-/medium-sized device. According to Marsden and Steel (1996), each active observatory with modest-sized instrumentation is needed. As recently working or prepared NEO searching projects with powerful telescopes network produce a lot of new Near-Earth Objects, a role of follow-up programs will increase.

As results of our NEO program, many minor planets precise positions were recorded (see Table I), with rising ratio of NEOs among them. Especially for efficiency illustration of the NEO program, the number of observed objects from the *NEO Confirmation Page* only (e.g. before their provisional designation), as well as the number of their reported precise positions, in 1999 and 2000, are shown in Figure 1. The term "observed objects" denotes only different objects number, not repeated observations of the same object.

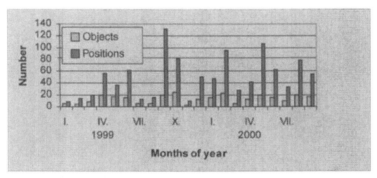

Figure 1. Number of observed NEOs, as well as number of reported positions, from *NEO Confirmation Page*

Together with increasing rate of NEOs follow-up observations, as the higher quality result we suggest the case of two Near-Earth-Objects identification: an Aten asteroid 1999 SN5 being the previously designed object 1998 RO1 (*M.P.E.C. 1999-T10, M.P.C. 36378*), and 1999 LS7 = 2000 LE2 (*M.P.E.C. 1999-L37, M.P.E.C. 2000-L20*).

Acknowledgements

This work was supported by the Slovak Scientific Grant Agency VEGA (Grant No. 1/7067/20) and by The Planetary Society Gene Shoemaker NEO Grant for 1998.

References

Kalmančok, D., Pittich, E. M. and Zigo, P.: 1994, 'Old telescope with new technology', In *Proceedings of the 3th International Workshop on Positional Astronomy and Celestial Mechanics* (A. L. Garcia Ed.), Cuenca, Spain, pp. 386–390.

Kalmančok, D., Kornoš, L., Svoreň, J. and Zigo, P.: 1995, 'Astrometry at the Astronomical Observatory in Modra', *Meteor Reports of the Slovak Astronomical Society (in Slovak)* **16**, pp. 57–62.

Marsden, B. G. and Steel, D. I.: 1996, 'Astrometry of Near–Earth Objects using small telescopes', *Earth, Moon and Planets* **74**, pp. 85–92.

Nakano, S.: 1996, 1997, 1998, 1999, 2000, *Reports on Activity in Minor Planets Observations*, Private communication.

Zigo, P. and Kornoš, L.: 1997, 'CCD cameras ST-6 and ST-8 and their use for astronomical measurements of asteroids and comets', *Acta Astronomica et Geoph. Universitas Comeniana* **XVIII**, pp. 101–106.

MOON–EARTH SEPARATION PROBLEM IN THE DYNAMICS OF NEAR EARTH ASTEROIDS

PAWEŁ KANKIEWICZ (ciupo@amu.edu.pl)
*Astronomical Observatory of Adam Mickiewicz University, Słoneczna 36,
PL–60–286 Poznań, Poland*

The number of discovered Near Earth Asteroids has dramatically increased in the last years. New observational data provide more informations for the dynamical studies of these objects. An important problem in the numerical studies of NEA motion is the estimation of possible perturbing effects. It is difficult to decide which effect can be neglected in the integration. Our results often depend on 'subtle' perturbations caused by small or distant objects. One of significant sources of perturbations in the motion of NEA is the Earth–Moon system. Some authors take into account the barycentre of the Earth–Moon system as a single perturbing body, but in some specific cases the separation of these bodies should be taken into account (especially, in the event of close encounters).

The influence of the Earth–Moon system on the evolution of NEA is investigated in the following article. The latest observational results and various integration methods are applied. The nature of perturbations caused by this 'binary' system was described and compared with the results obtained by using the 'barycentre model'. Additionally, the efficiency of several numerical algorithms is compared.

In the numerical simulation, only selected part of NEA population was used (known as PHA – Potentially Hazardous Asteroids). The motion of 128 PHA was integrated numerically. Initial orbital elements were adopted from the Minor Planets Circulars. Two dynamical models were taken into account:

– The barycentre model: the Sun + 9 massive planets, including the Earth–Moon barycentre
– The Earth–Moon model: the Sun + 9 massive planets, with the Earth-Moon system

Initial coordinates and velocities of planets (and the Moon) are taken from JPL DE406 ephemeris. Two integration algorithms were used:

– B–S method (Bulirsch, Stoer, 1966)
– RMVS3 (Regularized Mixed Variable Symplectic) (Duncan et al., 1998)

The most important conclusions are:

– B–S and RMVS3 methods are the efficient tools in statistical analysis of the short–term (10^3 y) NEA evolution. Differences between statistics obtained by both methods are relatively small. However, RMVS3 method is faster than

B–S. More detailed analysis of the efficiency of this method was described by Michel and Valsecchi (1997).

— Maximum differences between the final coordinates Δr obtained with the application of 'barycentre' and 'Earth–Moon' models are in an acceptable agreement with results presented by Todorovic–Juchniewicz (1985). Application of the the 'Earth–Moon' dynamical model may change the final coordinates significantly after 10–100 y. Propagation of differences between final velocities Δv looks similar.

— The distribution of selected orbital elements (a, e, i) changes after use the 'Earth–Moon' model in the numerical integration. There are small differences in the distribution of values of a and e after about 100 y. The distribution of inclinations changes after 10^3 y.

— Application of the 'Earth–Moon' dynamical model is very important in problems of NEA evolution and cannot be neglected, especially in the case of long–time integrations ($T > 10^3$ y). Gravitational influence of the Moon changes the distribution of NEA and can perturb significantly motion of asteroids in case of close approaches.

Acknowledgements

The autor aknowledges the financial support of the polish KBN grant 2 P03D 001 19.

References

Duncan M. J. , Levison, H. F. and Lee, M.: 1998, 'A Multiple Timestep Symplectic Algorithm for Integrating Close Encounters', *Astronomical Journal* **116**, 2067.

Michel, P. and Valsecchi, G. B.: 1997, 'Numerical Experiments on the Efficiency of Second–Order Mixed–variable Symplectic Integrators for N–body Problems', *Celest. Mech. Dyn. Astron.* **65**, 355–371.

Todorovic–Juchniewicz, B.: 1985, 'The Influence of Subtle Dynamical Effects on the Motion of Comets and Minor Planets' (in Polish), *Postepy Astronomii* **33**, 3–4.

DETERMINATION OF MASS OF JUPITER AND THAT OF SOME MINOR PLANETS FROM OBSERVATIONS OF MINOR PLANETS MOVING IN 2:1 COMMENSURABILITY WITH JUPITER

OLGA M. KOCHETOVA AND YULIJA A. CHERNETENKO
Institute of Applied Astronomy, St. Petersburg, Russia

Key words: Jupiter's mass, asteroid's masses, observations of minor planets

Great number of observations of minor planets accumulated by now gives possibility to estimate once again the value of the Jupiter' mass from perturbations of the minor planets being in 2:1 commensurability with Jupiter as proposed by Hill (1873). 9310 observations of 28 minor planets were used in general solution for orbital elements of minor planets and mass of Jupiter. The minor planet (944) Hidalgo was also included in the list of selected minor planets due to its close approaches to Jupiter (less than 1 au). In so doing the perturbations from 9 major planets and 5 asteroids in accordance with coordinates and masses from DE200/LE200 ephemeris were taken into account. The relativistic terms due to the Sun and Jupiter (Brumberg, 1999) were included in equations of motion. The observations were corrected for gravitational deflection of light and for phase effect by Lommel–Zeeliger law of scattering. The influence of weighting observations on accuracy of orbit parameters and Jupiter' mass was estimated. The weights were assigned depending on residuals on different intervals (we considered observations since 1900). It turns out that best result is obtained in the case when the observations are considered with equal weights. The results are presented in Table I. The comparison of our results with those of other authors leads to conclusion that minor planets give value of Jupiter' mass in good agreement with the results of its determination from tracking data.

TABLE I

Values of the Jupiter' inverse mass

Object of study	Sun-Jupiter mass ratio		Author
Pioneer & Voyager	1047.3486 ± 0.0008		Campbell et al., 1985
Martian landers	1047.34830	0.00017	Pitjeva, 1997
Minor planets	1047.3488	0.0006	This paper

In the second part of present work the masses of some minor planets have been found from their gravitational perturbations on a number of smaller planets. The corresponding pairs of minor planets were chosen from those being in 2:1

Pretka-Ziomek et al./Dynamics of Natural and Artificial Celestial Bodies, 333–334, 2001.
© 2001 *Kluwer Academic Publishers.*

commensurability with Jupiter. Sampling of the pairs was governed by criterion: minimal distance between planets is less or equal to 0.1 au. Usage of less stringent requirement than usually used 0.05 au is here warrantable as the approaches of planets are regularly repeated due to commensurability 1:1 of their mean motions and mutual velocities of the bodies are rather small. Hoffmann (1989) pointed out the fruitful of that approach. Observations of all selected perturbed minor planets were used in general solution for their orbital elements, masses of perturbing minor planets and mass of Jupiter. The obtained results are given in Table II.

One can add to the list of perturbed minor planets being in commensurability 1:1 the perturbed minor planets having occasional close approaches to perturbing minor planets. Result 2) for Hygiea gives the example of such solution.

TABLE II

Results of mass determination of some asteroids

Perturbing minor planet	Mass $(10^{-11} M_{Sun})$		Perturbed minor planets	Number of obs.
10 Hygiea 1)	5.0 ± 1.5		48,120,357,767	1118
2)	4.6	0.9	48,120,357,767,829,2619	1456
31 Euphrosyne	3.2	1.7	212,319,381,643,667,1070,1107, 1298,1309,1373,2379,2439,2882, 2986,3011,3210	2205
48 Doris	2.8	1.2	90,229,401,664,828,986,1003, 1154,1177,1254,2009,2164,2269, 5193,6106,6112,6203,6642,8590	2800
52 Europa	2.9	1.6	92,401,489,491,595,635,1015, 1489,1571,1674,2197,2405,2582	1803
65 Cybele	0.72	0.40	526,979,1082,1261,1815	865
511 Davida	7.9	2.3	328,508,567,621,780,1678,2296, 2361,2394,2452,2894,4677,10920	1617

References

Brumberg, V. A.: 1999, *Trudy IPA RAN* **4**, 199–224 (in Russian).
Campbell, J. K. and S. P. Synnott: 1985, *Astron. J.* **90**, 364–372.
Hill, G. W: 1873, *Collected Mathematical Works* **I**, 105.
Hoffmann, M.: 1989, *Asteroids II*, 228–239.
Pitjeva, E. V.: 1997, In *Proceedings of IAU Colloquium 165* 251–256.

EPHEMERIS MEANING OF PARAMETERS OF ASTEROID'S APPARENT MOTION

NATALYA KOMAROVA
Pulkovo Astronomical Observatory, Russia

Abstract. It describes the important role of the Apparent Motion Parameters of moving celestial body for its fast identification by means of its several CCD positions obtained during a single night observations.

Key words: Asteroids, artificial Earth satellite, positional CCD observations, apparent motion parameters, fast orbit determination

The Apparent Motion Parameters (AMP) of a celestial body are its normal place α, δ, an angular topocentric velocity μ and acceleration $\dot{\mu}$, positional angle of motion ψ and curvature C of its apparent trajectory. These parameters refer to a fixed moment of UT. They are connected with the first and the second derivatives $\dot{\alpha}$, $\dot{\delta}$, $\ddot{\alpha}$, $\ddot{\delta}$ and celestial object's coordinates α, δ. For example, the formulae for the first order AMP are following:

$$\mu^2 = (\dot{\alpha} \cos \delta)^2 + (\dot{\delta})^2, \qquad \psi = \arctan((\dot{\alpha} \cos \delta)/\dot{\delta}) \tag{1}$$

It was shown that the AMP can be calculated by means of statistical treating of AES dense accurate coordinates obtained from a short topocentric arc of positional observations. An accuracy of AMP was sufficiently high for the investigation of a motion of observed Satellite, and Dr. A.A.Kiselev [5, 6] proposed the method of an osculating orbit determination with the use of the Apparent Motion Parameters. The coordinates α, δ and the first order AMP μ, ψ are needed for a circular orbit determination, all six parameters allow us to calculate an elliptic preliminary orbit of the observed celestial object. The last orbit essentially depends on the accuracy of AMP.

Historically the AMP method can be considered as a further development of the classic orbit determination method suggested by Laplace in 1780 yr for comets [1]. The Laplacian method requires six observational parameters, namely normal place α, δ and the first and the second derivatives of spheric coordinates $\dot{\alpha}$, $\dot{\delta}$, $\ddot{\alpha}$, $\ddot{\delta}$. Both methods are the Direct Methods of calculation of preliminary celestial body orbit. They were investigated in Leningrad State University and Pulkovo observatory in 1970-90 yrs [1, 2, 5, 6] but their practical realization was limited due to the difficulties of the fast measurements and astrometric reductions of the traditional photographic observations. Now the epoch of supremacy of CCD technique is coming to Astronomy and an obtaining dense intensive coordinate sets have no limits as to accuracy of positions and their numbers distributed on a short topocentric arc.

Pretka-Ziomek et al./Dynamics of Natural and Artificial Celestial Bodies, 335–337, 2001.
© 2001 *Kluwer Academic Publishers.*

The system of Direct Method equations allows also to calculate the AMP with a rigid formulae depending on orbital elements of celestial body and topocentric position of an observer only [2]. Really the values α, δ, μ, ψ, $\dot{\mu}$, C, the first and the second derivatives of spheric coordinates α, δ and a distance d, radial velocity and radial acceleration can be easily taken out from simple solution of the system of Direct methods equations. This exact solution for a strict description of an apparent motion of any celestial body was obtained for the first time in astronomical practice.

The AMP and coordinate derivatives are great importance in ephemeris service and in identification of celestial bodies during the real time of their CCD observations. At Pulkovo observatory Dr.Victor L'vov and his colleagues developed the special ephemeris EPOS software with the use of AMP ideology [2, 4, 7]. With EPOS Software we test any positional asteroid observations and identify the observed celestial bodies especially with "old" orbital elements calculated several years ago on the base of a few observational positions. In this process we actively exploit the calculated AMP and their observed values. First of all we consider the sky area determined by telescope field of view where the CCD observations were made at given moment of UT. We find all asteroids which could be seen in it.

Our ephemeris calculations give us not only positions of asteroids registered in EPOS orbital catalogues but their AMP or the first derivatives of coordinates also. Comparison of these values allows us to reliably identify "old" objects with just discovered asteroids which got the new MPC nominations. In this problem we also take into account a coincidence of known orbital elements for "old" asteroid and a circular orbital elements calculated for a "new" asteroid by the AMP method with the use of LAPLACE Software created by Dr.V.Komarov in 1990 yr [3].

Analysis of some asteroid nominations gave the several new identifications in addition to the MPC service. It is safe to say that the Apparent Motion Parameters can be successfully used in the solution of the problem of celestial body identification and for its fast initial orbit determination in the modern CCD epoch.

Our research was supported by the Russian Astronomy Program's grant No 1.8.4.1.

References

1. Bykov, O. P.: 1978, 'The determination of the osculating orbit of the Earth satellite with the Laplace's method by means of photographic observations on a short arc', *Proc. of Astr.obs. of Leningrad University* **34**, pp. 156–164.
2. Bykov, O. P.: 1989, 'Determination of celestial bodies orbits with the Direct Methods', *Problemy issledovanija Vselennoi* **12**, pp. 328–356.
3. Bykov, O. P. and Komarov, V. V.: 1995, 'On the determination of a circular asteroid orbit with the use of its single CCD-observation (alpha, delta, alpha-dot, delta-dot, UT)', *Astronomical and Astrophysical Transactions* **8**, pp. 323–324.
4. Bykov, O. P. and L'vov, V. N.: 1997, 'The information status of dense series of topocentric positions of celestial bodies derived from CCD observations', *Baltic Astronomy* **6**, pp. 359.

EPHEMERIS MEANING OF ASTEROID'S APPARENT MOTION PARAMETERS 337

5. Kiselev, A. A and Bykov, O. P.: 1973, 'The determination of the satellite orbit by a single photograph with many satellite trails', *Sov. Astron. Journal* **50**, pp. 1298–1308.
6. Kiselev, A. A and Bykov, O. P.: 1976, 'The determination of a satellite elliptical orbit with the use of parameters of the satellites apparent motion', *Sov. Astron. Journal* **53**, pp. 879–888.
7. L'vov, V. N., Smekhacheva, R. I. and Tsekmejster, S. D.: 1999, 'EPOS - the Ephemeris Software Package for Solar System Bodies Research, User's Guide', Pulkovo, St.Petersburg.
8. Panova, G. V. at al.: 1959, 'The first Pulkovo catalogue of positional observations of the AES 1957 Beta 1', *Bull. of AES optical observations* **6(6)**, pp. 1–5.

PRIMORDIAL DEPLETION IN THE INNER REGION OF THE ASTEROID BELT, $A < 2.2AU$

FRANCISCO LÓPEZ-GARCÍA
Departamento de Astronomia y Geofisica, Facultad de Ciencias Exactas Universidad Nacional de San Juan, Argentina, E-mail: flgarcia@casleo.gov.ar

ADRIAN BRUNINI
Facultad de Ciencias Astronómicas y Geofísicas, UNLP, Paseo del Bosque s/n, 1900 La Plata, Argentina E-mail: abrunini@fcaglp.fcaglp.unlp.edu.ar

We report some evidences of orbital stability in the inner Solar System using numerical simulations. The studied objects are real asteroids and test particles. In the first case the elements were obtained from the Ephemerides of Minor Planets (EMP) and in the second were at random, in all cases $a < 2.2$ AU. The numerical integrations were performed using symplectic integrators, these algorithms are perfectly suited to study long time span with low eccentricity. The starting elements of fictitious asteroids, 1,000 test particles, were at random, with $a < 2.2AU$, $e < 0.15$, $i < 2°$ and ω, Ω, M $(0, 2\pi)$. The test particles are perturbed by the Sun and planets but not within themselves. We include the full gravitational effects since Mercury until Saturn. The initial positions and velocities of the planets as well as their masses and radius were obtained from the Nautical Almanac, 2000.0. In all the numerical simulations we used a step size of 0.015 yr not only to guarantee numerical precision but also to account for all close encounters with Mars, Earth and Venus. All test particles were investigated after each time step and were removed from the simulation those particles whose eccentricity become one or have entered the influence sphere of the inner planets or have close approaches to the Sun, less than 0.07 AU. We integrated the orbits for 100 Myr. This study is similar to recent research realized by Mikkola and Innanen (1995), Holman (1997), Evans and Tabachnik (1999) to study the stability of test particles in the Solar System. We must be careful when we numerically study the dymanical evolution of the inner Solar System because the orbital period of Mercury is ~ 88 days and this obligate us to use small time step. We found that only those asteroids from well defined narrow regions in semimajor axis, associated to the ν_{16} secular resonance and the 5:1 mean motion commensurability with Jupiter, can reach Mars crossing orbits in time scales comparable to the time scale of formation of the inner planets. This implies that only a small fraction of objects in this region could have been contributors to the accretion of the inner planets. Secular resonances with the inner planets and mean motion commesurabilities with both the inner and the outer planets play a key role in the primordial depletion of this region. To end the integrations some (a few) test particles (less than 12%) have close encounters within the Hill's sphere of Earth and Venus. On the other hand, the number of close encounter within the Hill's sphere of Mars is nearly 40%. For each close encounter we stored the mini-

Pretka-Ziomek et al./Dynamics of Natural and Artificial Celestial Bodies, 339–340, 2001.
© 2001 *Kluwer Academic Publishers.*

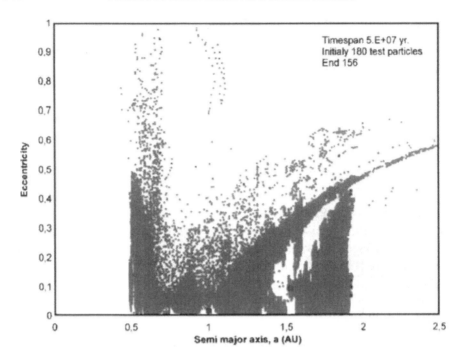

mum distance and the relative velocity to the planet, in such a way we may compute also the intrinsic probability of collision (or the mean interval between collisions), which was doing following Nakamura and Yoshikawa (1995). We found for collisions onto Mars $\Delta t = 3.x 10^7 yr$. which is one order of magnitude larger than the mean interval between collisions which was computed by Olson-Steel (1987). In the region between Venus and Earth the majority of test particles are ejected, only a few of them has stable orbits. Between Mercury and Venus almost all test particles show unstable orbits. Near to the terrestrial planets test particles are rapidly ejected and survive those whose eccentricities are very small (stable circular orbits).

Acknowledgements

FLG thanks the Director of Casleo for the use their facilities and appreciate the Mrs. L Navarro and Mr. J.L. Giuliani for their helping in the preparation of the manuscript.

References

Evans, N.W. and Tabachnik, S.: 1999, *Nature* **Vol. 399**, p. 41.
Holman, M.J.: 1997, *Nature* **Vol. 387**, p. 785.
Mikkola, S. and Innanen, K.: 1995, *M.N.R.A.S.* **277**, p. 497.
Nakamura, T. and Yoshikawa, M.: 1995, *Icarus* **116**, p. 113.
Ollson-Steel, D.: 1987, *M.N.R.A.S.* **227**, p. 501.

NUMERICAL SIMULATION OF THE MOTION OF SMALL BODIES OF THE SOLAR SYSTEM BY THE SYMBOLIC COMPUTATION SYSTEM "MATHEMATICA"

VADIM TITARENKO, LARISA BYKOVA and TATYANA BORDOVITSYNA
Applied Mathematics and Mechanics Institute, Tomsk, Russia

Key words: numerical simulation, asteroid motion, orbit determination, long computer word.

The problem of the investigation of close encounters of a small body with a large planet using long computer words (Wolfram, 1991) have been considered on the example of asteroid 1991 VG which has a lot of close encounters with the Earth, one of them was 0.0031 a.u. in December, 21, 1991. Together with the classical equations of motion (x_t) we have used the regularized equations (x_s) (Bordovitsyna et al., 1998).

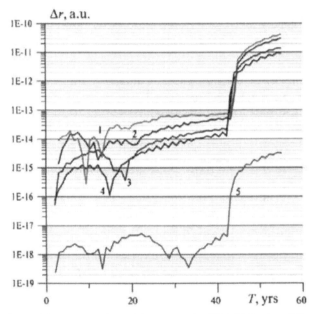

Figure 1. Estimations of the accuracy in position vector for 1991 VG using different algorithms of prediction of motion.

The estimations of the accuracy in position vector for various lengths of computer word defined in decimal digits L for various orders N of numerical methods and for two forms differential equations (x) are given on the Figure 1. For curve 1: $L = 19$, $N = 18$, $x = x_t$. For curve 2: $L = 19$, $N = 18$, $x = x_s$. For curve 3: $L = 25$, $N = 18$, $x = x_t$. For curve 4: $L = 25$, $N = 18$, $x = x_s$. For curve 5: $L = 28$, $N = 26$, $x = x_s$. The estimations show that the loss in accuracy takes place in all the cases.

Pretka-Ziomek et al./Dynamics of Natural and Artificial Celestial Bodies, 341–342, 2001.
© 2001 *Kluwer Academic Publishers.*

TABLE I

The computing accuracy of the matrices.

| Digits | Condition number of the initial matrices | | | |
| | 10^6 | | 10^{18} | |
	$\|\Delta I_1\|$	$\|\Delta I_2\|$	$\|\Delta I_1\|$	$\|\Delta I_2\|$
19	$2 \cdot 10^{-19}$	$1 \cdot 10^{-19}$	$4 \cdot 10^{-12}$	$2 \cdot 10^{-17}$
18	$6 \cdot 10^{-19}$	$1 \cdot 10^{-19}$	$1 \cdot 10^{-7}$	$3 \cdot 10^{-16}$
17	$2 \cdot 10^{-18}$	$1 \cdot 10^{-19}$	$1 \cdot 10^{-6}$	$1 \cdot 10^{-15}$
16	$4 \cdot 10^{-11}$	$2 \cdot 10^{-13}$	17	$3 \cdot 10^{-8}$

However the use of a long computer word and a numerical high order method with the regularized differential equations allows to put errors to insignificant digits.

The problem of fitting initial parameters by using a long computer word have been considered for 2349 Hathor. In the frame of linear square method of solving of the observational equation system $A\Delta q = b$ two algorithms for calculation of estimations Δq are compared. One of them is the Gaussian elimination method for calculation matrix Q^{-1}, $Q = A^T A$ and second is the method of singular decomposition connected with calculation of pseudo-inverse matrix $A^{\#}$ (Lawson and Hanson, 1974). The estimations of the computing accuracy of the matrices Q^{-1} and $A^{\#}$ using computer word with various digital position are given in Table I for the ill-conditioned and well-conditioned cases. Here $\|\Delta I_1\| = \|Q^{-1}Q - I\|$ and $\|\Delta I_2\| = \|A^{\#}A - I\|$, I is a identity matrix.

One can also see that in the case of ill-conditioned matrices the use of the singular decomposition with any length of computer word gives better results than the use of the Gaussian elimination method.

References

Bordovitsyna, T. V, Avdyushev, V. A. and Titarenko, V. P.: 1998, 'Numerical Integration in the General Three–Bodies Problem', *Research in Ballistics and Contiguous Problems of Dynamics*, **2**, Tomsk, 164–168 (in Russian).

Lawson, C. L. and Hanson, R. J.: 1974, *Solving Least Squares Problems*. Prentice-Hall Inc., Englewood Cliffs, N. J.

Wolfram, S.: 1991, *Mathematica, a System for Doing Mathematics by Computer*, Addison-Wesley Publishing Company, Redwood City.

THE PREDICTION OF THE MOTION OF THE ATENS, APOLLOS AND AMORS OVER LONG INTERVALS OF TIME

IRENEUSZ WŁODARCZYK (irek@entropia.com.pl)
Astronomical Observatory of the Chorzów Planetarium, 41-501 Chorzów 1, P.O. box 10, Poland

If we take into account two starting orbits which differ only by error of calculation of any orbital element, then after some time differences in mean anomaly between these neighbors orbits growth rapidly.

Equations of motion of 930 Atens, Apollos and Amors (AAA) were integrated 300,000 years forward using RA15 Everhart method (Everhart 1974). The Osterwinter model of Solar System was used (Oesterwinter and Cohen 1972). As the starting point orbital elements of the asteroids were taken from public-domain asteroid orbit database *astorb.dat* March 2000 update at *http://asteroid.lowell.edu*.

The differences in mean anomaly between unchanged and unchanged orbits were calculated. The changed orbits were constructed by adding or subtracting to the osculating orbital elements one after the other errors of determination of orbital elements.

As an example of calculations was orbit of Anteros - Amors type object, with the following errors of present author's orbital elements determinations: in mean anomaly, $dM = 0.^{\circ}00024$, in semimajor axis, $da = 0.00000004\,AU$, in eccentricity, $de = .0000007$, in argument of perihelion, $d\omega = 0.^{\circ}00057$, in longitude of ascending node, $d\Omega = 0.^{\circ}00051$, in inclination, $di = 0.^{\circ}00004$.

When the differences in mean anomaly were grater then 360° then calculations were stopped. In almost all cases after about 1000 years in forwards integration and after 2500 years in backwards integration differences in mean anomaly between neighbors orbits growth rapidly. Almost the same situation was when we have subtracted one after the other these errors of determination of orbital elements. It denotes, that it is impossible to predict behavior of asteroids outside this time. This time I have named *time of stability*. Thus time of stability of Anteros is 1000 years for forwards integration and 2500 years for backwards integration. Change of any orbital element leads to almost the same results. Therefore for other asteroids changes in semimajor axis were calculated only. For example time of stability for Toutatis is equal 250 years. The same result has obtained prof. Sitarski using another method of integration of equations of motion and applied another model of Solar System (Sitarski 1998).

To calculate times of stability I have changed only semimajor axes by the error of theirs determination. For simplicity I take for all these asteroids the same value of this error as for Anteros - 0.00000004 AU (6 km). The calculated times of stability for these AAA are very short (<1,000 years - 33%,<10,000 years - 89%,<30,000 years - 98%). Histogram of averaged times of stability for 916 AAA

Pretka-Ziomek et al./Dynamics of Natural and Artificial Celestial Bodies, 343–345, 2001.
© 2001 *Kluwer Academic Publishers.*

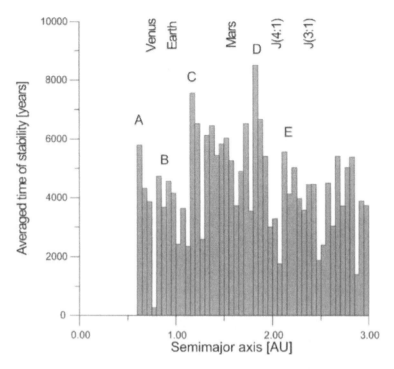

Figure 1. Histogram of calculated times of stability for 916 AAA asteroids.

asteroids, which started semimajor axes are shorter than 3.0 AU is presented on Figure 1.

This behavior of averaged times of stability of AAA is almost the same as the behavior of *survival times* of Evans and Tabachnik for testing particles in this same region on theirs Fig.1 (Evans and Tabachnik, 1999). They have integrated equations of motion of 1000 test particles which were in the beginning distributed on concentric rings with starting values of the semimajor axes between 0.1 and 2.2 AU. If the orbits of these test particles have entered the sphere of influence of planet, or have approached to the Sun closer than 10 solar radii, or have become hyperbolic then they have stopped theirs calculations. These times they have named *survival times*.

Between: Mercury and Venus (region A on Figure 1), Venus and Earth (B), Earth and Mars (C), Mars and region with mean motion resonance 4:1 with Jupiter (D) we observe grater values of *survival times* of Evans and Tabachnik as well as *times of stability* of present author. Furthermore both these times are much shorter in the regions of planets as well as in regions with mean motion resonances 4:1 and 3:1 with Jupiter - J(3:1), and J(4:1) on Figure 1, respectively.

References

Evans, N. and Tabachnik, S.: 1999, *Letters to Nature* **399**, 41.
Everhart, E.: 1974, *Cel. Mech.* **10**, 35.
Oesterwinter, C. and Cohen, C.: 1998, *Cel. Mech.* **5**, 3177.
Sitarski, G.: 1998, *Acta Astron.* **48**, 547.

ORBITAL MOTION IN OUTER SOLAR SYSTEM

J. KLAČKA
Institute of Astronomy, Faculty for Mathematics and Physics, Comenius University
Mlynská dolina, 842 48 Bratislava, Slovak Republic

M. GAJDOŠÍK
Ostredská 20, 821 02 Bratislava, Slovak Republic

1. Introduction

Motion of a point mass in gravitational fields of the Sun and of the galactic disk is important from the point of view of orbital evolution of cometary orbits. One of such numerical studies may be found, e.g., in Prętka and Dybczyński (1994). Since we want to find fundamental features of possible motions, we use time-averaged equations for orbital elements (e, i, ω) – conclusions may be applied to real situations for semimajor axes up to the order of 10^4 AU.

2. Time-averaged Equations for Orbital Elements

The dynamical model discussed in this paper is given by Eqs. (1) in Prętka and Dybczyński (1994); we will use abbreviation $k = 4\pi G\varrho$.

The case $k = 0$ corresponds to a Keplerian motion. If k is smaller than μ/r^3, we can consider the term $k\,z$ as a disturbing acceleration to two-body problem.

The significant time-averaged evolutionary equations for orbital elements are:

$$\left\langle \frac{de}{dt} \right\rangle = \frac{5}{4}k\sqrt{\frac{a^3}{\mu}}(\sin i)^2[\sin(2\omega)]e\sqrt{1 - e^2}; \left\langle \frac{da}{dt} \right\rangle = 0$$

$$\left\langle \frac{di}{dt} \right\rangle = -\frac{5}{8}k\sqrt{\frac{a^3}{\mu}}[\sin(2i)][\sin(2\omega)]\frac{e^2}{\sqrt{1 - e^2}}$$

$$\left\langle \frac{d\omega}{dt} \right\rangle = -\frac{5}{2}k\sqrt{\frac{a^3}{\mu}}\left\{ \frac{(\sin i)^2 - e^2}{\sqrt{1 - e^2}}(\sin \omega)^2 - \frac{1}{5}\sqrt{1 - e^2} \right\}. \tag{1}$$

Eqs. (1) show that it is useful to define a dimensionless variable τ

$$\tau \equiv \frac{5}{2}k\sqrt{\frac{a^3}{\mu}}t; \sqrt{1 - e^2} \cos i = \sqrt{1 - e_0^2} \cos i_0, \tag{2}$$

we have used also the fact that equations for eccentricity and inclination define the second relation of Eqs. (2), where the index 0 denotes initial quantities. The second relation of Eqs. (2) corresponds to conservation of the z – component of angular momentum – the component perpendicular to the galactic plane.

Eqs. (2) reduce Eqs. (1) to two equations:

$$\left\langle \frac{di}{d\tau} \right\rangle = \frac{1}{2}[\sin(2\omega)](\sin i) \frac{(1 - e_0^2)(\cos i_0)^2 - (\cos i)^2}{\sqrt{1 - e_0^2} \cos i_0}$$

$$\left\langle \frac{d\omega}{d\tau} \right\rangle = \frac{[1/5 - (\sin \omega)^2](1 - e_0^2)(\cos i_0)^2 + (\sin \omega)^2 (\cos i)^4}{\sqrt{1 - e_0^2}(\cos i_0)(\cos i)}. \qquad (3)$$

3. Mathematical Treatment

Eqs. (3) form a complete set of differential equations – initial conditions ω_0, i_0 must be added, of course. We may interpret solutions of Eqs. (3). At first

$$(\cos i_0)^2 = \frac{4}{5}(1 - e_0^2) \wedge \left(\omega_0 = \frac{\pi}{2} \vee \omega_0 = \frac{3\pi}{2} \right) \rightarrow \left\langle \frac{di}{d\tau} \right\rangle = \left\langle \frac{d\omega}{d\tau} \right\rangle = 0. \quad (4)$$

Eq. (4) defines one type of orbits – stationary orbits.

When considering periodic changes, one would await that $\dot{\omega}$ may be positive as well as negative. One obtains, on the basis of the second of Eqs. (3)

$$\dot{\omega} < 0 \iff \frac{1}{5}(1 - e^2) < (\sin \omega \sin i)^2 \left\{ 1 - \left(\frac{e}{\sin i} \right)^2 \right\}. \qquad (5)$$

Eq. (5) states that if the case $\dot{\omega} < 0$ may occur, the quantity ω can change periodically only in a small interval given by the values of e and i – oscillations in ω; the same holds for the quantity $\sin b = \sin \omega \sin i$. This is the second type of orbits. (The consequence of Eq. (5) is $e < \sin i$, and, $(\sin b)^2 > 1/5$.)

The final type of orbits corresponds to the case when the condition $\dot{\omega} > 0$ is permanently fulfilled: relation (5) can never hold during the orbital motion. In this case ω increases monotonically from 0 to 2π, b oscillates about zero.

The last two cases should correspond to the types of orbits found in Prętka and Dybczyński (1994) in a numerical way – by numerical integration of Eqs. (1). The two types of possible orbits are separated by the conditions: i) always $\dot{\omega} > 0$ holds (($\sin b)^2 < 1/5 \iff \dot{\omega} > 0$, always); ii) oscillations in ω (oscillations in ω $\iff (\sin b)^2 > 1/5$). We see that the results do not depend on semimajor axis and eccentricity (Prętka and Dybczyński (1994)), but on eccentricity e and $e/\sin i$.

4. One Consequence

The consequence of Eq. (4) and periodical solutions of Eqs. (3) is that the relation $a^3 P^2 = f(e_0, i_0, \omega_0)$ between semimajor axis a and period P of the change of the time-averaged orbital elements exists. This relation is interesting in comparison with the Kepler's third law: $a^3/T^2 = \mu/(4\pi^2)$, T is the period of revolution.

Acknowledgements

Authors want to thank to LOC for support. The work was supported by the Scientific Grant Agency VEGA (1/7067/20).

References

Prętka, H. and Dybczyński, P. A.: 1994, 'The galactic disk influence on the Oort cloud cometary orbits', in *Dynamics and Astrometry of Natural and Artificial Celestial Bodies* (K. Kurzyńska, F. Barlier, P. K. Seidelmann and I. Wytrzyszczak, Eds.), Astronomical Observatory of A. Mickiewicz University, Poznań, Poland, pp. 299–304.

SOLAR WIND AND MOTION OF METEOROIDS

JOZEF KLAČKA (klacka@fmph.uniba.sk)
Institute of Astronomy, Faculty for Mathematics and Physics, Comenius University, Mlynská dolina, 842 48 Bratislava, Slovak Republic

Abstract. The effect of solar wind on the orbital evolution of meteoroids is discussed. It is shown that results presented in recently published papers are incorrect. Non-radial component of solar wind is discussed from the qualitative point of view. It is shown that the orientation of non-radial component is opposite in comparison with the direction used in papers dealing with orbital evolution of meteoroids.

1. Introduction

Klačka (1994) has discussed the effect of solar wind particles on the motion of small interplanetary dust particles (meteoroids). He has pointed out that non-radial component of the solar wind has an opposite direction than the direction used in papers of the field of interplanetary matter. He has also tried to obtain some quantitative results following from consideration of the correct non-radial direction.

The newest papers deal with the effect of the solar wind on the motion of meteoroids not in a correct way. Even if the existence of non-radial component of the solar wind is neglected, the results presented by Banaszkiewicz *et al.* (1994) do not correspond to reality. Newer paper dealing with the orbital evolution of meteoroids (Cremonese *et al.* 1997) takes non-radial direction of solar wind into account, also. However, this direction is still taken in the incorrect way (see also references in Cremonese *et al.* 1997).

2. Radial Direction of the Solar Wind

Let us neglect non-radial component of the solar wind. Long-term orbital evolution of meteoroids due to the effect of radial solar wind is characterized by differential equations for orbital elements and these equations are of the same form as that for the Poynting-Robertson effect (P-R effect; for special form of interaction between meteoroid and solar electromagnetic radiation – see Klačka 2000). Simultaneous action of the P-R effect and solar wind yields for the change of semimajor axis a and eccentricity ε

$$\begin{aligned}\left\langle \frac{da}{dt} \right\rangle &= -(1+\eta)(2+3e^2)\frac{\alpha}{a(1-\varepsilon^2)^{3/2}} \\ \left\langle \frac{de}{dt} \right\rangle &= -(1+\eta)\frac{5}{2}\varepsilon\frac{\alpha}{a^2\sqrt{1-\varepsilon^2}},\end{aligned} \quad (1)$$

Pretka-Ziomek et al./Dynamics of Natural and Artificial Celestial Bodies, 351–353, 2001.
© 2001 *Kluwer Academic Publishers.*

352 JOZEF KLAČKA

if the same notation as in Banaszkiewicz *et al.* (1994) is used. Main parts of Eqs. (45) and (46) in Banaszkiewicz *et al.* (1994) do not correspond to our Eqs. (1).

3. Non-radial Direction of the Solar Wind – Qualitative Discussion

Using literature on solar physics, Klačka (1994) has pointed out that non-radial component of the solar wind brings down the effect of inspiralling of interplanetary dust particles toward the Sun generated by the P-R effect. However, according to (Cremonese *et al.* 1997, see also references in this paper) the non-radial component of the solar wind raises the effect of inspiralling of interplanetary dust particles toward the Sun generated by the P-R effect.

We can present another argument in favor of the non-radial direction of solar wind used in Klačka (1994). This argument is presented in Figure 3.20 (p. 84) in Hundhausen (1972). The important property of the non-radial component deals in fact that it is a function of distance from the Sun (e. g., Figure 3.21 (p. 87) in Hundhausen (1972)). This fact was not considered in Klačka (1994). However, the dependence on distance is not known, and, thus, no precise calculations can be reliably done. The only thing one can done is to make some other estimates (in comparison to Klačka 1994), using as a motivation the result presented in Figure 3.21 (Hundhausen 1972, p. 87). Supposing that the non-radial component depends on the heliocentric distance as $1/r$, the inspiralling toward the Sun due to the solar electromagnetic radiation will be stopped at several solar radii. This artificial example shows the importance of knowing exact dependence of the non-radial component on the heliocentric distance.

4. Conclusion

We have shown that even the approximation of radial solar wind is not correctly treated in papers dealing with orbital motion of meteoroids. We have shown that solar wind's real orientation of non-radial component is opposite in comparison to that considered in Cremonese *et al.* (1997 – see also references in this paper). Correct consideration of this non-radial component, together with its dependence on heliocentric distance, will help us in our better understanding of the evolution of small particles in solar system.

Acknowledgements

The author wants to thank to the Organizing Committees for the financial support which enabled his participation at the conference. The paper was supported by the Scientific Grant Agency VEGA (grant No. 1/7067/20).

References

Banaszkiewicz, M., Fahr, H. J. and Scherer, K.: 1994, 'Evolution of dust particle orbits under the influence of solar wind outflow asymmetries and the formation of the zodiacal dust cloud', *Icarus* **107**, 358–374.

Cremonese, G., Fulle, M., Marzari, F. and Vanzani, V.: 1997, 'Orbital evolution of meteoroids from short period comets', *Astron. Astrophys.* **324**, 770–777.

Hundhausen, A. J.: 1972, *Coronal Expansion and Solar Wind*, Springer-Verlag, Berlin-Heidelberg, 238 pp.

Klačka, J.: 1994, 'Interplanetary dust particles and solar radiation', *Earth, Moon and Planets* **64**, 125–132.

Klačka, J.: 2000, 'Electromagnetic radiation and motion of real particle', *Icarus* (submitted; astro-ph/0008510).

ON THE STABILITY OF THE ZODIACAL CLOUD

JOZEF KLAČKA
Institute of Astronomy, Faculty for Mathematics and Physics Comenius University, Mlynskd dolina, 842 48 Bratislava, Slovak Republic

MIROSLAV KOCIFAJ (kocifaj@astro.savba.sk)[*]
Astronomical Institute of the Slovak Academy of Sciences Dúbravská cesta 9, 842 28 Bratislava, Slovak Republic

1. Introduction

The Poynting-Robertson effect (P-R effect) (Robertson, 1937), (Klaĉka, 1992) is generally considered to be the real effect which influences the motion of interplanetary dust particles (IDPs) – meteoroids, in the Solar System. The most general case of the validity of the P-R effect requires that Eq. (120) (or, Eq. (122) for the moving particle) in (Klaĉka, 1992) holds.

The P-R effect causes inspiralling of met eoroids toward the Sun. As a consequence, the zodiacal cloud, a cloud of met eoroids moving in the inner part of the Solar System, is not stable unless new meteoroids from some parts of the Solar System are injected. Comets, asteroids or Edgeworth-Kuiper belt objects, e.g., (Gustafson *et al.*, 1987), (Lein-ert and Grün, 1990), (Lion *et al.*, 1996) seem to be the sources of meteoroids for zodiacal cloud.

It was already suggested that the motion of real meteoroid may significantly differ from the motion corresponding to the P-R effect-general equation of motion of IDP in terms of optical properties was presented by (Kiaçka and Kocifaj, 1994). However, the paper did not present any quantitative calculation. The possible (in-) stability of the zodiacal cloud and correct physical principles were discussed in (Klacka, 1994).

The exact numerical calculations for the real cosmic dust particle of the identical shape as the particle from NASA collection archived under identification code U2015 B10 (Clanton *et al.*, 1984) are of the main interest of this paper. Motion of the particle under the action of solar gravity and solar electromagnetic radiation is analyzed. Rapid rotation of the particle about defined axis of rotation is considered. Numerical solution of the interaction of radiation with the particle was based on so-called Discrete Dipole Approximation (Draine and Flatau, 1994).

2. Results

Figure 1 depicts time evolution of perihelion distance for meteoroids ejected from comet Encke (initial orbital elements are identical with those of comet Encke) for a

---
[*] This e-mail address is available for all problems and questions.

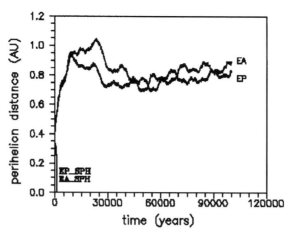

Figure 1. Time evolution of perihelion distances of meteoroids ejected from comet Encke.

special type of orientation of rotation axis. The index 'A' ('P') denotes that particle was ejected at aphelion (perihelion) of Encke's orbit. Lower left corner depicts time evolution for spherical particle (EP-SPH, EA-SPH), which corresponds to the P-R effect – rapid decrease of perihelion distance is evident.

Comparison of the real particle motion with the motion for the PR effect leads to significant conclusions. While the P-R effect yields spiraling toward the Sun (decrease of perihelion distance with time), motion of real meteoroid is much more complicated. Inspiralling toward the Sun during some time interval is followed by spiraling outward the Sun. This process is affected by starting geometry of the solved system particle-Sun, i.e. by selected particle rotation axis, current particle orientation and its position relatively to the Sun, and by initial orbital conditions.

3. Conclusion

The results are important for application to the stability of the zodiacal cloud. While the P-R effect (valid for spherical particles) yields instability of the zodiacal cloud due to monotonous inspiralling of the particles toward the Sun, our numerical simulations show that particle nonsphericity could radically modify its resulting motion in the Solar System. There were found situations when particles are characterized by the relatively stable trajectories. This fact can significantly contribute to explanation of the stability of the zodiacal cloud.

Acknowledgements

The authors want to thank to the Organizing Committees for the financial support which enabled our participation at the Conference. The paper was partially supported by the Scientific Grant Agency VEGA (grant No.1/7067/20 and 2/7175/20).

References

Clanton, V. S., Gooding, J. L., McKay, D. S., Robinson, G. A., Warren, J. L. and Watts, L. A. (Eds.): 1984, 'Cosmic Dust Catalog', (*Particles from collection flag U2015*), NASA, Johnson Space Center, Houston, Texas, **5**, No.1., 10.

Draine, B. T. and Flatau, P. J.: 1994, 'Discrete-dipole approximation for scattering calculations', *J. Opt. Soc. of America A* **11**, 1491–1499.

Gustafson, B. A. S., Misconi, N. Y. and Rusk, E. T.: 1987, 'Interplanetary Dust Dynamics', *Icarus* **72**, 582–592.

Klaĉka, J. : 1992, 'Poynting–Robertson effect. I. Equation of motion', *Earth, Moon, and Planets* **59**, 41–59.

Klaĉka, J.: 1994, 'On the Stability of the Zodiacal Cloud', *Earth, Moon, and Planets*, **64**, 95–98.

Klacka, J. and Kocifaj, M.: 1994, 'Electromagnetic Radiation and Equation of Motion for a Dust Particle', In *Dynamics and Astrometry of Natural and Artificial Celestial Bodies* (K. Kurzyńska, F. Barlier, P. K. Seidelmann and I. Wytrzyszczak, Eds.), Astronomical Observatory of A. Mickiewicz University, Poznań, Poland, 187–190.

Leinert, C., and Grün, E.: 1990, In *Physics and Chemistry in Space - Space and Solar physics*, (R. Schwenn and E. Marsch, Eds.), **Vol. 20**, Springer–Verlag, Berlin, 207–275.

Liou, J., Zook, H. A. and Dermott, S.F.: 1996, 'Kuiper Belt Dust Grains as a Source of Interplanetary Dust Particles', *Icarus*, **124**, 429–440.

Robertson, H.P.: 1937, 'Dynamical Effects of Radiation in the Solar System', *Mon. Not. Roy. Astron. Soc.*, **97**, 423–438.

INTERACTION OF STATIONARY NONSPHERICAL INTERPLANETARY DUST PARTICLE WITH SOLAR ELECTROMAGNETIC RADIATION

MIROSLAV KOCIFAJ (kocifaj@astro.savba.sk)*
Astronomical Institute of the Slovak Academy of Sciences Dúbravská cesta 9, 842 28 Bratislava, Slovak Republic

JOZEF KLAČKA
Institute of Astronomy, Faculty for Mathematics and Physics Comenius University, Mlynská dolina, 842 48 Bratislava, Slovak Republic

1. Introduction

Interplanetary dust particles are different in size, chemical composition, shape and physical-optical properties. The light scattering by such particles predetermines changes of their motion in the space. This fact is expressed by well-known radiation pressure, which was notoriously based on assumption of very simple models of particle shape. Mie theory is for example applicable when approximate the cosmic dust grains by spheres. The only forward and backward scattering efficiencies are important in this case. However, any irregularity of the particle shape will produce certain momentum in perpendicular projections to the direction of light propagation. This may be caused also by inhomogeneity of particle chemical composition (or particle density) (Eremin and Ivakhnenko, 1998). Particle shape specificity (cavities,...) plays dominant role in formation of light scattering diagram (Mishchenko *et al.*, 2000).

2. Scattering Diagrams and Non-radial Momentum

An angular structure of scattered radiation field is expressed by phase function, which is basis for calculation of the non-radial components of the momentum (Kocifaj *et al.*, 1999). Let G_0 characterizes the radial component, and G_1 and G_2 correspond to transversal components of momentum. Both the ratios G_i/G_0 ($i=1..2$) equal to zero for spherical particles. In such a case the motion of particles will fit the trajectory predetermined by Poynting–Robertson (P-R) effect (Burns *et al.*, 1979; Klačka, 1992). However, the non-radial components of the momentum occur for irregular targets. The motion of nonspherical particles can significantly differ from the P-R effect if $G_i/G_0 \gtrsim 10^{-4}$ (Kocifaj *et al.*, 2000).

The extensive and systematic calculations on arbitrarily shaped and composed particles (Draine and Flatau, 1994) were performed to evaluate the possible effect

---

* This e-mail address is available for all problems and questions.

Pretka-Ziomek et al./Dynamics of Natural and Artificial Celestial Bodies, 359–361, 2001.
© 2001 *Kluwer Academic Publishers.*

of radiation on the particle momentum. Especially, an attention was concentrated to porous as well as strongly absorbing rotating micro–sized particles. Although a compact spherical grain does not generate the non-radial momentum components, a porosity is responsible for their rising. It was shown that the porosity in 40% of particle volume can produce a value G_i/G_0 up to 0.01 for spheres, 0.02 for ellipsoids, 0.04 for rectangular targets, and 0.09 for really shaped grains. The asphericity factor of real particle shape was chosen to be about 1.4 – valid in average for interstellar cosmic dust (Hildebrand and Dragovan, 1995). In general, the higher degree of particle porosity, the probably greater significance of the non-radial momentum components. On the other hand, the higher content of strongly absorbing elements in the particle body, the smaller importance of the non-radial momentum components. Considering the iron grains in the calculation model (refractive index of which is about $1.27 - 1.37i$) the examined ratios decline approximately in one magnitude.

3. Conclusion

The greater the non-radial momentum components the more different trajectory of the particle could occur in comparison with the well–known P-R effect. The best candidate for such a motion are porous, non-absorbing particles. This effect will be less dominant for compact and strongly absorbing particles.

Acknowledgements

The authors want to thank to the Organizing Committees for the financial support which enabled our participation at the Conference. The paper was supported by the Scientific Grant Agency VEGA (grants Nos. 1/7067/20, and 2/7151/20). The authors are also very grateful to Bruce Draine and Piotr Flatau for their freeware DDSCAT code.

References

Burns, J. A., Lamy, P. L. and Soter, S.: 1979, 'Radiation forces on small particles in the solar system', *Icarus* **40**, 1–48.

Draine, B. T. and Flatau, P. J.: 1994, 'Discrete-dipole approximation for scattering calculations', *J. Opt. Soc. of America A* **11**, 1491–1499.

Eremin, Yu. A. and Ivakhnenko, V. I.: 1998, 'Modeling of light scattering by non-spherical inhomogeneous particles', *J. Quant. Spectrosc. Radiat. Transfer* **60(3)**, 475–482.

Hildebrand, R. H. and Dragovan, M.: 1995, 'The shapes and alignment properties of interstellar dust grains', *Astrophys. J.* **450**, 663–666.

Klačka, J.: 1992, 'Poynting–Robertson effect. I. Equation of motion', *Earth, Moon, and Planets* **59**, 41–59.

INTERACTION OF INTERPLANETARY DUST PARTICLE WITH SOLAR RADIATION 361

Kocifaj, M., Kapišinský, I. and Kundracík, F.: 1999, 'Optical effects of irregular cosmic dust particle U2015 B10', *J. Quant. Spectrosc. Radiat. Transfer* **63(1)**, 1–14.

Kocifaj, M., Klačka, J. and Kundracík, F.: 2000, 'Motion of really shaped cosmic dust particle in Solar System', In *Proc. of the Conference on Electromagnetic and Light Scattering by Nonspherical Particles*, Halifax, Canada (accepted).

Mishchenko, M. I., Hovenier, J. W. and Travis, L. D.: *Light scattering by nonspherical particles. Theory, measurements, and Applications.*

THE PHOTOGRAPHICALLY OBSERVED METEORS OF (PEGASIDS?) STREAM ASSOCIATED WITH COMET 18P/PERRINE-MRKOS

L. NESLUŠAN (ne@ta3.sk)

Astronomical Institute, Slovak Academy of Sciences, 059 60 Tatranská Lomnica, Slovakia

In a preliminary analysis of dynamics of 77 theoretical streams associated with the short-period comets moving in the orbits distant (more than 0.2 AU) from that of the Earth and having orbital period shorter than 10 years, we found that the particles of theoretical streams of 21 comets approached the Earth's orbit nearer than 0.2 AU and the particles of 14 of these comets could be identified with the actually observed meteors.

In this paper, a more detail analysis of another cometary stream previously detected is given. Specifically, we deal with the stream of comet 18P/Perrine-Mrkos. The used method of stream modeling was described in our previous paper (Neslušan L., 1999, Astron. Astrophys. 351, 752).

Only 5 observed meteors can be identified with the modeled particles. (Another 3 meteors are questionable as the members of the stream.) The characteristics of the identified meteors as well as the mean parameters of the diffuse stream are given in Table I (orbital elements) and Table II (radiants and velocities).

Looking at Table I or II, we can state that the stream of comet 18P has been active from October 20-th to November 6-th.

Comparing the resultant stream characteristics with those of streams known till now, one can notice a certain similarity between the found stream and the Pegasids meteor stream presented by Cook in 1973 in his Working List of Meteor Streams. However, an undoubted identification of both streams is questionable because the differences between the corresponding parameters are large enough. The radiant of found stream is not situated in constellation Pegasus, but in constellation Cygnus, near star τ Cyg (23° apart from the position given by Cook). The Pegasids are active from October 29-th to November 12-th.

Since the 18P/Perrin-Mrkos stream consists of ouly 5 observed meteors, one can ask whether such a low numbered stream is real. This question can most probably be answered positively. Such suggestion is supported by the fact that the stream meteors have been observed by 3 stations and in 5 various years. To confirm the reality of the stream more, we applied the method of selection of meteors from the database to appropriate phase space, i.e. we considered the mean orbit of stream (Table I) as the starting (normally parent body) orbit in the selection procedure. The diffuse stream was also distinguished using this procedure.

Pretka-Ziomek et al./Dynamics of Natural and Artificial Celestial Bodies, 363–364, 2001.
© 2001 *Kluwer Academic Publishers.*

TABLE I

The orbital elements of actually observed meteors associated with comet 18P/Perrine-Mrkos, which are identified with the modeled particles released by this comet. date + PSNO-date of meteor detection, its Publi- cation serial number, and Author or station code as presented in the IAU MDC database; π — longitude of perihelion. The angular elements are given to the same equinox as in the original IAU MDC database, i.e. 1950.0

date + PSNO	q [AU]	e	ω [°]	Ω [°]	i [°]	π [°]
10/20.7662/1979–102E	0.967	0.667	202.0	206.3	16.7	48.3
10/22.0680/1965–022F	0.973	0.614	199.6	208.2	12.7	47.8
10/25.2176/1957–241P	0.972	0.585	200.3	211.4	12.5	51.7
11/06.1620/1974–324F	0.975	0.666	196.7	223.0	12.4	59.7
11/06.2450/1958–331P	0.989	0.621	185.6	223.2	17.8	48.8
mean data of stream:						
—	0.975	0.631	196.8	214.4	14.4	51.3

TABLE II

The geophysical data of meteors listed in Table I. α, δ – equatorial coordinates (eq. 1950.0) of radiant; v_g, v_h – geocentric and heliocentric velocities; $L_\odot$ – solar longitude

date + PSNO	α [°]	δ [°]	v_g [kms^{-1}]	v_h [kms^{-1}]	$L_\odot$ [°]
10/20.7662/1979–102E	315.9	39.0	13.78	38.43	206.3
10/22.0680/1965–022F	316.4	32.0	11.49	37.81	208.2
10/25.2176/1957–241P	319.4	33.4	11.12	37.47	211.4
11/06.1620/1974–324F	324.9	33.7	11.74	38.53	223.0
11/06.2450/1958–331P	301.5	43.9	13.22	38.06	223.2
mean data of stream:					
—	315.6	36.4	12.27	38.06	214.4

Acknowledgements

The author thanks to Dr. Hans Rickman from the Uppsala University, Sweden, for the providing the computer code for the numerical integration of orbits. The work was supported, in part, by VEGA - the Slovak Grant Agency for Science (grant No. 5100). The numerical calculations were performed using computational resources (Origin, 2000) of the Computing Centre of the Slovak Academy of Sciences.

A SKETCH OF AN ORBITAL–MOMENTUM–BASED CRITERION OF DIVERSITY OF TWO KEPLERIAN ORBITS

L. NESLUŠAN (ne@ta3.sk)
Astronomical Institute, Slovak Academy of Sciences, 059 60 Tatranská Lomnica, Slovakia

If a database of meteors contains not only the data on their radiant and geo-centric velocity, but also their orbits, then the meteors of a particular shower can be separated from the database evaluating the similarity of orbits of individual shower-members. The Southworth-Hawkins' D-criterion has most frequently been utilized in this evaluation. This criterion is based on a mathematical quantification of diversity of two cone-section curves in space. Though it has appeared to be quite well applicable, some authors have objected that no dynamical aspects of evolution of a given orbit into other are respected within the criterion. The objection cannot be accepted absolutely, because we usually solve an inverse problem in the practice: to study the dynamical evolution of a meteor stream separated on the basis of considered criterion. A knowledge of the dynamics is aim of study and it cannot be included in the initial assumptions.

Nevertheless, the general principles known within the celestial mechanics can be considered in the initial assumptions, therefore a certain respecting of dynamical evolution in the similarity criterion is possible. In this paper, we suggest utilizing the orbital momentum to evaluate the dynamical evolution of a meteoroid from its initial orbit into the observed one. To determine the momentum of a meteoroid, we need to know its mass. This is unfortunately no known quantity, usually. So, we have to constrain to the orbital momentum per a unity of mass in our establishing of new criterion.

We abbreviately call the new criterion as C. The appropriate C-discriminant can be calculated by relation

$$C = \sqrt{(c_{xa} - c_{xb})^2 + (c_{ya} - c_{yb})^2 + (c_{za} - c_{zb})^2}$$ (1)

and represents, in fact, the magnitude of unit-mass-orbital-momentum difference between two Keplerian orbits. Quantities c_{xa}, c_{ya}, c_{za} (c_{xb}, c_{yb}, c_{zb}) are components of orbital momentum vector of initial (final) orbit divided by mass of the assumed body.

To estimate the applicability of the suggested criterion for an optimal separation of all shower meteors from a database, we applied it upon the IAU MDC database of photographical meteor orbits, from which two most numerous showers, Perseids and Geminids, were separated. We utilized so called "method of selection" for this purpose, which should provide, by the appropriate theory, the optimal separation.

Pretka-Ziomek et al./Dynamics of Natural and Artificial Celestial Bodies, 365–366, 2001.
© 2001 *Kluwer Academic Publishers.*

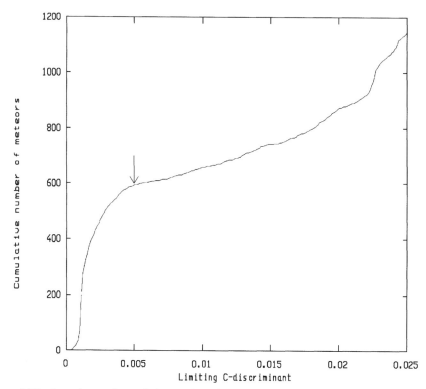

Figure 1. The dependence of cumulative number of Perseids separated from the IAU MDC catalogue on the limiting value of C-discriminant. The value of C for the definitive separation in the "break point" (pointed with an arrow) equals $C_b = 0.0050$ for this shower.

The break point in the dependence of cumulative number of selected Perseids on C-discriminant can be seen in Figure 1 (the break point is pointed out with an arrow). Qualitatively, the illustrated behavior is very similar to that obtained on the basis of D-discriminant within our earlier work. The same conclusion is valid for the Geminids. This means that C-criterion is as sensitive as its D equivalent to a change of orbit. The limiting values of C for the selection are equal to 0.0050 for Perseids and 0.0011 for Geminids.

Assuming the limiting values of C, we separated 594 Perseids and 199 Geminids, however. That means that the method used actually results the expected optimal numbers of separated meteors of these two showers, approximatively, if Southworth-Hawkins D-discriminant is replaced with newly established C equivalent. From this point of view, the orbital-momentum-based criterion appears to be a more appropriate criterion than its D counterpart.

Acknowledgements

The work was supported, in part, by VEGA – the Slovak Grant Agency for Science (grant No. 5100).

ON MODELING OF SMALL CELESTIAL BODY FRACTURE IN PLANET ATMOSPHERES

ILIA SEMENOV (icadks@online.ru) and VICTOR KOROBEINIKOV
Institute for Computer Aided Design, Russian Academy of Sciences, 2nd Brestskaya 19/18,
123056 Moscow, Russia

Abstract. The disruption of celestial body such as meteoroids, cometary fragments and asteroids is investigated by mathematical methods. The gas dynamic forces, inertial force and radiation act on the body. The main results of the paper is related to the creation of stress-strain and phase transition models and simulation of the celestial body fracture and breaking up during its flight in an atmosphere.

Key words: mathematical modeling, mechanic of fractures

1. Mathematical Model

The velocity of the body and its ablation during the first stage of the flight in an atmosphere are determined on the basis of the solution of the system of equations of the physical theory of meteors. The body is assumed to have form of body of revolution during its flight and breaking up. The initial height corresponds to the body position where Knudsen number is small enough. The Cauchy problem for the system of ordinary differential equations was solved by the Runge-Kutta method. We suppose flow around cosmic body (CB) is axial-symmetric and introduce cylindrical or spherical system of coordinates with origin inside the body, and axis of symmetry directed along the trajectory. The thermoelastic problem for a homogeneous isotropic body is solved by numerical and analytical methods. The quasi static approximation is used. Pressure distribution on the flying CB is known. The basic system of equations for thermoelastic behavior of CB material includes the Lame equation for displacement and the heat equation for temperature. Stress tensor components can be calculated by values of displacements by using equations of elastic body motion and deformation. On the body surface there are boundary conditions. To solve the boundary value problem for the equation of elasticity theory with appropriate boundary conditions we apply analytic methods which represents the unknown values (displacements, stresses etc.) by series using Legendre polynomials for spherical body with separation of partial solutions corresponding to the thermal and deceleration terms of the equation or by means of using ordinary polynomials for the case of cylindrical body. Numerical algorithms was worked out for axisymmetrical geometry and calculations were made.

Pretka-Ziomek et al./Dynamics of Natural and Artificial Celestial Bodies, 367–368, 2001.
© 2001 *Kluwer Academic Publishers.*

2. The Results of the Calculations

The local fracture can arise inside the body as well as near its surface. During motion along the trajectory, the inner fractured domain of the body will be increased and expanded toward body's boundary. The whole body is supposed as fractured one when the domain of crushed material occupies an inner part of the body and includes segments of the body's surface, i.e. fractured zone occupies place extended from one side of the body's surface to an opposite side. It was found that thermal stresses are not essential for considered cases. Four criteria of fracture have been used in the calculations: by maximum shear stresses; by maximum octahedral stresses; the Mohr criterion; the criterion of equivalent normal stresses on an octahedral area element (the modified Mohr criterion). By using of an energy equation and the mathematical catastrophe theory methods the energy criterion was also considered. The developed methods were applied to simulate the flight and disruption of icy bodies (fragments of comet heads), stone CB and metallic meteorites. Comparisons with observation data were made for real events. Carbonaceous chondrite variant of Tunguska CB was calculated for cylindrical geometry. After body fracture the breaking up stage of the body's material begins. The medium considered after mechanical fracture is supposed to be a mixture of a gas and small solid particles, where both the gas phase and the solid phase consist of a number of components in general case. This process was calculated by using equations of two phase gasodynamics.

The authors express their gratitude to Russian Fund of Fundamental Researches and INTAS for financial supports.

References

Grigorian, S. S.: 1979, 'About Moving and Fracture of Meteoroids in Planet Atmosphere', *Kosmicheskiye Issledovaniya* **17(6)**, pp. 875–893 (in Russian).

Korobeinikov, V. P., Gusev, S. B., Chushkin, P. I. and Shurshalov, L. V.: 1992, 'Flight and Fracture of the Tunguska Cosmic Body into the Earth's Atmosphere', *Computers and Fluids* **21(3)**, pp. 323–330.

Korobeinikov, V. P.: 1994, 'Heating and Fracture of Flying Large Cosmic Bodies', In *Proceedings of the 3rd International Conference, Integral Methods in Science and Engineering*, London, England.

Korobeinikov, V. P., Vlasov, V. I. and Volkov, D. B.: 1996, 'Numerical Study of Celestial Body Entering into Planets Atmospheres', *CFD Journal* **4(4)**, pp. 463–476.

APPROXIMATING TRAJECTORIES OF OBSERVED BOLIDES WITH ACCOUNT FOR FRAGMENTATION

N.G. BARRI, V.P. STULOV (stulov@inmech.msu.su) and L.YU.TITOVA
Institute of Mechanics, Lomonosov Moscow State University, Michurinskii Ave., 1, Moscow 119899, Russia

Abstract. A result of operating bolide networks in Europe and Northern America is a great number of photographed light segments of bolide trajectories. However, only in a few cases the analysis and extrapolation of these trajectories has made it possible to find the meteorites. It seems likely that this is because almost all meteoric bodies are fragmented in the atmosphere and then small-sized fragments burn out in dense layers.

On the basis of analyzing the bolide luminosity, the role of fragmentation was studied theoretically in (Popova, 1997). Using a radiating radius method, the author of (Popova, 1997) showed that, for a number of Prairie Network bolides (PN-bolides), the luminosity appreciably exceeds the value which can be anticipated for the bolide dynamic mass determined on the basis of the bolide trajectories. Accordingly, it was assumed that in these cases the dynamic trajectory corresponds to the largest fragment, while the light curve corresponds to the whole cloud of fragments, whose total mass may be much greater than the mass of one fragment.

The present paper is an attempt to determine the role of fragmentation on the basis of an observed trajectory without taking the bolide light curve into account. The trajectory is calculated in the "velocity - altitude" variables. For fitting the observed trajectory and the theoretical solutions obtained for different models of meteor body motion with ablation and fragmentation (Zubareva et al., 1999; Kulakov and Stulov, 1992), the method of least squares is used.

1. Principle of Selection of PN-bolides for the Analysis

From the tables (McCrosky *et al.*, 1979), we selected those bolide trajectories for which the highest excess of the photometric mass over the dynamical mass (coefficient k) was indicated in (Popova, 1997). In accordance with this rule, the trajectories of the following seven bolides (McCrosky *et al.*, 1979) were chosen: 38737* (k = 2,5), 39404 (12,1), 39406 (6,8), 39434 (14,8), 40151 (3,3), 40405 (5,0), and 40590 (Lost City, 2,5). As in the quoted works, we indicate the range of possible values of k (in (Kulakov and Stulov, 1992), the values obtained by two different methods are given). The mean value of k was used in calculations. It should be noted that all the values of k are greater (sometimes, much greater) than unity.

2. Trial Functions in the Method of Least Squares

It is assumed that fragmentation of a meteoric body starts inside a luminous segment of its trajectory. Accordingly, as the trial functions, we will use the analytical solutions (Zubareva *et al.*, 1999) obtained on the basis of the sequential fragmentation model. These solutions are:

$$y = \frac{3}{4} \ln \frac{\alpha}{2} + \frac{y_0}{4} - \frac{3}{4} \ln \left(v^{-2/3} - 1 \right),$$ (1)

$$\alpha = \frac{1}{2} c_d \frac{\rho_0 h_0 A_e}{M_e \sin \gamma}, \qquad y_0 = \ln \left(\frac{\rho_0 V_e^2}{\sigma_t} \right).$$

Here, α is the ballistic coefficient and y_0 is the altitude of fragmentation. The trajectory equation is written in dimensionless variables $y = h/h_0$, $v = V/V_e$; where h, V are the altitude and the velocity of the motion, h_0 is the homogeneous-atmosphere altitude, and V_e is the velocity of the body entry into the atmosphere. The other constants are as follows: c_d is the drag coefficient, ρ_0 is the atmospheric density at $h = 0$, A_e is the initial maximum midsection of the body, M_e is the body mass, γ is the angle between the trajectory and the horizon, and σ_t is the body tensile strength.

The parameter y_0 was calculated beforehand. For all bolides, we assumed: $\sigma_t = 10^7$ dyn·cm^{-2}; which is near the strength of carbonaceous chondrite. The parameter α was determined from the condition of the least square deviation of function (1) from the observational data. The minimum of the function:

$$Q(\alpha) = \sum_{i=1}^{n} [v_i - v(y_i, \alpha)]^2,$$ (2)

was determined. Here $v(y_i, \alpha)$ is the function inverse to (1), and v_i, y_i are the bolide velocity and altitude at the points of observation (McCrosky *et al.*, 1979). The value of α was found from the equation:

$$\sum_{i=1}^{n} [v_i - v(y_i, \alpha)] \left(1 + \frac{\alpha_1}{2} e^{-\frac{3}{4} y_i} \right)^{-\frac{5}{2}} e^{-\frac{4}{3} y_i} = 0,$$ (3)

$$\alpha_1 = \alpha \left(\frac{\rho_0 V_e^2}{\sigma_t} \right)^{1/3}.$$

This equation was solved by the Newton method.

3. Results of the Calculations and the Conclusions

In addition to the method described above, the observed trajectory was approximated by an elementary function which describes the motion of a single non-fragmented body

$$y = \ln \alpha - \ln \left(- \ln v \right).$$ (4)

For comparison, below we reproduce the data obtained in (Kulakov and Stulov, 1992) for the same bolides. These data were obtained on the basis of a single body model with the account of ablation. All the parameters obtained in the calculations are given in Table I.

TABLE I

No	PN-bolide	(1)	(4)	article (Kulakov and Stulov, 1992)	
		α	α	α	β
1	38737*	35,7	35,4	19,7	2,22
2	39404	44,7	48,3	30,3	1,10
3	39406A	24,7	27,3	15,7	2,15
4	39434	24,3	28,0	17,6	1,42
5	40151A	20,5	22,2	15,8	0,89
6	40405	58,3	46,2	28,1	2,24
7	40590	15,9	19,3	12,0	1,01

The values of the ballistic coefficient α obtained make it possible to reconstruct the relations $y = y(v)$ along the bolide trajectories and to compare them with the observational data. In almost all cases (except for PN40405), the results of (Kulakov and Stulov, 1992) ensure the best approximation of the observed trajectory, particularly at its lower segment. Probably, it is because ablation is neglected in formula (1). The main result of this study is that the shape of a luminous segment of a bolide trajectory does not display obvious influence of meteoric-body fragmentation.

The work received financial support from the RFBR (project 98-01-00864).

References

Kulakov, A.L. and Stulov, V.P.: 1992, 'Determination of parameters of meteoric bodies from observational data', *Astron. Bull*, **Vol. 26, No 5**, pp. 67–75 (in Russian).

McCrosky, R.E., Shao, C.-Y. and Posen, A.: 1979, 'Bolides of the Prairie Network. 2. Trajectories and light curves', *Meteoritics*, **Issue 38**, pp. 106–156 (in Russian).

Popova, O.P.: 1997, *Determining parameters of large meteoric bodies from observational data*, Ph. D. theses, Institute of Geospheres Dyn (in Russian).

Zubareva, E.N., Podkovkina, N.V. and Stulov, V.P.: 1999, 'Simulation of motion of a destroyed meteoric body in the atmosphere', *Bull. of Moscow State Univ, Math. and Mech.*, **No 3**, pp. 58–62 (in Russian).

ORBIT DETERMINATION OF THE GEOSYNCHRONOUS SPACE DEBRIS KUPON SATELLITE USING ITS CCD OBSERVATIONS WITH 2-METER TELESCOPE AT THE TERSKOL PEAK

S.P. RUDENKO, V.K. TARADY, A.V. SERGEEV and N.V. KARPOV

International Centre for Astronomical, Medical and Ecological Investigations, Golosiiv, 03680, Kiev-127, Ukraine

Key words: orbit determination, geostationary satellite, CCD observations

The CCD and photometric observations of the geostationary Kupon (97070A) satellite were performed using the astronomical complex of the two-meter Ritchey-Chretien-Coude telescope (Tarady, 2000) located at the Terskol peak in the Northern Caucasus at 3127 m altitude. The observations were performed at the different stages of satellite operation, namely, at launch, transition, controlled and libration orbits (Table I).

TABLE I

Summary of CCD observations of Kupon satellite

Time span	Orbit type	Number of nights	Number of observations
Nov. 12–13, 1997	Launch	1	8
Nov. 13 – Dec. 3, 1997	Transition	13	84
Jan. 17 – Mar. 17, 1998	Controlled	29	213
Apr. 18, 1997 – Jun. 22, 2000	Libration	26	212
Totally		69	517

The small size of the CCD camera field ($6' \times 6'$) required the use of the Guide Star Catalogue in astrometric processing of star and satellite coordinates. The root-mean square (rms) errors of the satellite spherical coordinates are 0.5–1.0". The satellite orbit was derived from all the collected observations using the software (Rudenko, 1995) elaborated to process CCD positional observations. The rms errors of the satellite coordinates are 50–200 m at 2–100 day orbital arcs without orbit corrections.

The regular monitoring of Kupon allowed us to detect another geosynchronous satellite Arabsat 1C (92010B) on January 18, 1998 at 4–5' distance from Kupon. Arabsat 1C was moved from 31° E to 55° E about 1–2 days before our "discovery". The regular CCD observations and orbit determination of both satellites allowed us to predict and observe five close encounters of Kupon and Arabsat 1C at 1.3–4.2 km

Pretka-Ziomek et al./Dynamics of Natural and Artificial Celestial Bodies, 373–374, 2001.
© 2001 *Kluwer Academic Publishers.*

which happened during February – March, 1998 (Table II). Such close encounters are dangerous for communication satellites since they can lead to their collision.

TABLE II

Close encounters between Kupon (97070A) and Arabsat 1C (92010B) in 1998

No.	Time span of data used	Encounter instant (UTC)	Distance [km]
1	25–31 Jan.	Feb. 3, 16:13	4.2±1.0
2	28 Feb. – 1 Mar.	Mar. 3, 15:12	1.3±0.3
3	3–7 Mar.	Mar. 4, 15:10	2.3±0.5
4	3–7 Mar.	Mar. 5, 15:00	3.1±0.7
5	3–7 Mar.	Mar. 6, 14:51	4.2±0.8

The Kupon satellite became a space debris object after an on-board failure took place on March 18, 1998. However, its tracking was continued after this event. Using the observations of Kupon obtained on April 18 – July 28, 1998, we derived that the satellite would librate between 55.0° and 94.8° E with the libration period equal to 769 days. On April 19, 2000 Kupon returned to the longitude from which it escaped. The processing of our data, obtained in April, 2000, showed that the minimum longitude of Kupon (54.96° E) was reached at 02:10 UTC on April 19, 2000. A series of encounters at 12–20 km distances between Kupon and Insat 2R (former Arabsat 1C) was predicted and observed by us in April, 2000.

Most of the results presented in the paper were derived almost in the real time. The results of observations were usually available in 3–12 hours after the last observation was obtained. Orbits derived were available in 12–24 hours.

References

Rudenko, S.: 1995, 'Software for the analysis of photographic observations of geosynchronous satellites', In *Proceedings of the Workshop: Accurate orbit determination and observations of high Earth satellites for geodynamics*, (A. Elipe and P. Paquet, Eds.), Walferdange, Luxembourg, pp. 63–70.

Tarady, V.K., Sergeev, A.V., Karpov, N.V., Rudenko, S.P., and Kulik, I.V.: 2000, 'The astronomical complex at the Terskol peak for observations of minor bodies of the Solar System', In *Near-Earth astronomy and the problems of investigations of small bodies of the Solar System*, Collected papers, Moscow, pp. 163–167 (in Russian).

MONITORING OF VARIATIONS OF THE UPPER ATMOSPHERE DENSITY

ANDREY I. NAZARENKO[1], VASILIY S. YURASOV[2], PAUL J. CEFOLA[3], RONALD J. PROULX[4] and GEORGE R. GRANHOLM[5]

[1]*Center for Program Studies, 84/32 Profsoyuznaya ul, Moscow 117810, Russia,*
E-mail: nazarenko@iki.rssi.ru
[2]*Space Research Center "Kosmos", Magadanshaya str. 10-59, Moscow 129345, Russia,*
E-mail: vyurasov@chat.ru
[3]*The Charles Stark Draper Laboratory, Inc., 555 Technologu Square, Cambridge, MA, 02139, USA,*
E-mail: cefola@draper.com
[4]*The Charles Stark Draper Laboratory, Inc., 555 Technology Square, Cambridge, MA, 02139, USA,*
E-mail: rproulr@draper.com
[5]*United States Air Force,*
E-mail: ggranholm@draper.com

Abstract. The concept of development of the system for monitoring of upper atmosphere density variations is considered.

Key words: satellite motion, upper atmosphere density variations

1. Introduction

One of typical problems of space surveillance is the increase of the accuracy of orbit determination and prediction for LEO satellites. The main source of errors in orbit determination and prediction are now the errors related with accounting the atmospheric drag disturbances. The latter errors, in their turn, are caused by the errors in atmospheric density calculation. The modern models of the atmosphere [11], [6], [7], [8], [12] are known to take into account the most essential effects of atmospheric density variations. However, even these models can provide only 10% accuracy under quiet conditions and 20-30% -under solar and geomagnetic disturbed conditions [9]. These models reflect only average tendencies in changes of atmospheric density. They describe only the "climate" in the upper atmospheric layers. But for high-precision prediction of LEO satellite motion the current and predicted "weather" data are necessary. The forecasting of "weather" in the upper atmosphere is a much more difficult problem. The "weather" in the upper atmosphere is formed and varies as a result of energy transfer from the Sun to the interrelated "magnetosphere - thermosphere - low atmosphere" system. The space surrounding the Earth is a highly dynamic environment, that responds sensitively to changes in the electromagnetic radiation, particle flows and magnetic fields coming from the Sun. Depending on the type of such changes, the near-Earth space environment responds with time delays of minutes to days.

The monitoring of the upper atmosphere density variations based on using the satellite orbital data can become the main component of the atmospheric "weather" service now. For determination of atmospheric density variations it is proposed to

Pretka-Ziomek et al./Dynamics of Natural and Artificial Celestial Bodies, 375–384, 2001.
© 2001 *Kluwer Academic Publishers.*

376   ANDREY I NAZARENKO ET AL.

use the orbital data estimates for low-altitude space objects, taken from the Space
Surveillance Systems (SSS) [4]. The number of LEO objects reaches some hundreds at this time. The majority of them are space debris. Their orbital elements
are updated in near-real time several times per day by the Russian and US Space
Surveillance Centers.

2. Methodical Problems

This atmospheric density tracking process includes:

1. The procedures for accumulating the atmosphere density observations from the usual output data of Space Surveillance Systems.
2. The procedure for constructing the atmospheric density variations model.
3. The procedure for estimating the true ballistic factors of satellites employed.
4. The procedures for forecasting the atmospheric density at future times.

The mathematical statement of the problem of estimating the atmospheric density
variations can be formulated in the general form as a problem of minimization of
the functional:

$$\min_{X, \delta\rho} F(X, \delta\rho(h, \varphi, \lambda, t)), \tag{1}$$

where:

$$F(\cdots) = \sum_{i=1}^{n} \sum_{j=1}^{m_i} \delta Z_{ij}^T K_{ij}^{-1} \delta Z_{ij}, \tag{2}$$

$$\delta Z_{ij} = (Z_{ij} - f_j(x_i, \rho_m, \delta_{WS}(h, \varphi, \lambda, t))), \tag{3}$$

$X = (x_1, x_2, \ldots, x_n)$, x_i is the vector of parameters of motion for j^{th} space
object (SO);

Z_{ij} are the measurements for the ith SO corresponding to time t_{ij};

$f_j(\ldots)$ is the functional dependence of measured parameters on motion parameters;

K_{ij} is the covariance matrix of measurements errors;

ρ_m is the density calculated from the atmospheric model;

$\delta\rho(h, \varphi, \lambda, t)$ is the density variation function;

h is altitude φ is latitude, λ is longitude, t is time.

Because of impossibility of using the satellite drag data to allocate the functional
dependence of density variations on latitude and longitude, we shall use in formula (2) for density variations the approximating linear-in-altitude function of the
following kind:

$$\frac{\delta\rho}{\rho_m}(h, t) = b_1(t) + b_2(t)(h - 400)/200. \tag{4}$$

Besides, the multi-dimensional problem of minimization of the generalized functional (1, 2) is reduced, in practice, to more simple one based on functional minimization for each of objects and on organizing the iterative processes of updating the values of approximating parameters of variations and ballistic factors. In our approach we are trying to measure the density variation term $\delta\rho/\rho_m$ at some particular altitude and time for the given "true" ballistic factor k_i and estimated ballistic factor $\hat{k}_{ij}$ which is obtained from the orbit determination as:

$$\frac{\delta\rho}{\rho_m}(h_{ij}, t_{ij}) = \left(\frac{\hat{k}_{ij}}{k_i}\right) - 1. \tag{5}$$

The "true" ballistic factor for each satellite is assumed to be constant and is given by:

$$k_i = \left(\frac{C_D A_x}{2m}\right)_i \tag{6}$$

where C_D is the drag factor, A_x is the cross-sectional area perpendicular to satellite's motion, and m is the mass. Subscript i relates here to the ith satellite, where the satellites are numbered from 1 to n. Some chosen objects, which have constant and precisely known ballistic factors are considered as "standard" satellites. Typically, a "standard" satellite should have a shape close to spherical one, and its size, mass, and dimensionless aerodynamic drag coefficient are supposed to be well known. The remaining objects are termed as "non-standard" satellites. For non-standard satellites the true ballistic factors can be approximated by averaged ones taken over a sufficient time.

In this case our task is reduced to using the density variation measurements for constructing the piecewise constant linear models. The "piecewise constant" term means here, that each linear model is constant with respect to time over its particular span and depends on the altitude only. The linearity of functions allows to solve the problem by using the least-square method. We "attribute" each observed ballistic factor to some particular time and altitude, group the ballistic factors into 3-hour spans and construct the linear models of atmospheric density variations for each time span.

Once the density variation models are calculated, the original atmospheric model plns corrections can be used to estimate the orbit and ballistic factor for any LEO space object. The orbital elements should be rather accurate in such a case, since we are no longer trying to fit the observations to an incorrect model. Besides, since we took into account the limitations in the density model, we are left with the errors in the ballistic factor only. We should, therefore, be able to gain more information on the "true" ballistic factor of a target object and, possibly, the information on its decay rates, mass or shape as well.

The legitimacy of such an approach was confirmed by a series of experiments [10], [1]. These experiments have shown that the accounting of density variations allowed to reduce the level of errors in satellite orbit predictions during geomagnetic storms more than twice.

Some major aspects, related to practical implementation of the proposed technique for estimating the atmospheric density variations, have been investigated under support and with contribution of the Draper Laboratory Company [3], [2], [5]. The fields of investigations in these works and their basic results are considered below.

2.1. STUDY OF THE CORRECTNESS OF DENSITY VARIATION REPRESENTATION AS AN ALTITUDE POLYNOMIAL AND THE IMPROVEMENT OF ALGORITHMS FOR UPDATING THE DENSITY VARIATIONS

The comparison of modeled and calculated values of density fluctuations was carried out for evaluating the quality of the solution of the problem stated. The values of variations were calculated for altitudes of 200, 400 and 600 km at the times of determination of b_{1j} and b_{2j} factors for three solar activity levels over the 50-day interval and for three versions of input data modeling. (Version 1 corresponds to "variations with measurement errors and ballistic factor's variability", Version 2: to "variations without measurement errors, but ballistic factor's variability", and Version 3 - to "pure variations"). For illustration, Figure 1 presents the modeled and calculated values of density variations for the Version 1 of modeling for the 600 km altitude. The generalized statistical data, which characterize the accuracy of determination of variations by using the proposed technique for a medium solar activity level, are presented in Table I. The data for Version 3 in Table I allow to judge, how valid is the choice of a linear function for approximating the altitude dependence of atmospheric density variations. It is seen that the maximum values of RMS errors for this version do not exceed 3%, the mathematical expectancies are also 3% or less. The largest errors, as it should be expected, take place at

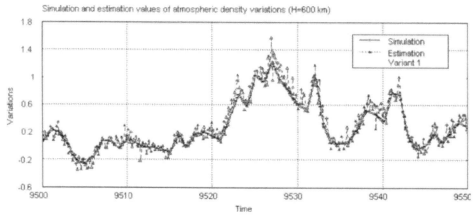

Figure 1. Simulated and estimated values of the atmosphere density variations at 600 km for the medium level of solar activity.

TABLE I

The relative errors of density variation determination

H,	Version 1		Version 2		Version 3	
km,	Mean	RMS	Mean	RMS	Mean	RMS
200	−0.010	0.047	−0.009	0.040	−0.027	0.024
400	0.006	0.041	0.006	0.039	0.007	0.006
600	−0.034	0.093	−0.036	0.088	−0.015	0.013

extreme points of the approximation interval and correspond to maximum values of atmospheric density variations (the medium solar activity level).

The data for the second version characterize the influence of the variability of space objects' ballistic factors on the accuracy of estimations obtained. The maximum values of RMS errors in density variation estimates are 8.8%, and the mathematical expectation is 3.6% in this case. Thus, the modeled variability of ballistic factors has resulted in nearly 3-fold increase of estimation errors.

The results corresponding to the first version characterize the degree of influence of the error in ballistic factor determination and their variability on the accuracy of estimations. These results are seen to insignificantly differ from the second version results.

2.2. IMPROVEMENT OF THE ALGORITHMS FOR BALLISTIC FACTOR ESTIMATION

For estimating the density variation we should accurately know the "true" ballistic factor k_i and its variability in the form of σ_i^2 for each of n satellites. However, this is usually not the case for non-standard satellites, which comprise the majority of space objects used in the atmosphere correction service. Instead, we must use some kind of estimation process to find the tim-averaged versions of these quantities, which are expressed as $\bar{k}_i$ and $\bar{\sigma}_i^2$. The fundamental assumption we rest upon in the derivations is that the modeled density correctly describes, on the average, the true density:

$$E[\rho_i - \rho_{m,i}] = E[\delta\rho_i] = 0. \tag{7}$$

Here E is the expectation operator and the terms written inside the brackets are considered to be random variables. Using Eq. (5), we can see that:

$$E\left[\frac{\hat{k}_{ij}}{k_i} - 1\right] = 0 \Rightarrow E[\hat{k}_{ij}] = k_i. \tag{8}$$

We have updated here the methods for obtaining $\bar{k}_i$ and $\bar{\sigma}_i^2$ [3]. In particular, the procedure for eliminating systematic errors as a function of altitude was offered.

The distribution of both standard and non-standard satellites over the maximally possible range of altitude is preferable from the viewpoint of providing the better accuracy. It was found that the recommended time interval for ballistic factors updating, which provides satisfactory accuracy, is about of solar rotation period, or 27 days.

2.3. STUDY OF THE POSSIBILITIES OF USING THESE ALGORITHMS WITH THE REAL DATA

The modeling results have shows that each 3-hour span should contain 60–90 ballistic factor estimations at least. To ensure such a quantity of measurements it is proposed:

- to choose at least 50-60 frequently tracked objects, which have sufficiently diverse inclinations, eccentricities, and perigee heights between 200 and 600 km;
- to include all known standard LEO satellites into this group.
- to organize updating of orbits of selected satellites over all observable revolutions;
- to organize accumulation of El-sets of the selected group of satellite;
- to renew the group of satellite in accordance with their decay.

2.4. STUDY OF AMPLITUDE-PHASE CHARACTERISTICS OF ERROR IN BALLISTIC FACTOR ESTIMATIONS

In using expression (5) the natural question arises: to which time instant and altitude is the $\delta\rho/\rho_m$ estimation "attributed"? In "attributing" the estimates to a particular altitude one should take into account the fact, that the atmospheric density sharply decreases with altitude. Therefore, it is reasonable to choose the minimum altitude of given satellite's orbit as basic one. The inaccuracy of such an altitude determination is assumed to be insignificant hereafter.

The question of the "attributing" of the $\delta\rho/\rho_m$ estimate is more complicated. The estimate of k was obtained as a result of processing the measurements over some time interval (t_{min}, t_{max}). Here t_{min} and t_{max} are, respectively, the beginning and end of the measurement interval used for orbit updating. The estimate k reflects the drag of the given satellite throughout the time interval, i.e. it represents an averaged estimate. It is determined by modeling that the amplitude phase distortions of atmospheric drag estimates, obtained by the least squares method and Kalman filter, are similar. It was also found that the optimum time of attributing the drag estimates differs from the instant of the last measurement. The smaller the time interval, the more time-localized the estimate k. However, for a smaller data measurement-processing interval the accuracy of estimates of variations can become unsatisfactory (see Figure 2). The dependence of the optimum attributing time and estimate accuracy on the mean drag level was found. For the specified accuracy of measurements and heli-geomagnetic conditions the recommended data

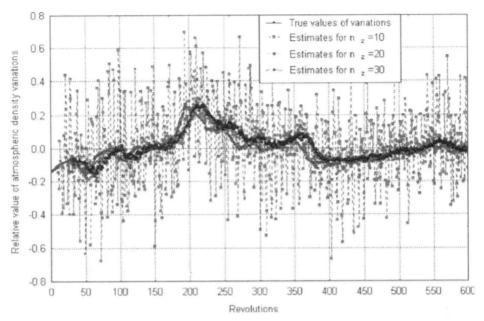

Figure 2. Estimation of variations for $n_z = 10, 20, 30$ revolutions.

pracessing interval was found to be equal to about 30 revolutions. The accuracy of estimates can be slightly improved by increasing the processing interval. However, the increase of the "lag" in estimates of variations can essentially lower their significance from the viewpoint of using them during the operative processing of measurements from ground sensors and increasing of the prediction accuracy of LEO satellites.

3. The Monitoring System Concept

The concept of a system for monitoring the upper atmosphere density variations, based on using the satellite orbital data, is presented in Figure 3. To use the obtained variations in practice it is necessary that, first, the monitoring be implemented continuously; second, the accuracy of estimations be maximum, and, third, the time lags in estimation of variations be minimum. To satisfy all these requirements, a number of technical and organizational problems must be solved.

First, the most important thing is the solution of the problem of monitoring system's information maintenance with the current orbital data. For these purpose it is expedient to join the information capabilities of US and Russian SSS. The results of some joint activities testify that these systems supplement each other from the information viewpoint. Potentially, the joint operation of these systems will allow to observe each LEO space object over all day-time revolutions. However, in order that this potential capability be implemented into practice, it is required

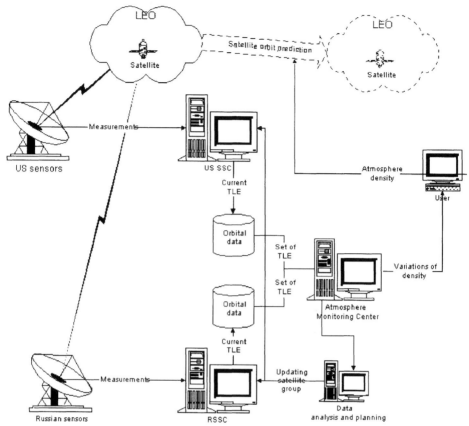

Figure 3. System of monitoring of the upper atmosphere density variations.

for the Russian SSS to solve the problems of accumulation and assuring the operative access of users to the obtained data. As to the US SSS it is expedient to change the adopted routine of orbits updating for the chosen objects and to organize the process of observation and updating these orbits over all observable revolutions.

Second, it is necessary to provide the access of users to the orbital data for estimating density variations with minimum time delay. This problem can certainly be solved at the modern level of telecommunication technology. However, some special arrangements are necessary to provide combining of the orbital data from the US and Russian Space Surveillance Systems in such a manner.

Third, the regular activity on the analysis of the structure and updating of LEO satellite constellation, used for estimating density variations, is required, namely: 1) the standard satellites must be chosen; 2) it is necessary to provide that the orbits of chosen space objects to represent the whole altitude range; 3) in accordance with decaying space objects, it is necessary to supplement the constellation with new objects and to organize collection and processing of the information; 4) in case of detecting some "anomalous" objects in a constellation (for example, active ones) it is necessary to exclude them from a constellation or to execute selective data processing. All this activity should be done simultaneously with estimations of atmospheric density variations at the Atmosphere Monitoring Center.

Fourth, it is necessary to correlate the format of presented orbital data. Apparently, the Two Line Elements (TLE) format could be used in this case. The procedures for correct recalculation of the Russian data in the TLE have already been completed.

Apparently, the problems accompanying practical implementation of a system for monitoring density variations are not exhausted with the above list. However, one should not think that the solution of all these problems is an indispensable condition for system implementation. Many of them influence the efficiency of obtaining the estimates and their accuracy only.

4. Conclusions

The approach to increasing the accuracy of orbit determination and prediction for LEO satellites is the organization of the upper atmosphere monitoring function. This one should be an analog of the weather service for the lower atmosphere. The upper atmosphere monitoring, based on the available satellite atmospheric drag data for all catalogued LEO satellites offers a low cost approach to this capability. These data are operationally updated in the US and Russian SSS as a result of regular satellite observations. Due to support from the Draper Laboratory Company side, very successful activity on perfecting the technique for monitoring of the upper atmosphere and testing its model was accimplished in 1997–1999.

References

1. Anisimov, V.D., Bass, V.P., Commissarov, I.N., Kravchenko, S.N., Nazarenko, A.I., Rychov, A.P., Sych, V.I., Tarasov, Y.L., Fridlender, O.G., Schackmistov, V.M. and Yurasov, V. S.: 1990, 'Results of research of the aerodynamic characteristics and upper atmosphere density with help of passive satellites 'PION", *Supervision of Artificial Celestial Bodies , USSR Academy of Sciences Astrocouncil*, **#86**.
2. Cefola, P.J., Nazarenko, A.I. and Yurasov,V.S.: 1999, 'Refinement of Satellite Ballistic Factors for the Estimation of Atmosphere Density Variations and Improved LEO Orbit Prediction', Paper AAS 99-203, Presented at *AAS/AIAA Space Flight Mechanics Meeting, Feb. 1999*.
3. Cefola, P.J., Nazarenko, A.I. and Yurasov, V.S.: 1999, 'Neutral Atmosphere Density Monitoring Based on Space Surveillance System Orbital Data', Presented at *AIAA/AAS Astrodynamics Specialist Conference, August 1999*.
4. Gorokhov, Yu.P. and Nazarenko A.I.:1982, 'Methodical questions of building the model of atmosphere parameters fluctuations', *Nablgyudeniga Isk. Sputn.*, *No 80*, Astronomical Council of the USSR Acad. Sci., Moscow.
5. Granholm, G.R., Proulx, R.J., Cefola, P.J., Nazarenko, A.I. and Yurasov V.S.: 1999, 'Near-real Time Atmospheric Density Correction Using NAVSPASUR Fence Observations', Paper AAS 00-179, Presented at *AAS/AIAA Space Flight Mechanics Meeting, Feb. 1999*.

6. Hedin, A.E.: 1987, 'MSIS-86 Thermospheric Model', *Journal of Geophysical Research*, **vol. 92**.

7. Jacchia, L.G.: 1977, 'Thermospheric temperature, density and composition: new models', *SAO Special Report*, **#375**.

8. Justus, C.G., Jeffris, W.R., Yung, S.P. and Johnson, D.L.: 1995, 'The NASA/MSFC Global Reference Atmospheric Model-1995 Version (GRAM-95)', *NASA Technical Memorandum 4715*.

9. Marcos, F.A.: 1990, 'Accuracy of Atmosphere Drag Models at Low Satellite Altitudes', *Adv. Space Research*, **10**.

10. Nazarenko, A.I., Amelina, T.A., Gukina, R.V., Kirichenko, O.I., Kravchenko, S.N., Markova, L.G., Tumolskaya, N.P. and Yurasov, V.S.: 1988, 'An Estimation of Efficiency of Various Ways of Atmosphere Density Variations Accounting at Satellite Motion Prediction During Geomagnetic Storms', *Supervision of Artificial Celestial Bodies*, **#84**, USSR Academy of Sciences Astrocouncil.

11. COSPAR: 1972, 'International Reference Atmosphere 1972 (CIRA-72)', Akademic verlag, Berlin 1984, 'The upper atmosphere of the Earth. Model of density for ballistic maintenance of the Earth artificial satellite flights', GOST 25645.115-84., Moscow, Publishing House of the Standards.

INFLUENCE OF THE LOCAL TROPOSPHERE ON GPS MEASUREMENTS

KRYSTYNA KURZYŃSKA and ALINA GABRYSZEWSKA

Astronomical Observatory, A. Mickiewicz University Słoneczna 36, 60-286 Poznań, Poland,
E-mail: kurzastr@amu.edu.pl

Abstract. It is demonstrated that no global model of the wet tropospheric zenith delay is able to reproduce the value obtained directly on the basis of aerological sounding with comparable accuracy. The spectacular artifacts resulting from the global model application are apparent seasonal variations of the GPS station altitude with respect to the reference ellipsoid. The method for the construction of a local wet tropospheric zenith delay model, based on a sufficiently representative sample of local aerological data, is proposed. Its systematic error does not exceed that of instantaneous local wet tropospheric zenith delay determination from the aerological data (0.5 cm) and reduces the apparent seasonal variations of altitude. The model is very convenient in every day use: it is sufficient to multiply the value of the actual refractivity at the ground level by a certain factor determined for a given month.

Key words: GPS – wet tropospheric zenith delay – local model – station apparent altitude variations

1. Introduction

An important motivation for writing this paper has been an analysis of changes in the coordinates of the Astrogeodynamical Station in Borowiec (Poland) defined with the help of GPS technique. They were determined on the basis of the coordinates calculated by Figurski (1998) with the help of BERNESE 4.0 program using measurements from 26 GPS stations. An intriguing effect has been a significant decrease of the coordinate H in summer (Lehmann and Figurski 1999). Today the accuracy of GPS measurements is very high so these changes of the altitude can be explained only as due to incorrect account of the effects coming from the atmosphere (Gendt and Beutler 1995). To check if our supposition is true we have compared several models of the wet tropospheric zenith delay (WTZD) determination as well as the CODE (Center of Orbit Determination in Europe) data with the instantaneous values of WTZD obtained on the basis of local aerological data. In fact, the aerological data were collected at the Aerological Station in Poznań, twenty kilometers from Borowiec, but, because no significant changes of properties of the higher atmospheric layers occur over such short distances, it is possible to use these data for Borowiec too. The subject of the study is the wet part of the tropospheric zenith delay only, as many authors (e.g. Davis et al. 1985) have already proved that its dry (hydrostatic) part is determined accurately enough. Also, we do not consider the mapping function for non-zero zenith distances because the

Pretka-Ziomek et al./Dynamics of Natural and Artificial Celestial Bodies, 385–394, 2001.
© 2001 *Kluwer Academic Publishers.*

precision of determination of the tropospheric delay at any zenith distance depends first of all on the accuracy of the zenith delay determination and there are very accurate mapping functions (e.g. Niell 1996) which can be used.

The organization of the paper is as follows. First we describe the way of the WTZD calculation on the basis of aerological data and the way of statistical error determination in the models considered further on. Next the insufficiency of the conventional as well as CODE global models of WTZD is proved. Finally, we propose the optimum, we hope, method of local WTDZ model construction.

2. Calculation of Wet Tropospheric Zenith Delay on the Basis of Local Aerological Data

The actual wet tropospheric delay can be determined only on the basis of actual local aerological data. We had at our disposal the aerological data from Poznań Aerological Station where soundings were made regularly twice a day (at noon and at mightnight) up to the end of 1989. Each sounding was made at the air pressure levels corresponding approximately to the heights 1.5, 3.0, 5.6, 7.2, 9.1, 11.7, 13.6, 16.2, 18.5, 20.6 and 23.9 km and gave the information about the air temperature and the dew point temperature at these levels. The accuracy of the measurements was ± 1 m for height and ± 0.2 C for temperature. The water vapour pressure was calculated from the dew point temperature. From all the series of sounding we have chosen a ten-year period to make a vast representative sample.

The WTZD is described by the integral (Davis et al. 1985)

$$S = 10^{-6} \int_0^{h_t} dh N(h),$$
(1)

where N and h are the wet part of refractivity and the height about the ground level, respectively, and h_t denotes the wet troposphere height. We have found that the optimum model functional dependence of the refractivity on height is of the form

$$N(h) = N(0) \exp(-ah - bh^2 - ch^3),$$
(2)

where $N(0)$ is the wet part of the refractivity at the ground level and a, b and c are some constants obtained by the Least Square Method applied to fit Eq. (2) to particular sounding data. Eq. (2) describes very well the data of all the soundings, even in the extreme cases, such as inversion of temperature, small or large temperature gradient and the cases when the decrease of water vapour pressure with height highly deviates from the exponential function. Examples of such fits are given in Fig. 1. The advantage of Eq. (2) is that it is free from any constrains either of a constant gradient of temperature or a concrete model of water vapour pressure distribution in troposphere.

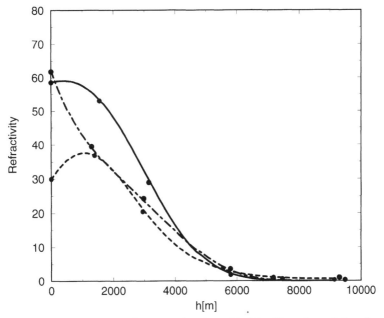

Figure 1. Examples of fits of refractivity as a function of height (2) to the real data for a typical case (dot-dashed line), a case with temperature inversion (dashed line) and a case with a very small temperature gradient within low-lying layers of atmosphere (solid line). The black circles denote experimental points determined from chosen aerological soundings.

From Fig. 1 it is immediately seen that in many cases an exponential function simpler than Eq. (2) does not describe the real instantaneous atmosphere accurate enough. We calculate S by numerical integration with such an integration step that the final accuracy is better than 0.1 cm. The final accuracy of the calculation of the tropospheric zenith delay depends not only on the integration step but also on the accuracy of the refractivity determination. The surface temperature is measured with an accuracy ± 0.1 C and the precision of determination of refractivity on the Earth surface is better than that determined from aerological data. We have estimated the resultant error of WTZD determination to be ± 0.5 cm.

Our ten-year sample contains more than 7300 individual soundings (both at night and during the day). For all these soundings, we have calculated WTZD to which we further refer as instantaneous WTZD and denote as S_{inst}. We use these values as the reference true values of WTZD and compare them with the values obtained with the help of various models considered. In order to determine the systematic and stochastic errors of particular models we perform a month statistics calculating the differences

$$\Delta S_m(i) = S_m(i) - S_{inst}(i), \tag{3}$$

where the index m distinguishes between different models and the integer variable i runs over all the days and nights of a given month during all 10 years ($n \approx 600$). The mean taken over the n-element sample

$$\overline{\Delta S_m} \equiv \frac{1}{n} \sum_{i=1}^{n} \Delta S_m(i) \tag{4}$$

is a measure of the systematic error and the square root of the empirical variance

$$\sigma_m \equiv \sqrt{\frac{1}{n} \sum_{i=1}^{n} \left[\Delta S_m(i) - \overline{\Delta S_m}\right]^2} \tag{5}$$

is a measure of the stochastic error.

3. Global Models of Wet Tropospheric Zenith Delay

There are many simple global models and formulas for the calculation of WTZD (e.g. Hopfield 1971, Saastamoinen 1972, Lanyi 1984, Davis et al. 1985, Baby et al. 1988, Mironov et al. 1993; we do not consider the global models for observers having at their disposal no data concerning the instantaneous weather conditions at the place of observation, e.g. Collins and Langley 1999, Hay and Wong 2000). The values of WTZD obtained with the help of these models practically coincide (they differ no more then a few millimeters) so to perform a comparison with S_{inst} we have chosen a simple formula given by Saastamoinen (1972):

$$S_{gl} = 0.002277(1255.0/T + 0.05)e, \tag{6}$$

where e is water vapour pressure in mb, T is temperature in K.

Eq. (6) includes only the meteorological data at the ground level at the site of observation. We calculated with the help of Eq. (6) the values for the whole 7300 element ten-year sample and compared them with S_{inst}, Eq. (1). In Fig. 2 we present the result of the comparison for a winter month (a typical January) and a summer month (a chosen typical July). It is seen that the simple model (6) renders the character of variation of the instantaneous values of WTZD but, practically, in more than 90% the values provided by it are *lower* than those calculated on the basis of aerological data. The differences vary with seasons and are especially large in summer months. This observation is confirmed in Fig. 3, where the systematic and the stochastic errors of model (6) are given, calculated for all succeeding months in a year on the basis of data collected for individual days within a given month over all ten years. Though the systematic error is significantly lower than the stochastic one, it displays a distinct year-wave behaviour which we suppose is the reason for seasonal oscillations of the station altitude mentioned in the Introduction.

Presently, at our disposal are also the values of WTZD determined by the Center of Orbit Determination in Europe (CODE). The center publishes the values of total tropospheric zenith delay for each two hours on every day for every GPS

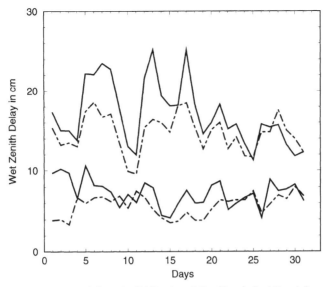

Figure 2. Month WTZD course of S_{inst} (solid lines) and S_{gl} (dot-dashed lines) for a chosen typical January (lower curves) and July (upper curves).

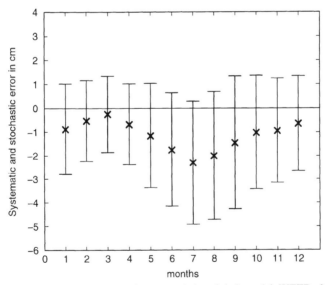

Figure 3. Systematic plus minus stochastic error of the global model WTZD determination in succeeding months.

station. We wanted to check if these values would fit S_{inst} better than S_{gl}, Eq. (6). Unfortunately, we are not able to compare directly the values given by CODE with those of S_{inst} because we do not have the aerological data for the years later than 1989. However, we can compare the CODE data with the values of S_{gl} which we already know to be certainly *lower* than the values of S_{inst}.

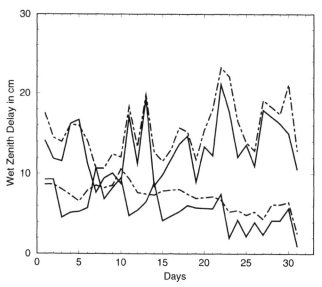

Figure 4. Month course of CODE WTZD (solid lines) and global WTZD, S_{gl} (dot-dashed lines) for January 1998 (lower curves) and July 1998 (upper curves).

The CODE data taken for the comparison were chosen for every day at noon. In order to get WTZD we subtracted from the values of the total tropospheric zenith delay given by CODE its hydrostatic part given by the formula

$$S_h = 0.002277 P F(\varphi, H)^{-1}, \qquad (7)$$

where P denotes the air pressure on the Earth surface in mb and the function F is:

$$F(\varphi, H) = 1.0 - 0.0026 \cos(2\varphi) - 0.00028 H, \qquad (8)$$

where φ is the latitude of the observational station and H the height above geoid in km. A comparison of the CODE WTZD with the values of S_{gl} calculated with the help of Eq. (6) (in the present case we have assumed, exceptionally, the meteorological conditions at the site of observation for Borowiec Station and not for the aerological station) was made for the whole year 1998. Fig. 4 presents the results for January 1998 and July 1998, which were typical months from the point of view of the weather conditions. As follows, generally, the values of the CODE WTZD are *still* lower than the values of S_{gl}, so we are not entitled to expect that the CODE estimations can be better approximations of S_{inst} than S_{gl}.

4. Local Model of Wet Tropospheric Zenith Delay

Taking into account the result of the previous section, we can draw a conclusion that no global model is accurate enough to eliminate completely the influence of troposphere on GPS measurements. In such a situation we have decided to make an attempt at finding a local model of atmosphere in a similar way as it was made for optical astronomical observations (Kurzyńska 1987, 1988). To reach this goal two kinds of averages have been considered:

- the average over a given month within the whole ten-year period – the Month Average (MA),
- the average over the whole ten-year period – the Climate Average (CA).

We have averaged not the aerological data but the values of parameters a, b and c which determine the refractivity dependence on height. According to Eq. (2) such an approach implies that the refractivity value at the ground level $N(0)$ influences as a global factor the refractivity value at any height. Consequently, one can rewrite integral (1) in the form analogous to Eq. (6) as the product

$$S_m = N(0) K_m \qquad (9)$$

with a factor

$$K_m = \int_0^{h_t} dh \exp -a_m h - b_m h^2 - c_m h^3. \qquad (10)$$

The index m distinguishes between different ways of averaging of the parameters. The values of factor K for particular month averages as well as for the climate average are given in Table 1. For the sake of comparison, we quote also the value of K for the global model of WTZD, Eq. (6). The WTZD calculated on the basis of such averaged parameter sets will be denoted as S_{MA^*} and S_{CA^*}, respectively. In calculating both WTZDs for a given day we have assumed the same value of the refractivity $N(0)$ at the ground level as in the calculations of the instantaneous WTZD, S_{inst}.

In Fig. 5 the day by day variation of both S_{MA^*} and S_{CA^*} are given for a chosen typical January and July and compared with the variation of the instantaneous

TABLE I

Values of coefficient K in cm for the particular months local model, climate local model and global model of WTZD

January	.2066	July	.2455
February	.2082	August	.2399
March	.2017	September	.2237
April	.2167	October	.2183
May	.2291	November	.2146
June	.2389	December	.2114
Climate	.2258	Global	.2190

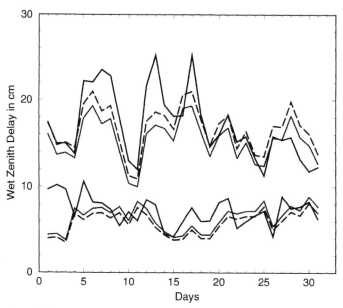

Figure 5. Month WTZD course of S_{inst} (solid lines), S_{MA^*} (dashed line) and S_{CA^*} (dotted lines) for a chosen typical January (lower curves) and July (upper curves).

WTZD, S_{inst}. It is seen that both models render the character of the variation of the instantaneous values of WTZD quite well, much better than S_{MA} and S_{CA}. Similarly as for the global model we calculated the systematic and stochastic errors for the S_{MA^*} and S_{CA^*} models using the whole ten-year material. In Fig. 6 the systematic error of both models for succeeding months is compared with that for the global model; the stochastic error is practically of the same value as for the global model (cf. Fig. 3). The value of systematic error for the S_{MA^*} model is comparable with that of the error of instantaneous WTZD determination from aerological data (±0.5 cm). As formulation of any other model with a smaller value of the stochastic error (about ±2.5 cm) can hardly be expected, the S_{MA^*} model can be accepted for application in analysis of routine measurements. The S_{CA^*} model, on the other hand, still displays a year-wave behaviour, though of a lower amplitude than the global one, and that is why we have rejected it.

5. Conclusions

After a comprehensive analysis we recommend a construction of a local model of WTZD determination following the rules presented in this paper. In our opinion, such a model could be shared by each group of GPS stations in IGS network lying close enough to have a similar type of climate. When constructing the model one should pay attention to truly representative character of the statistical sample of aerological data. The model is very convenient in every day use: it is sufficient

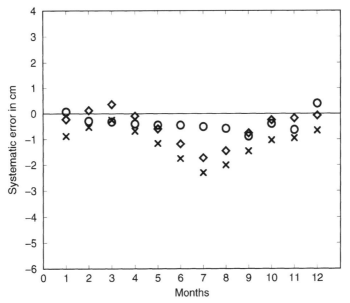

Figure 6. Systematic error of the MA^* model (circles), the CA^* model (diamonds) and, for comparison, the global model (crosses) of WTZD determination in succeeding months.

to multiply the value of actual refractivity at the ground level by a certain factor determined for a given month. In the case of GPS Borowiec Station considered the model reduces the apparent seasonal variations of altitude observed.

Acknowledgements

We thank Marek Lehmann and Gerd Gendt for discussions. The study has been supported in part by the Polish State Committee for Scientific Research (project 9 T12E 018 10).

References

Baby, H. B., Gole, P. and Lavergnat, J.: 1988, 'A model of the tropospheric excess path length of radio waves from surface meteorological measurements', *Radio Science* **23**, pp. 1023–1038.

Collins, P. and Langley, R. B.: 1999, 'Prediction for the WAAS user', *GPS World* **10, July**, pp. 52–58.

Davis, J. L., Herring, T. A., Shapiro, I. I., Rogers, A. E. E. and Elgered, G.: 1985, 'Geodesy by radiointerferometry: Effects of atmospheric modeling errors on estimates of baseline length', *Radio Science* **20**, pp. 1593–1607.

Figurski, M.: 1998, *WUT Analysis Raport*, **No. 938–990**.

Gendt, G. and Beutler, G.: 1995, 'Consistency in the troposphere estimations using the IGS network'. In *Proceedings IGS Workshop, Special topics and new directions* (Gendt G., Dick G., Eds.), Potsdam, pp. 115–127.

Hay, C. and Wong, J.: 2000, 'Tropospheric delay prediction of the Master Control Station', *GPS World* **11, January**, pp. 56–62.

Hopfield, H. S.: 1971, 'Tropospheric effect of electromagnetically measured range: Prediction from surface weather data', *Radio Science*, **6**, pp. 357–367.

Kurzyńska, K.: 1987, 'Precision in determination of astronomical refraction from aerological data', *Astron. Nachr.* **308**, pp. 323–328.

Kurzyńska, K.: 1988, 'Local effects in pure astronomical refraction', *Astron. Nachr.* **309**, pp. 57–63.

Lanyi, G.: 1984, 'Tropospheric delay affecting radio interferometry', *TDA Progress Report*, pp. 152–159.

Lehmann, M. and Figurski, M.: 2000, 'Local variations of selected GPS permanent stations', *Artificial Satellites* **35**, pp. 9–23.

Mironov, N., Linkwitz, K. and Bahndorf, J.: 1993, 'Wet component of tropospheric delay for microwaves from surface meteorological data', *Artificial Satellites* **28**, pp. 151–170.

Niell, A. E.: 1996, 'Global mapping functions for the atmosphere delay at radio wavelengths', *Journal of Geophysical Research* **101**, pp. 3227–3246.

Saastamoinen, J.: 1972, 'Atmospheric correction for the troposphere and stratosphere in radio ranging of satellites', *Geophysical monograph 15* (ed. Henrisken), pp. 245–251.

ORBIT DETERMINATION USING SATELLITE GRAVITY GRADIOMETRY OBSERVATIONS

ANDRZEJ BOBOJĆ
University of Warmia and Mazury, Institute of Geodesy ul. Oczapowskiego 1, 10-950 Olsztyn, Poland, E-mail: altair@uwm.edu.pl

ANDRZEJ DROŻYNER
University of Warmia and Mazury, Institute of Geodesy, ul. Oczapowskiego 1, 10-950 Olsztyn, Poland, E-mail: drozyner@uwm.edu.pl

Abstract. In the first part of this work we present a short description of the orbit determination method using measurements of the gravity gradient tensor components. These components are the functions of the gravity field coefficients and the gradiometric satellite position. It allows to make a system of the linear observation equations where the influence of the gravity field coefficients is neglected. Using the least squares adjustment method the normal equations can be obtained. Next, using these equations, the corrections to the initial state dynamic vector components are estimated. The corrected initial state dynamic vector enables determining of the more accurate satellite orbit by means of the numerical integration method.

The work second part contains the simulation results of the orbit improvement process using the simulations of the gravity tensor measurements. The various variants of the computation results are compared and some practical conclusions are given.

1. Introduction

The Satellite Gravity Gradiometry (SGG) is an observation technics, which enables determining of the second spatial derivatives of the geopotential. In this method the gravity tensor components (i.e. second spatial derivatives of the geopotential) are measured by means of the gradiometer, located on the satellite board (Rummel, 1988).

The gravity tensor can be presented in the 3×3 matrix form:

$$\mathbf{T} = \begin{bmatrix} T_{xx} & T_{xy} & T_{xz} \\ T_{yx} & T_{yy} & T_{yz} \\ T_{zx} & T_{zy} & T_{zz} \end{bmatrix}, \tag{1}$$

where: $\mathbf{T}$ — gravity tensor, $T_{kl}(k, l = x, y, z)$ – gravity tensor component, for example: $T_{xx} = \frac{\partial^2 V}{\partial x^2}$, $T_{xy} = \frac{\partial^2 V}{\partial x \partial y}$, $T_{yx} = \frac{\partial^2 V}{\partial y \partial x}$, etc.

The Satellite Gravity Gradiometry mission is planned in the nearest future (in the perspective up to five years).

Pretka-Ziomek et al./Dynamics of Natural and Artificial Celestial Bodies, 395–400, 2001.
© 2001 *Kluwer Academic Publishers.*

2. Research of the Application of the Satellite Gravity Gradiometry Observations to Determining the Earth Artificial Satellites Motion

The results of the future Satellite Gravity Gradiometry mission can be important for the Geosciences as well as in Space Physics and Celestial Mechanics. Thus the purpose of this work is to analyse the possibilities of the satellite orbit improving by means of the SGG measurements (i.e. gravity tensor components). This orbit improvement can be performed using the observation eqations in the following form:

$$T_{ij}^{obs} - T_{ij}^{c} = \frac{\partial T_{ij}^{c}}{\partial \mathbf{r}} \frac{\partial \mathbf{r}}{\partial (\mathbf{r}_0, \dot{\mathbf{r}}_0)} \begin{bmatrix} \Delta \mathbf{r}_0 \\ \Delta \dot{\mathbf{r}}_0 \end{bmatrix}. \tag{2}$$

In this equation: T_{ij}^{obs} – measured gravity tensor component (simulated using gravity field coefficients order and degree up to 360 - EGM96 model), T_{ij}^{c} – approximated gravity tensor component computed by means of the orbit improvement process using gravity field coefficients order and degree less than 360, $\frac{\partial T_{ij}^{c}}{\partial \mathbf{r}}$ – third spatial derivatives of the geopotential computed in the orbit improvement process, $\mathbf{r}$ – satellite position vector, $\frac{\partial \mathbf{r}}{\partial (\mathbf{r}_0, \dot{\mathbf{r}}_0)}$ – partial derivatives of the satellite position vector with respect to the initial dynamic state vector of the satellite. These derivatives are computed by means of the variational equations in the orbit improvement process, $\Delta r_0 = [\Delta x_0 \quad \Delta y_0 \quad \Delta z_0]^T$ – estimated correction vector to the components of the initial position vector of the satellite, $\Delta \dot{r}_0 = [\Delta \dot{x}_0 \quad \Delta \dot{y}_0 \quad \Delta \dot{z}_0]^T$ – estimated correction vector to the components of the initial velocity vector of the satellite.

Taking the observation equation in the presented form the observation equation system can be made. Then, using this observation equation system and the least squares method, the corrections to the components of the initial dynamic state vector of the satellite are estimated. It allows to correct the initial dynamic state vector of the satellite. Thus the improved satellite orbit is obtained by means of the Cowell numerical integration method.

In this work the performed researches consisted of several steps:

First Step: in this phase the measurements of the gravity tensor components were simulated using the EGM96 model (Earth Geopotential Model 96 order and degree up to 360) (Lemoine et al., 1998), (Metris et al., 1999).

Second Step: in which the corrections to the initial dynamic state vector were estimated using Satellite Gravity Gradiometry (SGG) observations. It was carried out in the TOP - SGG programm (Drożyner, 1995) by means of the least squares method algorithm. Next, the initial dynamic state vector was corrected.

In this phase of the research the various solution variants were obtained. These solution variants contained the corrected components of the initial dynamic state vector and their mean square errors. Determination of the various solution variants was possible thanks to the several modifications, which were carried out in the initial data of the orbit improvement process. These modifications included:

ORBIT DETERMINATION USING SGG OBSERVATIONS

- changes of the maximum order and degree of the gravity field coefficients, which were used in the estimation process,
- changes of the reference frame, where the measurements were simulated. These frames correspond the inertial and Earth – pointing satellite attitude. The first frame axes are parallel to the CIS (Conventional Inertial System at standard epoch J2000.0) reference frame axes. Its origin is located in the satellite mass center. The second frame called the orbital frame has the origin in the satellite mass center too. The Z axis of the orbital frame is always pointing along radial direction, the Y axis is perpendicular to the orbital plane, pointing along the angular momentum vector and the X axis there is in an orbital plane – completes the frame to the right – handed frame,
- taking different measured gravity tensor components and different total number of them to the orbit improvement process (in other words the different length of the orbital arc along which the gravity tensor components were simulated),
- taking three different orbits to the estimation of the initial state vector corrections. The taken orbits were almost circular (eccentricity is 0.001) and almost polar (inclination is 88 deg.). Their altitudes were equal about 160, 200 and 240 km,
- changes of the interval between the succesive gravity tensor measurements. These values were equal 30, 60 and 120 sec.,
- changes of the simulated observations by means of the random errors with standard deviations 10^{-1}, 10^{-2}, 10^{-3} and 10^{-4} E.U. ($1 E.U. = 10^{-9}$ sec.$^{-1}$). The errors had the normal distribution.

Third Step: in this last phase the analysis and the accuracy comparison of the computed variants were made. In that comparison the mean square errors (standard deviations) of the improved initial dynamic state vector components were taken into account.

3. Results and Conclusions

The examples of the accuracy comparison of the chosen solution variants (the solutions of the orbit improvement process) are showed on Figure 1 and 2. Figure 1 presents the mean square errors of the improved initial position vector components according to the lenght of the orbital arc or the total number of observations which were simulated along it. The initial data of the estimation process are the following: the 110×110 gravity field model (P:110, see Figure 1), full gravity tensor observation, interval between succesive gravity tensor observations (observation interval) equals 60 sec. and near circular improving orbit (altitude equals about 200 km, inclination 88 deg.). The relative best accuracy of the improved orbit results from this diagram for the one day orbital arc. Above this value the accuracies decrease. It may be caused by the cumulation computation errors effect.

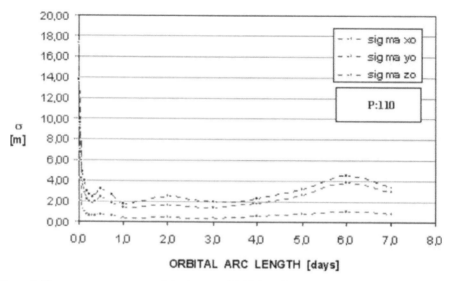

Figure 1. The mean square errors of the improved initial position vector components according to the orbital arc length. The points on the diagram present the obtained solution variants of the orbit improvement process.

On Figure 2, the mean errors of the improved initial position vector components are showed for seven successive computation variants. Here, the improving orbit, observation interval and orbital arc were changed. The 70 × 70 gravity field model was taken. An interesting property can be noticed on the presented figure. From comparing the first and the third variant accuracy it is clear that for increasing observation interval from 30 to 60 sec., the improved orbit accuracy decreases. This property is right for the following assumption: the total numbers of the used observations in both compared variants must be equal. This was achieved by the increase of the used orbital arc from 0.5 day (for the first variant) to 1 day (for the third variant).

Described solutions are only a part of all the obtained results. The comparative accuracy analysis of all obtained solutions of the orbit improvement process leads to the following conclusions:

1. The obtained solutions of the orbit improving simulation process indicate the practical realization possibility of this process using the real gravity tensor observations.
2. At the same geopotential model, the improved orbit precision increases in the case of growing gradiometric satellite orbital arc up to about one day. To the further growth of the solution accuracy:
   — the gravity field coefficient corrections must be determined,
   — the computation errors must be limited,
   — the input data accuracy and the significant figure number of the computation process must be increased.

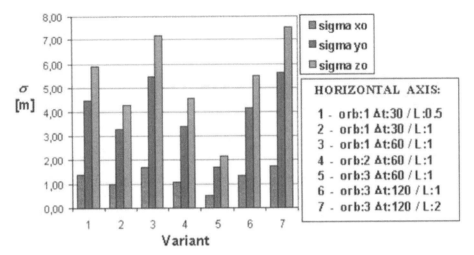

Figure 2. The mean square errors of the improved initial position components for the various solution variants with the different combinations of the improving orbit, observation interval and orbital arc length. The altitudes of the chosen orbits: about 160 km (orb:1), 200 km (orb:2) and 240 km (orb:3). The observation intervals: 30 sec. (Δt: 30), 60 sec. (Δt:60) and 120 sec. (Δt:120). The orbital arc lengths: 0.5 day (L:0.5), 1 day (L:1) and 2 days (L:2). The remaining initial data of the computation process: the 70 × 70 gravity field model and the full tensor observation simulation.

3. In case of observations simulation with respect to orbital frame the orbit improvement process solutions have a greater accuracy than the solutions with observations simulation with respect to inertial frame. It is recommended:
   — measurement of the full gravity tensor with respect to inertial frame,
   — measurement of the $T_{zz}(\frac{\partial^2 V}{\partial z^2})$ radial component with respect to orbital frame.
4. The random error modification of the simulated observations causes the following consequences:
   — for the 10^{-3} and 10^{-4} E.U. value of the random error standard deviation – there is no the clear change of the solutions accuracy,
   — for the 10^{-2} E.U. value of the random error standard deviation – the decrease of the solutions accuracy equals about 1.01 times,
   — for the 10^{-1} E.U. value of the random error standard deviation – the double decrease of the obtained solutions accuracy.
   The added errors had the normal distribution.
5. Increasing the improved orbit altitude, at the constant interval between the succesive gravity tensor measurements, leads to the greater precision of the obtained solution variants. The considered altitudes were equal about 160, 200 and 240 km.
6. Decreasing the interval between the succesive gravity tensor measurements, at the constant total number of the used observations, causes the increase of the improved orbit accuracy. Establishing the possible smallest value of the considered interval is an optimal solution.

400   ANDRZEJ BOBOJĆ

The other aspects of the Satellite Gravity Gradiometry will be research in the nearest future.

References

Colombo, O. and Kleusberg, A.: 1983, 'Applications of an Orbiting Gravity Gradiometer', *Bull. Géod.* **57**, pp. 83–101.

Drożyner, A.: 1995, 'Determination of Orbits with Toru/n Orbit processor', *Adv. Space Res.* **16**, No. 2.

Lemoine, F., Kenyon, S., Factor, J., Trimmer, R., Pavlis, N., Chinn, D., Cox, C., Klosko, S., Luthcke, S., Torrence, M., Wang, Y., Williamson, R., Pavlis, E. and Rapp, R.: 1998, 'The Development of the Joint NASA GSFC and the National Imagery and Mapping Agency (NIMA) Geopotential Model EGM96', *Report No.206861, July 1998.*

Metris, G., J. Xu, and Wytrzyszczak, I.: 1999, 'Derivatives of the Gravity Potential With Respect to Rectangular Coordinates', *Celest. Mech. and Dyn. Astron.* **71**, pp. 137–151.

Rummel, R., and Colombo, O.: 1985, 'Gravity Field Determination From Satellite gradiometry', *Bull. Géod.* **59**, pp. 233–246.

Rummel, R.: 1988, 'SGG Principles, State of the Art, Errors; Study on Precise Gravity Field Determination Methods and Mission Requirements', *Mid-Term Report, 1988.*

Sollazzo, C.: 1987, 'Satellite Gravity Gradiometer Experiment: Simulation Study', *MAS Working Paper* **No. 249**.

SEMI-ANALYTICAL MODELS OF SATELLITES MOTION FOR RUSSIAN SPACE SURVEILLANCE SYSTEM

ANDREY I. NAZARENKO

Center for Program Studies, 84/32 Profsoyuznaya ul, Moscow 117810, Russia,
E-mail: nazarenko@iki.rssi.ru

VASILIY S. YURASOV

Space Research Center "Kosmos", Magadanskaya str. 10-59, Moscow 129345, Russia,
E-mail: vyurasov@chat.ru

Abstract. The history of development of the semi-analytical satellites motion model for Russian space surveillance system (RSSS) is considered.

Key words: satellite motion theory, semi-analytical models

1. Introduction

Three types of integration methods of equations for celestial body motion are known. They are numerical, analytical and semi-analytical methods. The techniques of first two types are widely applied during long time. The methods of last type have attracted attention of the experts after the launch of the Sputnik satellite. These developments were caused mainly by requirements of space surveillance systems (SSS) such as accurate satellite catalogue maintenance [7], [8], [6]. The essential influence of perturbing factors of different nature makes it impossible to obtain the analytical solution of acceptable accuracy for all observed satellites. The impossibility of application of numerical methods is explained by their laborious and by the large run time while the number of software calling is about of 10^6 per day for RSSS. The averaging method [2], [5] developed by the Russian scientists was used as basic for the development of considered semi-analytical algorithms.

2. Basics of the Semi-analytical Models

Semi-analytical techniques combine the best feature of numerical and analytical methods to get the best accuracy and efficiency. The underlying approach is to separate the short-periodic perturbations from the long ones and secular effects so we can integrate numerically the equations for mean elements. The models of this type were developed in concerns of SSS in Russia by M. Lidov, A. Nazarenko, V. Yurasov, V. Boykov and in the USA by F. Hoots, J. Liu, P. Cefola, Mc-Claim, D.Danielson. These models are based on the classical methods of differential equations integration: the averaging Krylov-Bogolubov method and Hamilton-Jacobi canonical transformations method. We suppose that the most accurate and universal technique to develop semi-analytical models is the averaging method [2], [5]. Let's

Pretka-Ziomek et al./Dynamics of Natural and Artificial Celestial Bodies, 401–406, 2001.
© 2001 *Kluwer Academic Publishers.*

remind its main points. The task is to obtain the solution of a system of differential equations of a kind:

$$\dot{x} = \sum_j \varepsilon_j \cdot X_j(x, j),$$

$$\dot{y} = Y_0(x, y) + \sum_j \varepsilon_j \cdot Y_j(x, y), \tag{1}$$

where x, y are some vectors, and ε_j is small parameters related to the different perturbing factors. The functions X_j and Y_j are periodic functions of a variable y. The second-degree zonal gravitational coefficient c_{20} (J_2) $\approx 10^{-3}$ is usually the greatest one among other small parameters. Other perturbing factors (the anomalies of a gravitational field, the atmospheric drag etc.) have the order $\approx 10^{-6}$ in most cases.

The application of averaging method for an equation system with one "fast" variable is the simplest situation. Let's consider this case, as it is rather widespread. The use of the averaging method introduce to search such transformation of variables which would allow to separate fast variables y from slow x. As a result, the new set of equations is created which does not contain "fast" variables in right-hand-side part of equations:

$$\dot{\bar{x}} = \sum_j \varepsilon_j \cdot A_j(\bar{x}) + \sum_j \sum_k \varepsilon_j \cdot \varepsilon_k \cdot A_{jk}(\bar{x}) + \cdots,$$

$$\dot{\bar{y}} = f(\bar{x}) + \sum_j \varepsilon_j \cdot B_j(\bar{x}) + \sum_j \sum_k \varepsilon_j \cdot \varepsilon_k \cdot B_{jk}(\bar{x}) + \cdots \tag{2}$$

These equations can be integrated with a large time steps. Only few high accuracy semi-analytical models were elaborated worldwide because of their development is extremely laborious.

3. Milestones of Development of the Motion Models for RSSS

3.1. INITIAL STAGE (60-S)

Peculiarities of the situation at this stage were: the experience of orbit determination and forecasting had appeared during the flights of early Soviet satellites; simplest optical instruments were nsed for measurments ($\sim$100 sites); weak computers ($\sim$100,000 op/sec); thorough studies in satellite motion theory did exist (G. Duboshin, P. Elyasberg, M. Kislik, M. Lidov, I. Zhongolovich, Yu. Batrakov, T. Taratynova, D. Brouwer, Y. Kozai and others).

At this stage the numerical integration of satellite motion equations with simplest perturbations was mainly applied. The number of objects in the catalogue was equal to tens. It takes tens of minutes of computer time for one orbit updating.

The quantity of manual operations was in that time very great and there were large difficulties for new satellites detection. It takes weeks for the orbital elements determination of a new satellite. The simplest semi-analytical motion model for orbital elements in ascending node looks like:

$$x_N = x_{N-1} + \delta x(x_{N-1})_N,$$
$$t_N = t_{N-1} + T(x_{N-1})_N. \tag{3}$$

Here N is the revolution's number, t is the time equator plane interception, T is nodal period.

The main result of the motion model elaboration at this stage was a creation of analytical model of satellite motion. It was analogue to well-known US SGP-model. But this model was notable for more complete accounting of secular, long-periodical and short-periodical perturbations.

3.2. PERFECTING OF AUTOMATED SYSTEM (70-S)

At the second stage (1970–1980) the osculating elements at ascending node were used as the averaged ones. The perturbations of the first and second order were taken into account. General reasons for satellite motion models perfecting were: increase of the number of objects in space (approximately for 4–5 thonsand at the beginning of 80-s); putting into operation new radars; increasing of computing power; more accurate perturbation models; accumulation of the experience of new satellites detection; more accurate determination and forecasting of orbits.

Peculiarities for Russian space surveillance conditions were the following: limitation of territory, where measuring sensors were located, permanent deficit of computing power and high level of progress in mathematics and mechanics.

From the ratio "result/expenditures" point of view the satellite motion models are very effective SSS components in comparison with measuring and computing hardware. Therefore much attention was given for upgrading of these components. Main methodical and practical result of researches at the beginning of 70-s was an elaboration of the analytical and semi-analytical motion model based on an averaging method.

It was shown by Nazarenko that the averaged differential equations of second order may be written down as:

$$\frac{dx_N}{dN} = \delta^{(1)} x_N(2\pi) + \delta^{(2)} x_N(2\pi) - \frac{1}{2} \frac{\partial \delta^{(1)} x_N(2\pi)}{\partial x_N} \delta^{(1)} x_N(2\pi) + O(\varepsilon^3). \tag{4}$$

Here osculating elements in ascending node are chosen as mean elements. The values $\delta^{(1)} x_N(2\pi)$ and $\delta^{(2)} x_N(2\pi)$ are perturbations per revolution of the first and second order. The following difference equation corresponds to this differential equation:

$$x_{N+1} = x_N + \delta^{(1)} x_N(2\pi) + \delta^{(2)} x_N(2\pi). \tag{5}$$

For a fast variable appropriate differential and difference equation looks like:

$$\frac{dt_N}{dN} = T(x_N)_\Omega - 0.5 \cdot \delta T(x_N),$$
$$t_{N+1} = t_N + T(x_N)_\Omega,$$
$$T(x_{N+1})_\Omega = T(x_N)_\Omega + \delta T(x_N). \tag{6}$$

These equations were integrated analytically by approximate drag accounting and numerically for atmosphere dynamic model. Thus, the unified system of specialized motion models adapted in the best way to predict various satellite orbits was constructed. All these developments were done under a management of Nazarenko A.I.

Main results of intrusion of these technique and some other technique developed by the end of 70 s are: processing of all obtained measurements (about 10,000 per day); multiple increase of the number of objects in the catalogue; carrying out more accurate orbital elements determination on every revolution, where measurements were obtained; reducing of information processing time; increasing of the accuracy of orbits determination and forecasting. Quality of surveillance tasks had solving at that stage was nearly as good as at present.

3.3. FURTHER STAGES (AFTER 1980)

The transition to classical averaged orbital elements was hereinafter executed. Thus the number of the taken into account perturbing factors was increased essentially. At the beginning of 80-s the considerable step in further progress of semi-analytical motion models was made. These developments were made by V. Yurasov from SRC "Kosmos" [9], [10], [11], [12], and by V. Boykov from "Vympel" Corporation as well [3]. The main information about modern semi-analytical V. Yurasov's model is presented below.

4. Current Universal Semi-analytical Model by Yurasov

The universal semi-analytical model was developed and realized as mainframe software at the beginning of 1980 [9]. At first it was used for testing specialized algorithms and programs for space object (SO) motion prediction in the RSSS.

The universality of the model is understood here as its suitability and flexibility for precise orbit prediction for various heights, eccentricities and inclinations. The universality is achieved by choice of nonsingular orbital elements and taking into account all essential perturbations in the most general form.

A special representation of expressions for the right-hand-side of the averaged equations and short-periodical perturbations was developed. These expressions have the following structure in most cases:

$$\dot{x}_j = \sum_m F_m^{(j)}(a, e, i) S_m^{(j)}(\omega, \Omega, M, \varphi(M, \theta), \bar{r}_k),$$

$$\delta x_j = \sum_m f_m^{(j)}(a, e, i) s_m^{(j)}(\omega, \Omega, M, \theta, \bar{r}_k), \tag{7}$$

where $F_m^{(j)}(a, e, i)$, $f_m^{(j)}(a, e, i)$ are two very slowly-changing functions of the mean semimajor axis, eccentricity and inclination with do not experience the first-order secular perturbations with respect to c_{20}; $S_m^{(j)}(\omega, \Omega, M, \varphi(M, \theta), \bar{r}_k)$ are the slowly-changing functions; $s_m^{(j)}(\omega, \Omega, M, \theta, \bar{r}_k)$ are the fast-changing functions; $\bar{r}_k$ is third body position vector and m is the multi-index.

Piecewise constant approximations for calculation of the functions $F_m^{(j)}(a, e, i)$, $f_m^{(j)}(a, e, i)$ are used. This allows to reduce considerably the computation time by avoiding laborious calculations of complex functions at each integration step. The maximum order of geopotential expansion and third-body potential are set in model as parameters before compilation of source code and during run time. For an integration of the averaged equations the Runge-Kutta method of the fourth order was realized. In most cases the integration step size is adopted equal about of one day. However for reentry satellites the step size of a fraction of period is used.

The accuracy estimations of the universal semi-analytical model were carried out many times for various orbits [9], [10], [11], [12]. For example, the comparison of semi-analytical prediction by universal model with numerical integration by the Everhart method [4] was obtained by using JGM-02 with harmonics up to 70×70. It shows that maximum values of along track, radial and cross-track deviations have not exceeded 0.01 sec, 13 m and 13 m respectively over 30 day prediction interval.

A PC-version of this model is now exist [11]. It can be applied for processing the geostationary space objects measurement information [12] and this model is flow used in activities connected with the tracking of important space objects. Some examples of these works were the reentry determination of an orbital complex "Salyut-7"-"Cosmos-1686" [7], satellites "Cosmos-398" and FSW-1-5 [1].

References

1. Andrewschenko, V., Batyr, G., Bratchikov, V., Dicky, V., Veniaminov, S. and Yurasov, V.: 1996, 'Reentry Time Determination Analysis for "Cosmos 398" and FSW-1-5', *Proceedings of U.S.-Russian Second Space Surveillance Workshop*, Poznay', Poland.
2. Bogolyubov, N.N. and Mitropolskij, Yu. A.: 1961, *Asymptotic methods in the theory of nonlinear oscillations*, Gordon and Breach, New York.
3. Boykov, V.F., Machonin, G.N., Testov, A.V., Khutorovsky, Z.N. and Shogin, A.N.: 1988, 'High-speed algorithms of the predictions of satellites for the operative orbit estimation and living time of artificial satellites', *Observation of artificial celestial objects*, **#84**, USSR Academy of Sciences Astrocouncil, Moscow.
4. Everhart, E.: 1985, 'An efficient integrator that uses Gauss-Radan spacings', In *Dynamics of Comets: Their Origin and Evolution* (A. Carusi and G.B. Valsecchi, Eds.), Reidel Pubi. Co., 185–202.

5. Grebenikov, E.A. and Ryabov, Yu.A.: 1971, *Qualitative methods in celestial mechanics*, Nauka, Moscow.
6. Kuzmin, A.A.: 1993, 'Information Capabilities of the Domestic Space Surveillance Concerning Space Debris', *The technogenic space debris problem*, Moscow, Kosmoinform, 22–32.
7. Nazarenko, A.I.: 1991, 'Determination and Prediction of Satellite Motion at the end of the Lifetime', *Proceedings of the International Workshop "Salyut7/Kosmos-1686 Reentrv"*, ESOC, Darmstadt.
8. Votintzev, Yu.V.: 1993, 'Unknown Troops of Disappeared Super-State', *Militaryhistorical Journal*, #8–11.
9. Yurasov, V.S.: 1984, 'The application of universal semi-analytical method for satellite motion propagation in the atmosphere', *Observation of artificial celestial objects*, **#82**, USSR Academy of Sciences Astrocouncil, Moscow.
10. Yurasov, V.S.: 1994, 'The application of universal semi-analytical method for geostationary satellite motion propagation', *Proceedings of the "Conference Observation programs of deep space objects and celestial bodies of solar system"*, Sankt-Petersburg.
11. Yurasov, V.S.: 1996, 'Universal Semi-analytical Satellite Motion Propagation Method', *Proceedings of U.S.-Russian Second Space Surveillance Workshop*, Poznań, Poland.
12. Yurasov, V.5. and Moscovsky, A.A.: 1996, 'Geostationary Orbit Determination and Prediction', *Proceedings of U.S.-Russian Second Space Surveillance Workshop*, Poznań, Poland.

SEMI-ANALYTICAL METHOD TO STUDY GEOPOTENTIAL PERTURBATIONS CONSIDERING HIGH ECCENTRIC RESONANT ORBITS

PAULO HENRIQUE CRUZ NEIVA LIMA JR. (paulohc@cdt.br)
DEF-FACAP

SANDRO DA SILVA FERNANDES (sandro@ief.ita.cta.br)
IEF-ITA-CTA

RODOLPHO VILHENA DE MORAES (rvm@feg.unesp.br)
DMA-FEG-UNESP

Abstract. A dynamic system describing the orbital motion of artificial satellites with mean motion commensurable with the rotation of the Earth, including harmonics of high degree and order in the geopotential perturbations, is studied. An integrable kernel is searched through canonical transformations. One resonant angle is fixed and the dynamic system can be reduced to a two degree of freedom system. The coefficients of the Hamiltonian are computed analytically. Numeric simulations for highly eccentric orbits and 2:1 resonance are presented.

Key words: Resonance, Geopotential Perturbations, High Eccentricity

1. Introduction

Analytical computation of the perturbations due to the geopotential in the orbital motion of an artificial satellite, considering harmonics of high order and degree, expressed in terms of the mean anomaly, presents two major problems: a) developments in power of the eccentricity for high eccentric orbits and b) resonance among the enrolled frequencies.

Concerning to the first problem, in the classical Kaula's theory for the geopotential, the solutions can be used only for orbits with eccentricity $e < 0.6627\ldots$. Near this value the convergence of the series enrolled is very slow and, for orbits with eccentricities greater than this, the series do not converge. Hansen's coefficients has been successfully used to solve this problem (Wnuk, 1988) and this will be done here. Inasmuch, for the resonant case, due to the properties of the Hansen's coefficients (Lima Jr., 1998), it was possible to derive analytical expressions for the coefficients of equations obtained in this paper. Resonance introduces small divisors and special care must be taken to not compromise the validity of the solution.

As resonance, it will be considered here the commensurability between the satellite's mean motion frequency and the frequency of the Earth's rotational motion. MOLNYIA, KOSMOS, PROGNOZ, EXPLORER 28, ESA-GEOS are examples of any of several artificial satellites whose orbits have such characteristics: high eccentricity and resonance. A reduced Hamiltonian will be constructed, con-

Pretka-Ziomek et al./Dynamics of Natural and Artificial Celestial Bodies, 407–413, 2001.
© 2001 *Kluwer Academic Publishers.*

408   LIMA JR. ET AL.

taining only secular terms and periodic terms with commensurability between the frequencies of the motion. It this Hamiltonian will be not considered other long or short period terms than these. The influence of resonance in the orbital motion of artificial satellites has been investigated by this way by other authors (Lane, 1988; Sochilina, 1982; Grosso and Sessin, 1993; Ely and Howell, 1996; Vilhena de Moraes et al., 1995).

Following, it will be constructed a new Hamiltonian considering a given resonance. A canonical transformation will be performed reducing the degree of freedom of this Hamiltonian system. This final Hamiltonian system will be integrated numerically.

Thanks to recurrent formulas for Hansen's coefficients it will be possible to find two first integrals making the system of differential equation of motion an integrable system. Numerical simulations for orbits having eccentricities up to 0.9 shows the behaviour of the dynamical system near the region of the 2:1 resonance.

2. Equation of Motion in Canonical Form

Using Hansen's coefficients, the geopotential can be written as (Osório, 1973):

$$U = \frac{\mu}{2a} + \sum_{l=2}^{\infty}\sum_{m=0}^{l}\sum_{p=0}^{l}\sum_{q=-\infty}^{\infty} \frac{\mu}{a}\left(\frac{a_E}{a}\right)^l J_{lm} F_{lmp}(I) H_q^{-(l-p)(l-2p)}(e) \tag{1}$$

with

$$\varphi_{lmpq} = qM + (l-2p)\omega + m(\Omega - \Theta - \lambda_{lm}) + (l-m)\frac{\pi}{2} \tag{2}$$

Here $a, e, I, M, \omega, \Omega$ are the Keplerian elements, $\Theta = \omega_E t$ is the sidereal time, ω_E is the Earth's angular speed, λ_{lm} is the longitude of the semi-major axis of symmetry for the harmonic (l,m); $F_{lmp}(e)$ represent the inclination functions and $H_q^{-(l-p)(l-2p)}(e)$ are the Hansen's coefficients. Let us introduce the canonical set of variables (X, Y, Z, x, y, z) related with the classical Delaunay variables by the following canonical transformation:

$$X = L \qquad Y = G - L \qquad Z = H - G \tag{3a}$$

$$x = l + g + h \qquad y = g + h \qquad z = h \tag{3b}$$

Therefore, in the extended phase plane the equations of motion of an artificial satellite submitted to the geopotential perturbations can be put in Hamiltonian form with Hamiltonian $H = H(X, Y, Z, \Theta, x, y, z, \theta)$ where θ is a variable conjugated to Θ (Lima Jr., 1998).

The central part of the Hamiltonian, describes the motion of the satellite when the Earth is considered as spherical and with a homogeneous distribution of mass. The disturbing function is composed by the other terms. We can distinguish in the

SEMI-ANALYTICAL METHOD

disturbing function two kinds of terms: secular, related with odd zonal harmonics and periodic (short and long period), related with the tesseral and even zonal harmonics.

3. Resonant Dynamical System

In this section it will be constructed a new Hamiltonian H_{sr} containing as periodic part just the resonant terms extract from the Hamiltonian H. A first integral can be found for the canonical system generated by the new Hamiltonian H_{sr}. This first integral enable us to perform a Mathieu transform in order to reduce the order of the system.

The resonance considered here comes from the problem of the motion of an artificial satellite whose orbital period is multiple of the Earth's rotation period. Thus, if n stands for the orbital mean motion, the resonance condition can be expressed by $qn - \omega_E \theta = 0$. We will denote by $\alpha = \frac{q}{m}$, where q and m are integers, the commensurability of the resonance. Commensurability related with the tesserals causes the appearance of small divisors during the integration of the equations of motion, endangering the validity of the solution. Terms whose arguments have such commensurability are called resonant terms. The part of the Hamiltonian containing just secular and resonant terms will be called reduced Hamiltonian H_{SR}. Substituting the commensurability of the resonance α, and considering just secular and resonant terms we obtain an new Hamiltonian H_{SR}

The Hamiltonian system with Hamiltonian H_{SR} has the following first integral (Lima Jr., 1998):

$$\left(1 - \frac{1}{\alpha}\right) X + Y + Z = C \tag{4}$$

where C is a constant. Using this first integral let us perform the following Mathieu transformation:

$$X_1 = X \quad Y_1 = Y \quad Z_1 = \left(1 - \frac{1}{\alpha}\right) X + Y + Z \quad \Theta_1 = \Theta \tag{5a}$$

$$x_1 = x - \left(1 - \frac{1}{\alpha}\right) z \quad y_1 = y - z \quad z_1 = z \quad \theta_1 = \theta \tag{5b}$$

Since from equation (5), Z_1 is a constant, the degree of freedom of the system was reduced by one. This Hamiltonian system yields valid for any commensurability among the frequencies of the mean anomaly and of the sidereal time.

410   LIMA JR. ET AL.

4. One Resonant Angle

Fixing a value for one resonant angle , we can consider as short period terms all the periodic terms different from the fixed one (Lane, 1988; Lima Jr., 1998). Choosing α, all the possible frequencies that can assume the resonant frequency can be obtained when varying the values assumed by the coefficients l, m, and p, where $l \geq 2, m \geq 2, 0 < p < 1$. So, the frequency:

$$\frac{d\varphi_{lmp(\alpha m)}}{dt} = m\left(\alpha\frac{dx_1}{dt} - \frac{d\Theta_1}{dt}\right) + (l - 2p - m\alpha)\frac{dy_1}{dt} \tag{6}$$

containing fixed coefficients is the unique resonant frequency to be considered henceforth in the Hamiltonian. We will call this frequency as critical frequency. The frequency of the arguments of Ω and ω can enhance the effects of the small divisors. The coefficients l, m and p can vary, generating several frequencies for one given resonance. These frequencies are lightly different each from other, around the commensurability $\alpha\dot{x}_1 - \dot{\Theta}_1$.

Once defined a critical frequency, we will consider a new Hamiltonian H_C containing only secular terms and terms with the critical frequency. The order of the Hamiltonian system with Hamiltonian H_C can be reduced observing that for the this system the following relation is valid:

$$(k - m\alpha)X_1 - m\alpha Y_1 = C_2 \tag{7}$$

where $k = l - 2p$ and C_2 is a constant.

Performing the canonical transformation:

$$X_2 = X_1 \qquad Y_2 = (k - m\alpha)X_1 - m\alpha Y_1 \qquad \Theta_2 = \Theta_1 \tag{8a}$$

$$x_2 = x_1 + \left(\frac{k - m\alpha}{m\alpha}\right)y_1 \qquad y_2 = (\frac{1}{m\alpha}) \qquad \theta_2 = \theta_1 \tag{8b}$$

the Hamiltonian system is reduced to the following system:

$$\frac{dX_2}{dt} = -\sum_{p=0}^{\infty} B_{(2p+k)mp(\alpha m)}(X_2, C_2, C_1)sen\varphi_{(2p+k)mp(\alpha m)}$$

$$\frac{d\varphi}{dt} = m\alpha\frac{\mu}{X_2^3} - m\omega_E - m\alpha\sum_{j=1}^{\infty}\frac{\partial B_{2j,0,j,0}(X_2, C_2, C_1)}{\partial X_1} \tag{9a}$$

$$-m\alpha\sum_{p=0}^{\infty}\frac{\partial B_{(2p+k)mp(\alpha m)}(X_2, C_2, C_1)}{\partial X_1}\cos\varphi_{(2p+k)mp(\alpha m)} \tag{9b}$$

These equations can be integrated by quadrature. Once solved the reduced system, algebraic transformations enable us to get the time variation in terms of the Delaunay variables (and then in terms of the Keplerian elements). In fact, we can

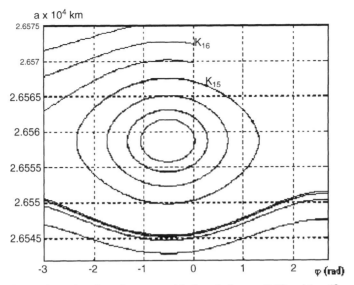

Figure 1. semi-major axis versus critical angle for $e = 0.90$ and $i = 4°$.

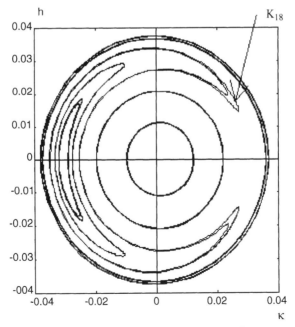

Figure 2. Phase plane for $e = 0.70$, $i = 50°$.

412 LIMA JR. ET AL.

return to the Delaunay variables inverting the transformations (9a), (9b), (8a), (8b), (3a) and (3b) and using the first integrals defined by equations (5) and (8).

5. Numerical Simulations and Conclusions

In what follows, some examples for critical angles near the $2 : 1$ resonance are exhibited. The coefficients m, p, k and α, related with the orbit of the satellite, are computed as follows: $\alpha = 1/2$, m is an even number, $k = 2$ and $p = 0, 1, 2$.

The following initial conditions were considered at the instant t_0: $a_0 = 26561$, $176\,\text{km}$, $e_0 = 0.70$, $e_0 = 0.90$, $i_0 = 4°$, $i_0 = 50°$.

It was considered here just the harmonics J_{20} and J_{22}. The curves were constructed considering a_0 for several values of δa.

In Figure 1, K_{15} was obtained for $a_0 + 5.3$ km , $\varphi = 0$ and K_{16} for $a_0 + 7.9$ km, $\varphi = 0$. The curves above the closed curves come from integrations with initial condition $\varphi = 0$ and the curves below the closed curves have initial conditions $\varphi < -2\pi$.

Figure 2 shows the phase plane for $e = 0.70$ and $i = 50°$.

The curve K_{18} is such that $a = a_0 + K_{18}$, $K_{18} = 8.5$ km, $\varphi = 0$.

In conclusion, imposing two hypothesis, two first integrals can be found for the equations of motion of satellites perturbed by the geopotential and with high eccentric resonant orbits. Also, the order of system was reduced, making it an integrable system. Numerical simulations can illustrate the range of possibilities given by this analytical method.

Acknowledgements

This work was supported by FAPESP under contract N°00/03185-1.

References

Ely, T.E.A. and Howell, C.K.: 1996, 'Long Term Evolution of Artificial Satellite Orbits Due to Resonant Tesseral Harmonics', *The J. Astron. Sci.* **Vol. 44,2**, pp. 167–190.

Grosso, P.R. and Sessin, W.: 1993, 'Resonance Effects on Satellite Orbits', *Advances in Astron. Sci.* **Vol. 2**, pp. 314–326.

Lane M.T.: 1988, 'An Analytical Treatment of Resonance Effects on Satellite Orbits', *Celestial Mechanics* **Vol. 42**, pp. 3–38.

Lima Jr. P H.C.L.: 1998, *Sistemas Ressonantes a Altas Excentricidades no Movimento de Satélites Artificiais*, Doctoral These, ITA, S.J.C., (in Portuguese).

Osório, J.P.: 1973, *Perturbações de Órbitas de Satélites no Estudo do Campo Gravitacional Terrestre*. Imprensa Portuguesa, Porto.

Sochilina, A.S.: 1982, 'On the Motion of a Satellite in Resonance with its: Rotating Planet', *Celest. Mech.* **Vol. 26**, pp. 337–352.

Vilhena de Moraes, R. , Fitzgibbon, K.T. and Konemba, M.: 1995, 'Influence of The 2:1 Resonance in The Orbits of GPS Satellites', *Adv. Space Res.* **Vol. 16, 12**, pp. 37–40.

Wnuk, E.: 1988, 'Tesseral Harmonic Perturbations for High Order and Degree Harmonics', *Celest. Mech.* **Vol. 44**, pp. 179–191.

COORDINATED ORBITAL CONTROL FOR SATELLITE CONSTELLATIONS AND FORMATIONS

GIOVANNI B. PALMERINI (g.palmerini@caspur.it)
*Scuola di Ingegneria Aerospaziale - Università di Roma "la Sapienza" via Eudossiana 18 – 00184
Roma, Italy*

Abstract. Preserving the desired orbital configuration under perturbing effects is an important issue
for multiple satellite missions concepts. Several guidance strategies have been proposed in the past,
aiming to reduce the ground station workload. The experience gained on this subject at the Università
di Roma is presented, outlining the approaches studied, their advantages and their drawbacks. A
comprehensive model, based on the work by McInnes on the potential function method, is introduced.

Key words: satellite constellations, orbital control

1. Introduction

Recent scientific or telecommunications missions make use of a distributed archi-
tecture, with the payload being divided among several platforms. Injection errors
and perturbing forces act on this system, possibly limiting its capability to satisfy
the requirements. The common aspect of all these multiple platform mission concepts is the need for a correct relative positioning of the satellites, to be obtained by
means of a dedicated control action. Performing guidance strategies are therefore
required to identify a configuration which is suitable for the mission concept, as
well as optimal from the point of view of energy consumption.

Different views already exist on this subject. The acknowledged basis is that
the on-board autonomy has to be pursued, in order to reduce the need for a heavy
ground-segment. Wertz *et al.* (1998) proposed the absolute station keeping, where
each platform belonging to the system is constrained in its box. This technique
is similar to the one currently used for single geostationary satellites, and is basically built on a quite accurate prediction of the orbit that each platform will fly
on once launched. The orbital control package becomes a standard asset, with a
correct modelling of perturbations which is good for all missions operating in
similar conditions, while accuracy of navigation data and thruster characteristics
fit the specific system. No intersatellite links are needed, with a simplest system
design.

The drawback of such a technique is that it does not address the peculiarities of
the constellation or formation, that really performs a mission as a unique payload.
Common drifts of all the platforms should be corrected by this method, with a clear
wasting of propellant and a reliability problem due to the frequent maneuvres.

Pretka-Ziomek et al./Dynamics of Natural and Artificial Celestial Bodies, 415–424, 2001.
© 2001 *Kluwer Academic Publishers.*

416 GIOVANNI B. PALMERINI

On the other hand, relative station-keeping techniques deal with spacing among satellites instead of looking for its position in an absolute frame. As mission requirements hold for the set of platforms, this approach can save control action, but it does require the knowledge of the current state of the system. Starting from these data, referred to a perturbed condition, the guidance strategy should evaluate the set of maneuvres needed to recovery a configuration satisfying the requirements. In the following, we will deal with the approaches worked on at the Università di Roma, offering some comments on the problems and opportunities faced in such investigation. Then, we will focus on the more promising of these approaches, which is able to offer, at least, a solution converging to the nominal status and easy to implement, while optimal solution envisaged by other methods are quite often very difficult to compute. This powerful technique, lent by dynamic systems theory and introduced by McInnes (1995), allowed us to mechanize a concept already presented and hopefully useful to obtain a simple control strategy.

2. Linear Optimization Method

A first approach to a co-ordinated station keeping strategy can be achieved by a detailed analysis of the satellite system behaviour in time. This method, mainly developed at LAAS and CNES (Lasserre *et al.* (1997)), starts from the classical linear system, to be considered in the neighbours of the current system state X

$$X' = AX + BU \tag{1}$$

Matrix A includes principal perturbations acting, and the control action U is performed when orbital parameters X are unable to satisfy the requirements. This linearization leads to a schedule of orbital maneuvres, also depending on thruster performances. A linear optimization provides the magnitude of the thrust actions that minimizes propellant consumption with respect to the configuration constraints. The advantage of such a technique is the ready availability of powerful software packages for linear optimization; the drawback is the strong dependence on the accepted timeline, limiting the analysis to a set of solutions, and possibly leaving aside the optimal one.

Some work has been carried on at the Università di Roma on this subject. Extensive analysis were performed in a test case of a low altitude formation, where the air-drag effect was a big player. Therefore, out-of-plane corrections have been divided from the more frequent in-plane ones and separately evaluated and executed off-line, as a process operated by the ground control centre. The in-plane dynamics, limited to orbital semi-axes and phases, can be obtained at the time $t + 1$ from the previous state at the time t by means of:

$$\begin{bmatrix} \Delta a_h \\ \Delta a_k \\ \Delta \theta_{hk} \end{bmatrix}_{t+1} = \begin{pmatrix} D_h \Delta t + 1 & 0 & 0 \\ 0 & D_k \Delta t + 1 & 0 \\ J_h \Delta t & J_k \Delta t & 1 \end{pmatrix} \begin{bmatrix} \Delta a_h \\ \Delta a_k \\ \Delta \theta_{hk} \end{bmatrix}_t +$$

$$
+ \begin{pmatrix} T_a \Delta t & 0 \\ 0 & T_a \Delta t \\ T_\theta \Delta t & T_\theta \Delta t \end{pmatrix} \begin{bmatrix} \Delta v_h \\ \Delta v_k \end{bmatrix} + \begin{bmatrix} d_h \\ d_k \\ 0 \end{bmatrix}
\tag{2}
$$

(referred to the simple case of two following platforms h and k) where J terms take into account the oblateness of the Earth

$$
J_h = -1.5 \frac{n_o}{a_o} [1 + 3.5 J_2 \frac{R_E^2}{a_o^2} (4cos^2 - 1)]
\tag{3}
$$

and D terms represent the effect of the drag, assuming a linear law for density ρ centered on the design semi-axis

$$
\rho = \rho_o - k^2(a - a_o)
\tag{4}
$$

$$
D_h = -\frac{A_h}{n_h} C_{D_h} (.5\rho_o \sqrt{\mu/a_o} - k^2 \sqrt{\mu a_o})
\tag{5}
$$

while T terms assess the effects of the thrust action on the parameters and d terms take into account the orbit decay due to the air-drag during the manoeuvre. The optimality problem to be solved is

$$
min \sum_t \sum_h \Delta v_{h,t}
\tag{6}
$$

with sums extended to all time steps (t) and to all platforms (h), to be completed with the constraints on the magnitude of the thrusters actions and on the amplitude of the parameter variations allowed for:

$$
|\Delta V_h| \leq \Delta V_h^* \quad |\Delta a_h| \leq \Delta a_h^* \quad |\Delta \theta_h| \leq \Delta \theta_h^*
\tag{7}
$$

Constraints can be considered to be different as the platforms belonging to the constellations carry different payloads, and therefore different tolerances should apply. Results clearly claimed for the superiority of the relative control, from both the aspect of the propellant consumption and of the frequency of the thrusting actions: the frequency has to be as limited as possible in order to reduce the reliability risk, due to possible thrusters' wrong pointing or early/late switching on and off.

3. The 'Mean Constellation' Approach

In 1993 Lamy and Pascal proposed the concept of mean constellation, as the configuration which is able to satisfy the requirements and is the cheapest one, in terms of propellant consumption, to achieve from current, perturbed status. This new orbital configuration should be the target of the maneuvre, to be executed at the time allowed for by thrusters characteristics or ground control monitoring

418   GIOVANNI B. PALMERINI

needs. In 1994 Graziani *et al.* proposed to use Lagrangian multipliers to evaluate the mean constellation parameters with the introduction of a cost function

$$\phi = \sum_{s=1,Nsat} \alpha_s \sum_{m=1,M} c_m (\Delta p_m^s)^2 \tag{8}$$

where p_m^s are the differences in the M (≤ 6) lagrangian parameters of interest between the current, perturbed configuration and the mean constellation, c_m terms take into account the cost of the correction of the $m - th$ parameter (they will be larger for parameters involving plane correction, as inclination or node) and α_s are related to the manoeuvering capability of the satellite s (they can be increased as the propellant available on that satellite decreases). Target configuration can be found by means of Lagrangian multipliers considered in the augmented penalty function:

$$\Phi = \phi + \sum_{i=1,I} \lambda_i g_i \tag{9}$$

where g_i are the constraints defined by the desired differences $\overline{p}$ among the platforms, referred to the orbital parameter of interest

$$g_i = \sum_m p_m^r - p_m^s - \overline{p_m^{r,s}} \tag{10}$$

and possibly including the foreseeable effect of perturbations. The mean constellation parameters p will be given by

$$\left[\frac{\partial \Phi}{\partial p_1^1} \cdots \frac{\partial \Phi}{\partial p_M^1} \cdots \frac{\partial \Phi}{\partial p_1^{Nsat}} \cdots \frac{\partial \Phi}{\partial p_M^{Nsat}} \frac{\partial \Phi}{\partial \lambda_1} \cdots \frac{\partial \Phi}{\partial \lambda_H} \right]^T = \underline{0} \tag{11}$$

At this time, once the target has been identified for each single platform, an orbital maneuvre can be evaluated, with respect to the thruster steering capabilities of the spacecraft and to their performances, as well as depending on the time interval allowed for the return to an operational status.

4. Comments on the Application of Previous Techniques

Previous approaches have been useful to validate the relative control strategy with respect to the absolute one, assessing its expected advantages. The analysis pointed out that a tuning of the codes is necessary in order to fit specific test cases, so that the software created for a constellation mission is not immediately portable to other projects, especially if orbital parameters are not really similar. On the other hand, multipurpose, already available optimization packages can be easily integrated in this process, helping in managing the number of variables to be considered in order to take into account the constraints. At the end, this technique can represent a good solution during operation phases, if the maneuvres have to be carried out in the traditional way, i.e. by means of telecommands from the ground control center.

However relative control strategies present several problems in case of constellations/formations composed by a large number of platforms, as the amount of work at the ground control station would be difficult to carry out. Of course, autonomous operations impose inter-satellite links, as every satellite should know the state of others, and this requirement translates in a remarkable task for the spacecraft designer. Moreover, also allowing for the additional hardware needed for the intersatellite link, the on-board implementation of these relative control strategies seems really difficult, as the related computation effort is not negligible and really awkward for typical spacecraft resources. New approaches are therefore required to pave the way for autonomous orbital control of large satellite constellations.

5. The Potential Function Method

A breakthrough in the station-keeping problem was represented by the sequence of papers by McInnes(1995). The strength of his approach is to make use of Lyapunov method in order to outline a strategy for control action which is proofed to be convergent to the desired solution. A potential function V, such that, according to the Lyapunov conditions, $V(x) > 0$, $\frac{dV}{dt} < 0$ ($\forall x \neq \underline{0}$), can be defined adding to the kinetic energy term a potential term, depending only on the position, which assumes minimal value in the configuration correctly spaced slots. In such a way, the thrust required to achieve a desired configuration can be obtained by equaling the total time derivative to a negative term

$$\frac{dV}{dt} = \nabla_x V \dot{X} + \frac{dV}{dt} = -kX^2 \tag{12}$$

and then, by substitution of the system dynamics equations for $\dot{X}$, solving for the accelerations appearing in the dynamics law. The method has been originally applied to the Euler Hill linearized problem, but can be also used, as remarked by McInnes, for non-linear problems. Figure 1 shows the convergence to the desired angular separation for the (classical) non-linear representation

$$\ddot{r}_s - r_s \dot{\theta}_s = -\mu/r_s^2 + a_{r,s} \tag{13}$$
$$r_s \ddot{\theta}_s + 2\dot{r}_s \dot{\theta}_s = a_{\theta,s} \tag{14}$$

of the problem of a constellation ($s = 1, N_{sat}$) to be equally spaced on a ring (the same case presented in the Euler-Hill form by McInnes). Transient behaviour, as reported in figure 1, is determined on the selection of k in Eq.(12) as well as on additional coefficients that can be included in the potential expression.

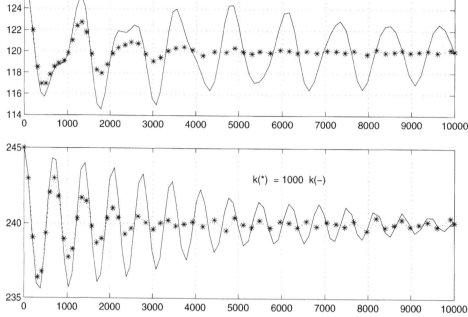

Figure 1. Angular distances (deg) vs. time (s) between platforms belonging to the same orbital plane for different values of the parameters - Non linear case, 3 satellites on the plane, i.c. 0, 115, 242 degrees.

6. Application of the Potential Function Method

Investigations on orbital control strategy made clear the need for a tool able to provide the distance of the state of the system from the unperturbed configuration, and to evaluate the amount of thrust needed to correct. This feature can be represented by a 'spring', centered in the nominal positions of the platforms and with moving extrema representing the satellite, which moves away from its position due to orbital perturbation or thruster off nominal behaviour. The idea, previously introduced by Palmerini and Graziani (1994) and reported as a sketch in fig.2, was to build a complete model of a constellation where distances among close platforms are represented by a set of springs. Additional springs represent the altitude of each platform. Desired distances are represented by the spring length at rest. The actual lengths of the springs vary during the orbital propagation due to the perturbations acting. The required control action should be therefore identified and evaluated as the set of manoeuvres needed in order to counteract these variations. Not all the connections have to be considered as active, but their selection can be made on the real system case, depending on which parameters have to be controlled, i.e. on the orbits selected for the satellites and on the constraints to be respected [as an example, altitude springs will be important for low altitude constellations,

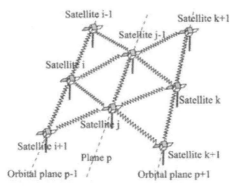

Figure 2. The constellation model with springs representing in-plane and out-of-plane 'links' to be considered for configuration keeping.

where the air drag is very significant]. As far as it concerns the inter-plane springs, the dynamic behaviour in terms of node differences and anomalies can be easily reported to distances among satellites by means of trigonometry relations. McInnes relevant work actually offers the possibility to implement this model, being sure that convergence to the target constellation will be achieved. The potential function can be simply chosen as

$$V(x) = \frac{1}{2} X^T K X \qquad (15)$$

while higher order springs should be possibly selected too.

This spring potential can be considered for the angular separation control in the case described in the previous paragraph, obtaining results quite similar (figure 3) to the ones presented by McInnes(1995) by selecting the potential as

$$\sum_{i,j \ i \neq j} \lambda_i \cos^{-2}(x_i - x_j + \pi) \qquad (16)$$

The cost of the manoeuvre in terms of propellant, depending on the law assumed for the potential function, as well as on on the accuracy required in station-keeping, can be evaluated by means of the sum of the squared acceleration, showing some possible advantage for the spring potential case (acceleration ratio smaller than unity in fig. 4).

A criteria to evaluate the convergence of the method can be obtained by means of the results presented by Kalman and Bertram back in 1960: if it is possible to consider a value of c such that

$$\dot{V}(X) \leq c^2 V(X) \qquad (17)$$

that c^{-2} can be considered as the time constant of the system, so that largest c do imply faster convergence.

Several different real world situations can be represented by selectively choosing the spring stiffness. Varying stiffness can implement the need for a correct

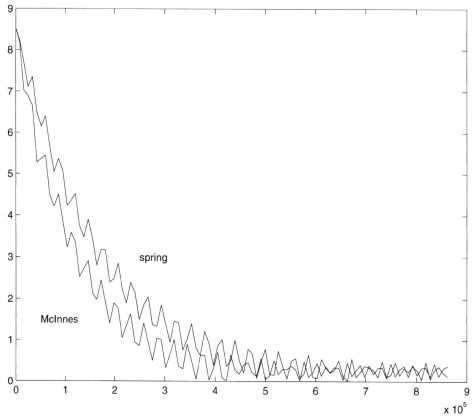

Figure 3. Differences in anomalies (deg) vs. time (s) for 3 satellites belonging to the same orbital plane [rms values]. Initial conditions are 5.3, 118.4, 241.1 degrees.

redistribution of the propulsive action among different satellites in order to maximize the system lifetime. Moreover, a threshold in the stiffness can be introduced, allowing for a better representation of the propulsion system behaviour.

An interesting subject is the control of the intersatellites distances in configuration conditions far from the designed ones. Potential function method has been applied to these topics by Colantoni (1999), dealing with the case of correction manoeuvres being executed on satellites close to the constellation plane crossing, when possible collision risk arises.

Future promising work will be focussed on the possible introduction of potential functions tailored in such a way to benefit of low-cost manoeuvres. It means that the steepness of the potential will be designed according to the propellant cost of the relevant correction, in order to drive the system through consumption reduced trajectories. An example of such a technique should be provided by the nodal out-of-plane corrections by means of Earth oblateness driven, in-plane manoeuvres, and should be obtained by adding this conservative perturbation potential to the global potential.

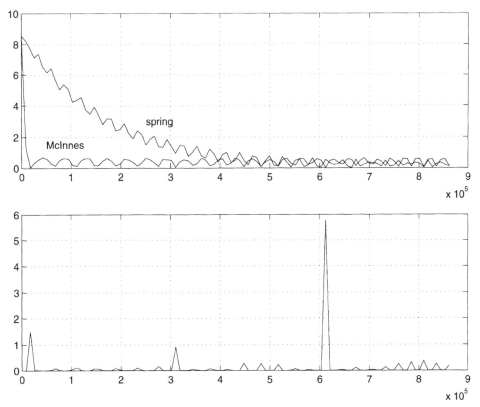

Figure 4. Differences in anomalies [deg - rms values] and ratio of the squared acceleration required [spring model vs. McInnes model] in a test case with i.c. equal to figure 3 case and different parameters for the control laws.

7. Conclusions

Relative orbital control strategies of multiple satellite systems are deemed to offer better performances in terms of amount and frequency of thrusting actions needed. The work carried on at the Università di Roma with a linear optimization approach confirmed these findings. However this technique presents strong computation requirements, and are therefore difficult to implement on board. A different and quite promising solution has been recently presented by McInnes. His potential function method approach, lent by Lyapunov work on dynamic systems theory, deserves a great interest for orbital control analysis. This technique has been investigated in this paper, and led to a concept of constellation/formation control model by means of simple, spring- representing, additional potential functions.

References

Cipollone, L.: 1998, 'Controllo Orbitale di una Costellazione di Satelliti', *Thesis in Aerosp. Eng.*, Università di Roma *La Sapienza*.

Colantoni, M.: 1999, 'Autonomous Orbit-Keeping of a Satellite Constellation with several Orbital Planes', Paper IAF-99-ST-99-W.2.04, *50th Congress of the International Astronautical Federation*, Amsterdam.

Graziani, F., Palmerini, G.B., Teofilatto, P.: 1994, 'Design and Control Strategies for Global Coverage Constellations', *Journal of the Brazilian Society of the Mechanical Sciences*, **XVI**, pp. 181–187.

Kalman, R.E. and Bertram J.E. : 1970, 'Control System Analysis and Design Via the Second Method of Lyapunov', *Transactions of the ASME - Journal of Basic Engineering*, pp. 371–393.

Lamy, A., Pascal, S.: 1993, 'Station Keeping Strategies for Constellations of Satellites', Paper AAS 93-306, *AAS/AIAA Astrodynamics Specialist Conference*, Victoria.

Lasserre, E., Dufour, F., Bernoussou, J., Brousse, P., Lefebvre, L. : 1997, 'Linear Programming Solution to the Homogeneous Satellite Constellation Station Keeping', Paper IAF 97AF-97-A.4.08, *48th Congress of the International Astronautical Federation*, Turin.

McInnes, C.R. : 1995, 'Potential Function Methods for Autonomous Spacecraft Guidance and Control', Paper AAS 95-447, *AAS/AIAA Astrodynamics Specialist Conference*, Halifax.

McInnes, C.R.: 1995, 'Autonomous Ring Formation for a Planar Constellation of Satellites', *Journal of Guidance, Control and Dynamics*, **18(5)**, pp. 1215–1217.

Palmerini, G.B., Graziani, F. : 1994, 'Coverage Keeping Strategies for Low-Earth Orbit Satellite Constellations', Paper IAF-94-M.4.292, *45th Congress of the International Astronautical Federation*, Jerusalem.

Wertz, J.R., Collins, J.T., Dawson, S., Koenigsmann, H.J., Potterveld, C.W. : 1998, 'Autonomous Constellation Maintenance', In *Space Technology Proceedings: Mission Design and Implementation of Satellite Constellations* (J.C. Van der Ha, Ed.), Kluwer Academic Publ.

THE SPACE DEBRIS EVOLUTION MODELING TAKING INTO ACCOUNT SATELLITE'S COLLISIONS

ANDREY I.NAZARENKO (nazarenko@iki.rssi.ru)
Center for Program Studies, 84/32 Profsoyuznaya ul. Moscow 117810, Russia

NICKOLAY N. SMIRNOV and ALEXEY B. KISELEV (ebifsun1@mech.math.msu.su)
Moscow M.V.Lomonosov State University, Vorobiovy Gory, Moscow 119899, Russia

This paper is aimed at developing the techniques for mathematical modeling of long–term orbital debris evolution within the continual approach framework. The Russian Space Debris Prediction and Analysis (SDPA) model [1], [2], [3] is used as a basis of this study. Under this approach the equations of evolution contain some source terms responsible for variations of quantities of different fractions of orbital debris population caused by fragmentation and collisions. Our efforts were concentrated at determining the source terms for the equations of evolution, at developing the numerical–analytical technique for integrating the equations of evolution, which makes it possible to obtain the results within the reasonable time interval using modern PCs.

Our investigation indicate, that the number of collisions of small–size particles (sizing smaller than 1 cm) between each other, as well as with larger objects, is much more, than the number of mutual collisions of cataloged objects (sizing larger than 10 – 20 cm). This result testifies to the necessity of taking into account mutual collisions of space debris of different sizes.

The space debris model (SDPA), applied in this study, allows to take into account a variety of sizes of colliding particles, a variety of masses of particles of the given size, a variety of possible impact velocities as well as a variety of altitudes, where the collision can take place. All listed circumstances allow us to take into account the fragmentation model published in paper by A.B. Kiselev [4].

The technique of space debris evolution modeling is developed, which takes into account the consequences of collisions of objects of the different sizes. This process technique includes the following principal components: the fragmentation model for high–speed impact, the technique for calculation of average collision consequences, the technique for calculation of the matrix of SO mutual collision probabilities and the technique for integration of evolutional equations. Our technique is adapted to implementation on an usual PC.

The estimation of the mean contribution of SO collision consequences into cataloged SD environment is made. The objects sizing larger than 0.1 cm at altitudes up to 2000 km are considered. It is found that the maximal contribution of collision consequences is reached in the altitude range of 800 – 1000 km with accounting of all mutual collisions and equals now 33% of the general contamination level of this high–altitude layer by particles sizing 0.25–0.5 sm.

Pretka-Ziomek et al./Dynamics of Natural and Artificial Celestial Bodies, 425–426, 2001.
© 2001 *Kluwer Academic Publishers.*

The topical directions of further researches are the decreasing a low boundary of the considered particles sizes by 1–2 orders of magnitude and the accounting of the collisions of technogeneous objects with micrometeorites.

It is expedient to consider some alternate fragmentation models for studying a possible scattering of collision consequences. The authors are ready to cooperate on this problem with all interested persons and organizations.

Acknowledgements

The research described in this paper was supported by the Russian Fund for Basic Investigations (Grant No. 98–01–00218).

References

1. Kessler, D., Cour–Palais, B.: 1978, 'Collision Frequency of Artificial Satellites: the Creation of Debris Belt', *Journal of geophysical research*, **Vol. 83, A6, June 1978**.
2. Kiselev, A.: 2000, 'The Model of Fragmentation of Space Debris Particles under High Velocity Impact', *Moscow Univ. Mechanics Bulletin* **No 5**.
3. Nazarenko, A.: 1993, 'A Model of Distribution Changes of the Space Debris', *The Technogeneous Space Debris Problem*, Moscow, COSMOSINFORM.
4. Nazarenko, A.: 1996, 'Aerodynamic Analogy for Interactions between Spacecraft of Different Shapes and Space Debris', *Cosmic Research* **Vol. 34, No3**.
5. Nazarenko, A.: 1997, 'The Development of the Statistical Theory of a Satellite Ensemble Motion and its Application to Space Debris Modeling', *Second European Conference on Space Debris*, ESOC, Darmstadt, Germany 17–19 March 1997.

INCLINATION CHANGE USING ATMOSPHERIC DRAG

WALKIRIA SCHULZ[1], ANTÔNIO F.B.A. PRADO[1] and RODOLPHO VILHENA
DE MORAES[2]

[1] *National Institute for Space Research – INPE, Brasil*
[2] *DMA-FEG-UNESP, Quaratingueta, Brasil*

Abstract. The study of atmospheric maneuvers for satellites is an important aspect of the exploration of planets surfaces and atmospheres. To take more advantage of the gains due to atmospheric maneuvers, we need to develop an optimal control law to maneuver the satellite. This control law should minimize the fuel consumption required by the velocity variation due to engines on board the satellite. With this in mind, the aim of the present work is to establish whether a performance gain can be achieved using aerodynamic forces combined with propulsive ones, when the objective is changing a satellite orbital plane around the Earth.

Key words: atmospheric maneuvers, artificial satellites

1. Transfer Orbit Description

The problem discussed here is concerned with orbit transfer maneuvers. In the special case considered, an initial and a final orbit around the Earth are completely specified. The problem is to find how to transfer the spacecraft between those two orbits in such a way that the fuel consumed is minimum. There is no time restriction involved here, and the spacecraft can leave and arrive at any point in the given initial and final orbits. The maneuver is performed using an engine that is able to deliver a thrust with constant magnitude and variable direction and with the use of the atmosphere to make a plane change. The mechanism, time and fuel consumption to change the direction of the thrust are not considered in this work. An impulsive engine maneuver is used for comparison with the strategy described above.

As the maneuver goes on, the trajectory followed by the satellite can be divided into two parts: outside the atmosphere the spacecraft is supposed to follow a Keplerian motion controlled only by the thrusts (whenever they are active); inside the atmosphere, thrusts are not firing and motion is governed by two forces: Earth's non-homogeneous gravity field and atmosphere drag force.

For propulsive maneuvers purposes, thrusts are assumed to have the following characteristics: *(i)* fixed magnitude; *(ii)* constant ejection velocity; *(iii)* free angular motion; and *(iv)* operation in on-off mode.

The solution is given in terms of the time-histories of the thrusts (pitch and yaw angles) and fuel consumed. For the propulsive part of the mission, any number of "thrusting arcs" (arcs with the thrusts active) can be used for each maneuver. Instead of time, the "range angle" (angle between the radius vector of the spacecraft and an arbitrary reference line in the orbital plane) is used as the independent variable.

Pretka-Ziomek et al./Dynamics of Natural and Artificial Celestial Bodies, 427–428, 2001.
© 2001 *Kluwer Academic Publishers.*

2. Conclusions

The expected advantage for the atmospheric maneuvers procedure leans on the fact that, depending on the aerodynamic characteristics of the vehicle (bank and attack angles), the atmosphere would take charge of the target plane change, saving a significant portion of the necessary fuel for the accomplishment of the complete orbital change. This expectation is valid, because the aerodynamic characteristics can be found and chosen before the entrance of the vehicle in the atmosphere.

However, our simulations show that this problem is highly dependent of such initial conditions as mass of the vehicle and orbital elements of the initial and final orbits. Thus, it can be advantageous to accomplish a descent to the terrestrial atmosphere in certain cases and completely disadvantageous in others. It is also necessary to take in consideration the fact that a vehicle, that will face the current heating conditions of one or more passages by the atmosphere, should be appropriately prepared with coatings or other protection systems. This adaptation can mean more mass due to the increment of insulating material and, consequently, a smaller quota of available mass on board for the storage of fuel. Thus, resulting in aeroassisted maneuvers which do not imply significant fuel savings.

Acknowledgements

The authors are grateful to FAPESP (Foundation to Support Research in São Paulo State) for the contract number 96/11205-5.

References

Mease, K. D.: 1988, 'Optimization of aeroassisted orbital transfer: current status', *Journal of the Astronautical Sciences*, **36**, 1/2, pp. 7–33.

Miele, A.: 1996, 'Recent advances in the optimization and guidance of aeroassisted orbital transfers', *Acta Astronautica*, **38**, 10, pp. 747–768.

Regan, F. J. and Anandakrishnan, S. M.: 1993, *Dynamics of atmospheric re–entry*, Washington, DC: American Institute of Aeronautics and Astronautics, Inc., 584 p.

Vinh, N. X.: 1981, *Optimal trajectories in atmospheric flight*, Amsterdam, The Netherlands: Elsevier, 402 p.

Walberg, G. D.: 1985, 'A survey of aeroassisted orbit transfer', *Journal of Spacecraft and Rockets*, **22**, 1, pp. 3–18.

AN OPTICAL SCANNING SEARCH FOR GEO DEBRIS

HIROAKI UMEHARA (ume@cr1.go.jp) and KAZUHIRO KIMURA
Kashima Space Research Center, Communications Research Laboratory (CRL) Hirai, Kashima, Ibaraki, 314-0012 Japan (http://www. crl. go.jp/ka/control/people/ume/index-e.html)

A geostationary orbit (GEO) is invaluable to communications, broadcasting, and so on. The number of space debris near GEO is growing, as is the amount of geostationary satellites. Surveys and orbit-determinations of GEO debris are necessary to avoid collisions with satellites (Arimoto et al., 1994). Up until now, the U.S. and Rnssia have taken the initiative in surveillance (National Research Council, 1995). In Japan, two high-resolution telescopes are being constructed to detect small unknown objects (Isobe et al., 1999). Many currently unknown debris will soon be discovered. Concerns exist about the operating expense associated with the detection and the orbit determination. A broad, systematic, and efficient method to search for debris is designed in this paper. To confirm the validity of the method, test observations were made using a telescope in our laboratory, CRL. Let the three aims be further specified. A broad search is performed by scanning GEO. To do a systematic search, surveyed ranges should be sorted according to orbital elements. Logging the surveyed regions of the horizontal coordinates is not vital. An efficient search is considered to be reduce the number of observations. We know that at least two night observations are necessary to accurately determine the orbit of GEO debris (Kawase, 2000). So, the same objects must be detected on two nights. A fixed point is observed in the equatorial-celestial coordinates in two or three nights. The point is located at $\bar{\alpha}$ [deg] right ascension, zero declination, and distance from the geocenter to the GEO. This observation corresponds to a GEO scan done on the ground. So, this is a broad observation. The telescope is rotated in the western direction at $(\Delta t/240)$-degree intervals, and CCD images are taken at Δt-second intervals. It is necessary to take two images of each object to distinguish it from dark CCD noise. As a result, we can evaluate approximate time derivatives of measured values. The geodetic longitude λ [deg] and latitude θ [deg] are expressed by

$$\theta(t) = -I_x c(t) - I_y s(t), \lambda(t) = \bar{\lambda}(t) + \frac{360}{\pi}\{e_x c(t) - e_y s(t)\},$$

$$\bar{\lambda}(t) = \lambda(t_\alpha) + \frac{D}{T}(t - t_\alpha), c(t) = \cos\frac{2\pi}{T}t, s(t) = \sin\frac{2\pi}{T}t, \tag{1}$$

where t [sec] is the time. The inclination vector (I_x [deg], I_y [deg]), the period T [sec], and the drift rate per period D [deg/day] are approximated. Only the eccentricity vector (e_x, e_y) is not given by this scan.

Notice that the pieces of debris with a definite range of the right ascension of ascending node are selected if the inclination of the detected debris is larger than the latitudinal angle of view. This scan, therefore, provides a systematic observation. Follow-up observations are necessary to determine (e_x, e_y). Doing the above scan, the same objects can be easily detected several nights later. In fact, $\theta(t)$ is calculated based on information from the above scan. Therefore, a search along the line of longitude with the expected $\theta(t)$ is sufficient to detect the same objects. This decreases operating costs. This method of detection, therefore, allows an efficient observation. Moreover, this procedure is appropriate for parallel measurement of many pieces of GEO debris.

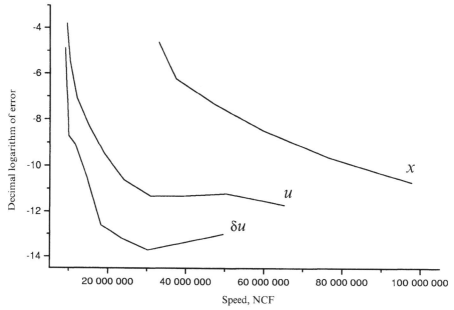

Figure 1. Coordinates of GEO objects. o: 10, +: 11, x: 12 Dec. 1999.

Figure 1 shows the result of a scan at $\bar{\alpha} = 90$ [deg]. The scan was repeated over the three consecutive nights. If an object was detected at a time t_0, the coordinates $(\alpha(t_0), \theta(t_0))$ and the right ascension (t_0) were measured. The same object was detected at the time $t_0 + \Delta t$. Assume that t is the time when the object passes through $\bar{\alpha}$. Then,

$$X_\alpha = X(t_0) + \frac{\Delta X}{\Delta \alpha}(\bar{\alpha} - \alpha_0), \quad t_\alpha = t_0 + \frac{\Delta t}{\Delta \alpha}(\bar{\alpha} - \alpha_0), \qquad (2)$$

where $\quad \Delta X = X(t_0 + \Delta t) - X(t_0), \quad X \in \{\lambda, \theta, \alpha\}$.

The coordinates $(\lambda_\alpha, \theta_\alpha)$ of the objects were plotted in the figure. Seven drifting objects were detected. We were able to estimate I_x, I_y, T and D. Follow-up observations and orbit determinations were successful.

We used a 35-cm telescope with F3.6 located at 140 degE and 36 degN. This telescope has a CCD of 1.6 million pixels, shutter with a 10-msec error from UTC, and the software for detecting and measuring GEO objects. The authors express their thanks to Dr. Seiichirou Kawase for assisting in orbit determination.

References

Arimoto, Y., Takaini, H., Hiromoto, N., Sawada, N., and Aruga, T.: 1994, *Journal of the Communications Research Laboratory*, 41[**3**] pp. 195–207.

National Research Council: 1995, *National Academy Press*, Washington, D.C.

Isobe, S. and Japanese Spaceguard Association: 1999, *Advances in Space Research* **23**, pp. 33–36.

Kawase, S.: 26-30 June 2000, *15th International Symposium "Spaceflight Dynamics"*, Biarritz, France, MSOO/43.

CO–LOCATION OF GEOSTATIONARY SATELLITES BY IMAGINARY–INTERACTION MANEUVERS

HIROAKI UMEHARA (ume@crl.go.jp)
Kashima Space Research Center, Communications Research Laboratory (CRL) Hirai, Kashima, Ibaraki, 314-0012, Japan (http://www.crl.go.jp/ka/control/people/ume/index-e.html)

Increased use of the geostationary orbit has led to intensive study of co-location methods. Dynamical features and optimal control of cluster-satellite systems have been studied in the CRL (Sawada and Kawase, 1997). The ultimate goal of this study is to realize *autonomous uncoordinated co-location* that will enable to be co-located without monitoring even if various organizations put satellites into orbit without considering the locations of other satellites. As a preliminary step, autonomous *coordinated* co-location is developed in this paper.

McInnes (1995) considered autonomous formation-keeping in a planar many-satellite system. In his model, a desired constellation of satellites is defined as a regular-polygon configuration with a geocenter. McInnes introduced *imaginary interaction* in the orbit control. Continuous maneuvers are performed in the longitudinal and radial directions as if repulsive interaction with frictional force were applied to the satellites. The desired constellation thus corresponds to the equilibrium state of an imaginary-interaction system. The system leads to the equilibrium state.

Here, the imaginary interaction is applied to eccentricity-inclination separation (e-i separation). Use of this separation is a convenient co-location strategy (Soop, 1994). Assume that the only maneuvers needed is the impulsive thrusts in the longitudinal direction. An n-satellite system with non-equal σ_i ($i = 1, 2, \ldots, n$) is considered, where σ_i means the effective cross-section to mass ratio of the i-th satellite. If σ_i is equal for all satellites, the maneuver schedule for the e-i separation is simple. All co-located satellites are influenced by common perturbations. Both the relative eccentricity-vectors and the relative inclination-vectors are translated parallelly. Simultaneous thrusts of equal strength maintain the e-i separation. However, near-misses may occur with non-equal σ_i. Figure 1(a) shows a simulated result of the sun-point-perigee strategy for all satellites. The trajectories of $\boldsymbol{e}_i$ are represented in the nine-satellite system, where $\boldsymbol{e}_i$ is the eccentricity vector of satellite i. Two vectors coincide with each other. The two corresponding satellites experience a near-miss at least twice a day. The multiple-maneuver strategy is one way to avoid near-misses. However, fuel consumption and operational cost are still high. Through the strategy described below, the vectors are separated with a single maneuver. Let the terminal points of $\boldsymbol{e}_i$ ($i = 1, 2, \ldots, n-1$) be initially positioned at the vertices of the regular $(n-1)$-polygon. Let the terminal point of $\boldsymbol{e}_n$ be located at the center of the polygon. Where does $\boldsymbol{e}_i$ jump to in response to an impulsive thrust? The direction of the vector shift depends on the thrust time. The distance of

Pretka-Ziomek et al./Dynamics of Natural and Artificial Celestial Bodies, 433–434, 2001.
© 2001 *Kluwer Academic Publishers.*

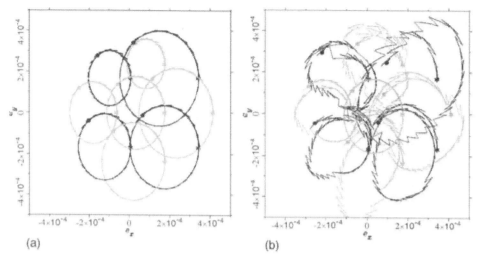

Figure 1. Eccentricity vectors: (a) sun-pointing-perigee strategy, (b) imaginary-interaction strategy.

the shift depends on the thrust strength. Only e_n is assumed to be controlled by the sun-point-perigee strategy. The other vectors are controlled by the $(n-1)$-particle dynamics on a ring with frictional repulsive interaction. The interaction is induced by the potential:

$$U(\phi_1, \phi_2, \ldots, \phi_{n-1}) = \sum_{i=1}^{n-2} \sum_{j=i+1}^{n-1} \frac{1}{\cos^2(\phi_{ij})}, \quad \phi_{ij} = [\phi_i - \phi_j + \pi]/2. \quad (1)$$

Here, ϕ_i is the angle between the e_x-axis and $e_i - \bar{e}$, where $\bar{e}$ is e_n after the sun-point-perigee thrust. The isotropic configuration corresponds to the equilibrium state, i.e., the minimum of U. The coincidence $\phi_i = \phi_j$ ($j \neq i$) results in divergence of U. Impulsive thrusts are performed as the eccentricity vectors jump to the equilibrium configuration. Figure 1(b) shows the result of using the imaginary-interaction strategy. The eccentricity vectors maintain their separation although the vector trajectories fluctuate. The terminal points form approximately a regular octagon at the same time as when two vectors coincide in Fig. 1(a).

This paper is a sequel of Umehara (2000) written in Japanese. The author thanks Dr. Seiichirou Kawase for initially giving me the rough idea for the imaginary-interaction guidance.

References

McInnes, R. C.: 1995, *J. Guidance, Control, and Dynamics* **18**, pp. 1215–1217.
Sawada, F. and Kawase, S.: 1997, *Proceedings of the 12th International Symposium on "Space Flight Dynamics"*, ESOC, Darmstadt, Germany, pp. 65–69.
Soop, E. M.: 1994, *Kluwer*, Sect.5.6.
Umehara, H.: 2000, *Journal of the Japan Society for Aeronautical and Space Sciences* **48**, pp. 118-121 (in Japanese).

EVOLUTION OF SUPER-GEOSTATIONARY ORBITS ON A TIME SPAN OF 200 YEARS

IWONA WYTRZYSZCZAK, Sł AWOMIR BREITER and BARTOSZ MARCINIAK
Astronomical Observatory of A. Mickiewicz University, Poznań, Poland

The evolution of a randomly generated population of 6000 super-geostationary orbits was studied on a period of 200 years. Equations of motion, in rectangular coordinates, were integrated by means of a sixth-order explicit symplectic integrator (Kinoshita et al., 1991). The perturbations were: the geopotential effects up to degree and order 4, the gravitational influence of Sun and Moon, and the direct solar radiation pressure without taking into consideration the shadow crossing.

The elements of super-geostationary orbits were randomly generated in the following way: the initial epochs were distributed over one Julian year starting from the epoch JD2000; the perigee h_p and apogee h_a altitudes were uniformly taken from the interval of $50 - 1000$ km above the stationary radius $A = 42164.16943$ km (the semi-major axes and eccentricities were computed according to the values h_p and h_a); the inclinations of the orbits were taken randomly from the range of $0°$ to $15°$, and the Keplerian elements Ω, ω and M of a satellite were drawn from the interval of $0°$ to $360°$.

The perigee height was checked at every step of integration. When it decreased below the security zone of 35 km above the geostationary radius, the integration was interrupted and the initial as well as final orbital elements were stored in a separate file.

From the sample of 6000 orbits, only 139 (2%) crossed the security limit of 35 km. From the qualitative point of view this result is a little less than the 5% ratio obtained by Breiter (1998) but is a consequence of taking into account the initial eccentricities from a more narrow interval.

Figures 1a and 2a present respectively the distribution of initial semi-major axes vs. eccentricities, and of perigee vs. apogee heights above the geostationary ring for all initial orbits. Figure 1b presents the distribution of initial semi-major axes against eccentricities of 'unsafe' orbits and Figure 2b – the distribution of their initial perigee and apogee heights.

There is a strict correlation between the semi-major axes of 'unsafe' orbits and their eccentricities. Even very low orbits can be 'safe' over the time span of 200 years if their eccentricities are small enough. For example, an orbit with a semi-major axis of 100 km above the geostationary radius may be 'safe' if its eccentricity is less than 0.001, whereas for an orbit with the semi-major axis of 500 km the eccentricity should be less than 0.01.

The fact that initial perigee heights of all 'unsafe' orbits are below 100 km does not testify the 'absolute safety' of higher orbits but rather is an artifact resulting

Pretka-Ziomek et al./Dynamics of Natural and Artificial Celestial Bodies, 435–436, 2001.
© 2001 *Kluwer Academic Publishers.*

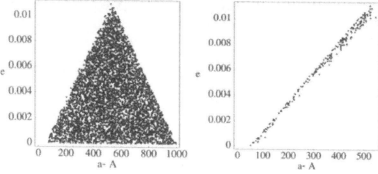

Figure 1. Distribution of initial elements of a) the tested orbits, b) the 'unsafe' orbits. Units are kilometers.

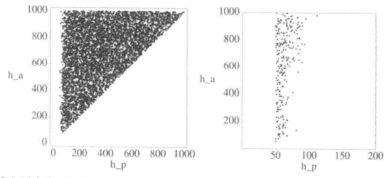

Figure 2. Initial distributions of perigee and apogee heights above the geostationary ring for a) all orbits b) those which crossed the safety region. Units are kilometers.

from the choice of the sample with small eccentricities (up to 0.01 - Figure 1a). No other significant correlations have been detected. More detailed analysis of a bigger sample of orbits is presented in a paper by Wytrzyszczak and Breiter (2000).

References

Breiter, S.: 1998, Testing the Safety of Storage Orbits Beyond the Geostationary Ring, *Artificial Satellites* **33**, 25–29.

Kinoshita, H., Yoshida, H. and Nakai, H.: 1991, Symplectic Integrators and their Application to Dynamical Astronomy, *Celest. Mech. Dynam. Astron.* **50**, 50–71.

Wytrzyszczak, I. and Breiter, S.: 2000, Long Term Evolution of Disposal Orbits Beyond the Geostationary Ring, *Advances in Space Research*, in press.

DYNAMICAL METHOD FOR THE ANALYSIS OF THE SYSTEMATIC ERRORS IN STELLAR CATALOGUES

J.A. LÓPEZ AND F.J. MARCO
Departamento de Matemáticas Univ Jaume I de Castellón, Spain
E-mail: lopez@mat.uji.es, marco@mat.uji.es

M.J. MARTÍNEZ
Departamento de Matemática Aplicada Univ Politecnica de Valencia, Spain
E-mail:mjmartin@mat.upv.es

Abstract. The recent introduction of the new reference frame and new star catalogues gives special interest to the study of the probable existence of systematic errors among them and between each catalog and the reference frame.

Several methods are currently used to approach to this last process. One of them is the analysis of the residuals given by the Observation minus Calculus errors of the minor planets. In this paper it is developed a method of analysis of the residual function for each asteroid. This method is exact up to second order and it allows the development of more accurate expressions to determine the signal and the initial constants. A determination from 1850-1996 ground based observations is done.

Key words: asteroids, reference frames, catalogues.

1. Introduction

One of main problems in astronomy is the determination of the errors in the reference catalogues. There are some methods to study these errors. The main dynamical methods for this purpose are based in meridian observations of the sun and the inner planets, in the analysis of the occultations of the stars by the moon and in observations of minor planets.

The idea of using minor planets to analyze the reference systems was proposed by Numerov(1935). Branham (1994) made an analysis of the equinox and the equator of FK5 catalog based in ground minor planets observations. His results for J2000 are $\Delta A = .''186 \pm .''086$. Batrakov (1999) obtained a correction $\Delta A = .''066 \pm .''041$. In this paper the differences among the catalogues based in the FK5 and the dynamical reference, based in the planetary theory, are studied using observed minor planets positions. We use observed positions taken from the M.P.C. (Mardens, 1999) of the Smitsonian Astrophysical Observatory. These positions are referred to the FK5 system, and the planetary theory VSOP87 Bretagnon (1988).

2. Functional Model

Let $\{\{(\alpha_i^r, \delta_i^r)\}_{i=1}^{n_r}\}_{r=1}^{N}$ be a set of observed positions, of the minor planet r at the epoch t_i^r referred to the FK5 system. Let $(\sigma_{r,1}^o, \ldots, \sigma_{r,6}^o)$ be the orbital elements of the asteroid r at the initial osculation epoch. From this set of elements we can obtain the orbital elements for each epoch t_i' by means of the Lagrange's planetary equations. These equations are integrated using a Gragh-Burlish-Stoer (Burlish, 1966) algorithm. The second member of the Lagrange's planetary equations is evaluated using the x,y,z coordinates given from the planetary theory VSOP87 for the planets (the earth- moon barycenter is considered instead of the earth), we also include four asteroids in a Keplerian orbit. The theoretical topocentric position of the asteroid at the epoch is obtained by means of a reduction to topocentric coordinates, in this process the planetary aberration is included.

The differences between the calculated and observed coordinates are due to errors in the reference from which the observed positions are reduced and also to errors in the initial osculating elements of the minor planet r. The work is divided in two tasks: first, the improvement of the orbital elements by means of an individual correction and, in the second place, a global correction of a reference system.

2.1. IMPROVEMENT OF ORBITAL ELEMENTS OF ASTEROIDS

One of the most popular methods for orbital elements corrections was developed by Brouwer (1961). An extension of this method was developed by Marco (1996). A new extension up to second order is proposed.

The variation in the calculated spherical coordinates $(\alpha(t_i), \delta(t_i))$ of the contribution of the asteroid errors $\Delta\vec{\sigma_o}$ to the individual residual O-C in right ascension and declination is given by:

$$\Delta\alpha^{calc} = \alpha(\vec{\sigma_o} + \Delta\vec{\sigma_o}, t_i) - \alpha(\vec{\sigma_o}, t_i) \tag{1}$$

$$\Delta\delta^{calc} = \delta(\vec{\sigma_o} + \Delta\vec{\sigma_o}, t_i') - \delta(\vec{\sigma_o}, t_i) \tag{2}$$

Up to second order in $\Delta\vec{\sigma_{r,o}}$ we can write

$$\Delta\alpha^{calc} = \sum_{i=1}^{6} \frac{\partial\alpha}{\partial\sigma_i^o} \Delta\sigma_i^o + \frac{1}{2} \sum_{i=1}^{6} \sum_{j=1}^{6} \frac{\partial^2\alpha}{\partial\sigma_i^o \partial\sigma_j^o} \Delta\sigma_i^o \Delta\sigma_{.j}^o \tag{3}$$

$$\Delta\delta^{calc} = \sum_{i=1}^{6} \frac{\partial\delta}{\partial\sigma_i^o} \Delta\sigma_i^o + \frac{1}{2} \sum_{i=1}^{6} \sum_{j=1}^{6} \frac{\partial^2\delta}{\partial\sigma_i^o \partial\sigma_j^o} \Delta\sigma_i^o \Delta\sigma_j^o, \tag{4}$$

where it is necessary to evaluate the following derivatives $\frac{\partial\alpha}{\partial\sigma_i^o}, \frac{\partial\delta}{\partial\sigma_i^o}, \frac{\partial^2\alpha}{\partial\sigma_i^o\partial\sigma_j^o}, \frac{\partial^2\delta}{\partial\sigma_i^o\partial\sigma_j^o}$. To this aim, we can write for the first and second order derivatives, the operators:

$$\frac{\partial}{\partial\sigma_k^o} = \sum_{i=1}^{6} \frac{\partial\sigma_i}{\partial\sigma_k^o} \frac{\partial}{\partial\sigma_i} \tag{5}$$

$$\frac{\partial^2}{\partial\sigma_k^o \partial\sigma_s^o} = \sum_{i=1}^{6} \frac{\partial^2\sigma_i}{\partial\sigma_k^o \partial\sigma_s^o} \frac{\partial}{\partial\sigma_i} + \sum_{i=1}^{6} \sum_{j=1}^{6} \frac{\partial\sigma_j}{\partial\sigma_k^o} \frac{\partial\sigma_i}{\partial\sigma_s^o} \frac{\partial^2}{\partial\sigma_i\partial\sigma_j} \tag{6}$$

To evaluate the quantities $\frac{\partial \sigma_i}{\partial \sigma_k^o}$, and $\frac{\partial^2 \sigma_i}{\partial \sigma_k^o \partial \sigma_s^o}$ we proceed as follows: let $\frac{d\vec{\sigma}}{dt} = \vec{f}(\vec{\sigma}, t)$ be the Lagrange's planetary equations, and from the derivative operators we get

$$\frac{\partial}{\partial \sigma_k^o}\left(\frac{d\vec{\sigma}}{dt}\right) = \frac{\partial}{\partial \sigma_k^o}\vec{f}(\vec{\sigma}, t) = \sum_{i=1}^{6}\frac{\partial \sigma_i}{\partial \sigma_k^o}\frac{\partial}{\partial \sigma_i}\vec{f} \tag{7}$$

$$\frac{\partial^2}{\partial \sigma_k^o \partial \sigma_s^o}\left(\frac{d\vec{\sigma}}{dt}\right) = \sum_{i=1}^{6}\frac{\partial^2 \sigma_i}{\partial \sigma_k^o \partial \sigma_s^o}\frac{\partial}{\partial \sigma_i}\vec{f} + \sum_{i=1}^{6}\sum_{j=1}^{6}\frac{\partial \sigma_j}{\partial \sigma_k^o}\frac{\partial \sigma_i}{\partial \sigma_s^o}\frac{\partial^2}{\partial \sigma_i \partial \sigma_j}\vec{f} \tag{8}$$

The temporal derivative is a Lie one and derivatives with respect to the initial elements are exterior ones so, we can swap them and we get for r,k,s=1,..,6

$$\frac{d}{dt}\left\{\frac{\partial \sigma_r}{\partial \sigma_k^o}\right\} = \sum_{i=1}^{6}\frac{\partial \sigma_i}{\partial \sigma_k^o}\frac{\partial f_r}{\partial \sigma_i} \tag{9}$$

$$\frac{d}{dt}\left\{\frac{\partial^2 \sigma_r}{\partial \sigma_k^o \partial \sigma_s^o}\right\} = \sum_{i=1}^{6}\frac{\partial^2 \sigma_i}{\partial \sigma_k^o \partial \sigma_s^o}\frac{\partial f_r}{\partial \sigma_i} + \sum_{i=1}^{6}\sum_{j=1}^{6}\frac{\partial \sigma_j}{\partial \sigma_k^o}\frac{\partial \sigma_i}{\partial \sigma_s^o}\frac{\partial^2 f_r}{\partial \sigma_i \partial \sigma_j} \tag{10}$$

through the integration of this ordinary differential equations system with the initial values $\sigma_i(t_o) = \sigma_i^o$, $\left\{\frac{\partial \sigma_i}{\partial \sigma_k^o}\right\}_{t=t_o} = \delta_{i,k}$ where $\delta_{i,k}$ is the Kroneker symbol, and $\left\{\frac{\partial^2 \sigma_i}{\partial \sigma_k^o \partial \sigma_s^o}\right\}_{t=t_o} = 0$, $i, j, k = 1, .., 6$ we get the involved quantities. To integrate them it is necessary the evaluation of the first, second and third derivatives of the disturbing potential. For this purpose we use the Simon (1988) method: $\frac{\partial^2 \Re}{\partial \sigma_i \partial \sigma_m} = \vec{V_i}.(\partial^2 \Re).\vec{V_m} + \vec{V_{i,m}}.\partial \Re$ where $\vec{V}$ is the ecliptic vector of position for the asteroid, $\vec{V_m}$, the derivative of $\vec{V}$ with respect to σ_m,, $\vec{V_{i,m}}$, the second derivatives, $\partial \Re$ and $\partial^2 \Re$ are, respectively, the gradient and the Hessian matrix of the disturbing potential with respect to the elements. Starting from these quantities we can get $\Re_{k,j,l} \equiv \frac{\partial^2 \Re_k}{\partial \sigma_j \partial \sigma_l}$, $\Re_{kjl} = \vec{V_{k,j,l}}.\partial \Re + \vec{V_{k,j}}(\partial^2 \Re)\vec{V_l} + \vec{V_{k,l}}(\partial^2 \Re)\vec{V_j} + \vec{V_k}(\partial^3 \Re \otimes \vec{V_l})\vec{V_j} + \vec{V_k}(\partial^2 \Re)\vec{V_{j,l}}$ where we use the notation: $\partial_i^2 \Re = \frac{\partial \Re}{\partial x_i}$ i=1,2,3 ; $\partial^3 \Re = [\partial_1^2 \Re, \partial_2^2 \Re, \partial_3^2 \Re]$; $\vec{V_{k,j,l}} = \frac{\partial V_{k,j}}{\partial \sigma_l}$

The differences between observed and improved calculated spherical coordinates of minor planet are $\Delta \alpha_k = \Delta_0 \alpha_k + \Delta_1 \alpha_k + \Delta_2 \alpha_k$ and $\Delta \delta_k = \Delta_0 \delta_k + \Delta_1 \delta_k + \Delta_2 \delta_k$, where the subindex in Δ represents the order in $\Delta \vec{\sigma_o}$.

The residual function is R($\Delta \vec{\sigma_o}$) = $\sum_{k=1}^{N}(\Delta \alpha_k^2 \cos^2 \delta_k + \Delta \delta_k^2)$, where N is the number of observations for each minor planet. The residual function is developped up to second order in $\Delta \vec{\sigma} = \vec{W}$. From the minimum condition for R we get the improved normal system:

$$S\vec{\Delta \sigma^o} = \vec{W}, \tag{11}$$

where the components of the matrix S and the vector $\overrightarrow{W}$ are

$$
(S)_{j,k} = \sum_{i=1}^{N} \left[\frac{\partial \alpha}{\partial \sigma_j^0}(t_i) \frac{\partial \alpha}{\partial \sigma_k^0}(t_i) - \Delta_0 \alpha_i \sum_{j=1}^{6} \frac{\partial^2 \alpha}{\partial \sigma_k^o \partial \sigma_j^o}(t_i) \right] \cos^2 \delta_i +
$$

$$
+ \sum_{i=1}^{N} \left[\frac{\partial \delta}{\partial \sigma_j^0}(t_i) \frac{\partial \delta}{\partial \sigma_k^0}(t_i) - \Delta_0 \delta_i \sum_{j=1}^{6} \frac{\partial^2 \delta}{\partial \sigma_k^o \partial \sigma_j^o}(t_i) \right] \tag{12}
$$

$$
(W)_j = \sum_{i=1}^{N} \left[\Delta_0 \alpha_i \frac{\partial \alpha}{\partial \sigma_j^0}(t_i) \cos^2 \delta_i + \Delta_0 \delta_i \frac{\partial \delta}{\partial \sigma_j^0}(t_i) \right] \tag{13}
$$

2.2. GLOBAL CORRECTION AT REFERENCE SYSTEM

The second task in this work is the analysis of bias in the reference system. Let the FK5 catalog system axis be defined by three small rotation angles $\epsilon_x, \epsilon_y, \epsilon_z$ (clockwise) around the axis OX,OY,OZ of the dynamical reference frame. In first order we have:

$$
\overrightarrow{X_{Cat}} = R_1(\epsilon_x) R_2(\epsilon_y) R_3(\epsilon_z) \overrightarrow{X_D}, \tag{14}
$$

where $\overrightarrow{X_D}$ is the position in the Dynamical reference and $\overrightarrow{X_{Cat}}$ is the position in the FK5 system.

The differences in the spherical coordinates are

$$
\Delta \alpha = \alpha_{Cat} - \alpha_D = \epsilon_x \tan \delta \cos \alpha + \epsilon_y \tan \delta \sin \alpha - \epsilon_z \tag{15}
$$

$$
\Delta \delta = \delta_{Cat} - \delta_D = -\epsilon_x \sin \alpha + \epsilon_y \cos \alpha \tag{16}
$$

These angles are connected with the Euler angles $\Delta A, \Delta L, \Delta \varepsilon$ by $\epsilon_x = -\Delta \varepsilon$, $\epsilon_y = \Delta L \sin \varepsilon$, $\epsilon_z = \Delta A - \Delta L \cos \varepsilon$

The residual function can be represented by $R(\epsilon_x, \epsilon_y, \epsilon_z) = \sum_{r=1}^{N} \sum_{i_r=1}^{n_r} \{\Delta \alpha_{r,i_r}^2 \cos^2 \delta_{r,i_r} + \Delta \delta_{r,i_r}^2\}$, where $\Delta \alpha_{r,i_r} = \alpha_i^r - \alpha(\overrightarrow{\sigma_{r,o}} + \Delta \overrightarrow{\sigma_{r,o}}, t_i)$ and $\Delta \delta_{r,i_r} = \delta_i^r - \delta(\overrightarrow{\sigma_{r,o}} + \Delta \overrightarrow{\sigma_{r,o}}, t_i)$ the normal system derived from the minimum of global residual function resolution gives the values for $\epsilon_x, \epsilon_y, \epsilon_z$.

3. Numerical Results

The method exposed in section 2.1 is applied to a set of positions of minor planets observations between 1850–1996. The initial values are given by integration from the I.T.A. tables (Batrakov, 1997), the positions from the M.P.C.s of the S.A.O. and the J2000 are selected for the osculation epoch elements. Table I contains the

DYNAMICAL METHOD FOR THE ANALYSIS OF THE SYSTEMATIC ERRORS

TABLE I
The elliptical elements at the osculating epoch J2000(FK5)

ast	a	e	i	Ω	ω	M
1	2.7664961	.0783723	10.58352	80.49370	73.9235	6.17659
2	2.7723227	.2296400	34.84632	173.19772	310.26548	352.96020
3	2.6680351	.2584429	12.96734	170.17240	248.03087	240.26946
4	2.3615351	.0900230	7.13407	103.95110	149.58936	341.02147
6	2.4248415	.2021553	14.76791	138.86464	238.94907	46.07502
7	2.3853371	.2303998	5.52420	259.87696	145.07568	53.40430
11	2.4530721	.0991558	4.62161	125.65836	194.29923	270.03568
18	2.2951431	.2181028	10.12863	150.57890	227.65986	68.26790
25	2.4001070	.2562072	21.57916	214.30869	90.40548	102.03851
39	2.7699114	.1115526	10.36845	157.24206	208.51464	232.76724
40	2.2675744	.0461007	4.25612	94.31649	268.94117	231.43949

TABLE II
Number of observations and errors before and after reduction

1	2	3	4	6	7	11	18	25	39	40
3804	4190	3389	3111	2179	1804	1789	1596	492	2305	1481
2"41	2"22	2"27	2"32	2"40	2"57	2"48	2"49	2"50	2"29	2"64
1"84	1"58	1"59	1"76	1"71	1"77	1"66	1"63	1"69	1"67	1"84

improved elements for these epoch of osculation. The mean quadratic errors are given in seconds before and after the improvement. They are contained in Table II: The mean quadratic errors are given in seconds before and after the improvement. They are contained in Table II:

The numerical results for the rotations angles are:

$$\epsilon_x = -."015 \pm ."011 \epsilon_y = ."009 \pm ."010 \epsilon_z = ."041 \pm ."008$$

and for the zero point of the FK5 we have $\Delta A = ."061 \pm ."034$

4. Concluding Remarks

The orientation of reference frame FK5 with respect to a dynamic reference has been determinated. The results may be compared with other determinations. They are compatible with the Equinox correction given by other authors. This method, together with a new reduction of astrometric plates with respect to the new astrometric catalogues, can be used to determine the orientation errors of some catalogues.

References

Batrakov, Yu.V. and Shor, V.A.: 1997, 'Ephemerides of Minor Planets for 1998', *Inst. Theoretical Astronomy St. Petersburg* Russia.

Batrakov, Yu.V., Chernetenko, Yu.A, Gorel, G.K. and Gudkova, L.A.: 1999, 'Hipparcos catalogue orientation as obtained from observations of minor planets', *Astron. Astrophys.* **Vol. 352**, pp. 703–711.

Branham, Jr. and Sanguin, J.: 1996, 'The FK5 Equator and Equinox', In: López,A. e.a (Eds.) *Proceedings of the III International Workshop on Positional Astronomy and Celestial Mechanics*, Cuenca (1994) Spain, pp. 429–435.

Bretagnon, P. and Francou, G.: 1988, 'Variations Seculaires des Orbites Planetaires. Thèorie VSOP87', *Astron. Astrophys.* **Vol. 114**, pp. 69–75.

Brower, D. and Clemence, G.: 1961, 'Methods of Celestial Mechanics', *Academic Press*.

Burlish, R. and Stoer, J.: 1966, 'Numerical treatment of ordinary differential equations by extrapolation methods', *Numerische Mathematik* **Vol. 8**, pp. 1–13.

Marco, F.J., López, J.A, Martinez, M.J.: 1996, 'A Time Dependent Extension to Brower's Method for Orbital Elements Corrections', *I.A.U. Symp.* **172**, pp. 199–202.

Marsden, B.G.: 1999, M.P.C. Electronic Version, Minor Planets Center. S.A.O.

Numerov, B.V.: 1935, *Journal des Observateurs XVIII* **Vol. 4**, pp. 57– (Translated from the original in Russian).

Simon, J.L.: 1988, 'Calcul des dèrivées premieres et secondes des équations de Lagrange par analyse harmonique', *Astron. Astrophys.* **Vol. 175**, pp. 303–308.

RESONANT EXCITATION OF DENSITY WAVES IN GALAXIES

EVGENY GRIV, MICHAEL GEDALIN and DAVID EICHLER
Department of Physics, Ben-Gurion University, Beer-Sheva 84105, Israel

Abstract. Linear kinetic theory is developed to describe the resonant Landau-type excitation of spiral density waves in a self-gravitating, rapidly and nonuniformly rotating, spatially inhomogeneous, and collisionless stellar disk of flat galaxies. This Landau excitation of waves is suggested as a mechanism for the formation of observable nonradial structural features, such as spiral arms, and the slow dynamical relaxation of disk galaxies, in a parameter regime of Jeans stability.

Key words: Galaxies, kinematics and dynamics, structure, instabilities and waves

1. Introduction

Comparatively recently a new type of unstable spiral density waves was found in collisionless N-body simulations of galactic disks by Sellwood and Lin (1989) and Donner and Thomasson (1994). True recurrent instabilities of small-amplitude collective oscillations developed in the numerical models spontaneously through nonlinear changes to the particle distribution function. The shape of the spirals corresponds to Toomre's (1964) critical wavenumber $k_T = \kappa^2/2\pi G\sigma_0$, where κ is the epicyclic frequency and σ_0 is the equilibrium surface density, and certainly is not shape predicted by the standard Lin–Shu–Kalnajs dispersion relation for linear density waves (Sellwood and Lin, 1989, p. 999; Donner and Thomasson, 1994, Fig. 7). Spiral amplitudes were greatly in excess of an expectation from the level of particle noise, and amplitudes do not scale with the number of particles. Therefore the new type of spontaneous spiral activity observed in N-body experiments cannot be explained by the simple Toomre (1990) and Toomre and Kalnajs (1991) "kaleidoscope of chaotic arm features" which are responses to the random density irregularities orbiting within the particulate disk. Also, it seems unlikely that the mechanism of groove modes suggested by Sellwood and Kahn (1991) explains recurrent spiral patterns in disk galaxy simulations. This is because growing aperiodically "negative mass instabilities" provoked by grooves in a self-gravitating medium are no more than the ordinary Jeans-type instabilities. The latter instabilities can be suppressed by random velocity spreads due to a finite "temperature" of the disk (Lovelace and Hohlfeld, 1978). In addition, the groove instability discussed first by Lovelace and Hohlfeld (1978) needs a special structure in the distribution function for its excitation. Such a structure, however, is hardly met in galactic disks. Moreover, the properties of recurrent spirals do not correspond to any of the known instabilities. The mode is driven by gradients

Pretka-Ziomek et al./Dynamics of Natural and Artificial Celestial Bodies, 443–452, 2001.
© 2001 *Kluwer Academic Publishers.*

444   GRIV ET AL.

in the phase space density of model particles at corotation resonance. According to Sellwood and Lin (1989), it therefore strongly resembles *Landau excitation* in plasmas (inverse Landau damping effect).

The discovery by N-body simulations of the new type of spiral activity seems to solve some basic problems of the dynamics of stellar systems. In particular, recurrent spiral patterns may be responsible for observed density wave spiral structures and dynamical relaxation of flat galaxies. In the present work, motivated by the N-body calculations, a Landau excitation of oscillations in a disk of mutually gravitating stars in flat galaxies is proposed as a driver of spiral arms. The problem is studied by applying plasma kinetic theory.[1] A local Maxwellian distribution is assumed since the beginning, so an instability reported here has nothing to do with the groove mode.

2. Equilibrium

Since Chandrasekhar's (1960) studies, in stellar systems pairwise star–star gravitational encounters are well recognized to have no influence on the system's evolution. Then the collisionless Boltzmann kinetic equation for the distribution function of stars $f = f(\vec{r}, \vec{v}, t)$ is given by

$$\frac{\partial f}{\partial t} + \vec{v} \cdot \frac{\partial f}{\partial \vec{r}} - \frac{\partial \Phi}{\partial \vec{r}} \cdot \frac{\partial f}{\partial \vec{v}} = 0, \tag{1}$$

where $\Phi(\vec{r}, t)$ is the total gravitational potential determined from the Poisson equation. As a further simplification, we consider the problem of stability of an infinitesimally thin disk (Lin and Shu, 1966; Lin et al., 1969; Shu, 1970). This is a valid approximation for perturbations with a wavelength that is greater than a typical thickness of the system.

We assume small perturbations about equilibrium such that $f = f_0(r, \vec{v}) + f_1(\vec{r}, \vec{v}, t)$ and $\Phi = \Phi_0(r) + \Phi_1(\vec{r}, \vec{v}, t)$, with $|f_1/f_0| \ll 1$ and $|\Phi_1/\Phi_0| \ll 1$ for all $\vec{r}$ and t. If a medium is only weakly inhomogeneous on the scale of the oscillation wavelength, in the approximation of the WKB method the perturbations of the equilibrium quantities are selected in the form of a plane wave (in the rotating reference frame we are using) $\aleph_m = \delta\aleph_m \exp(ik_r r + im\varphi - i\omega_* t)$, where $\delta\aleph_m$ is the slowly varying amplitude, $(1/k_r)(d/dr) \ln \delta\aleph_m \ll 1$, $\omega_* = \omega - m\Omega$ is a Doppler-shifted complex frequency of excited waves as seen by the moving star, and $\Omega(r)$ is the angular velocity of the differential galactic rotation at the distance r from the center. Evidently $\aleph_m$ is a periodic function of φ, and hence the azimuthal nonnegative mode number m must be an integer, and gives the number of spiral

---

[1] Even though the similarity between the gravitational and Coulomb interactions is well known and explored in stellar dynamics since 1960s (Bertin, 1980; Binney and Tremaine, 1987; Griv and Peter, 1996), plasmas are significantly different from gravitating systems. The main reason for expecting the difference is the sign of interaction: gravitating systems, because of the nature of the gravitation force, are always spatially inhomogeneous.

RESONANT EXCITATION OF DENSITY WAVES   445

arms. The rapidly varying part of $\aleph_m$ is absorbed in its phase ($|k_r|r \gg 1$). Also here k_r and $k_\varphi = m/r$ are the radial and azimuthal components of the wave vector $\vec{k}$, respectively, $k^2 = k_r^2 + k_\varphi^2$, and r, φ, and z are the galactocentric cylindrical coordinates. The axis of galactic rotation is taken oriented along the z-axis. In the framework of the linear theory, we can select one of the Fourier harmonics: $\aleph = \delta\aleph \exp(ik_r r + im\varphi - i\omega_* t)$ and $\delta\aleph = $ const. It is convenient to write the eigenfrequency ω_* in a form of the sum of the real part $\Re\omega_*$ and the imaginary part $i\Im\omega_*$; the existence of solutions with $\Im\omega_* > 0$ implies instability of oscillations, that is, the excitation of waves. Initially the disk is in equilibrium ($\partial f_0/\partial t = 0$). The function f_0 which describes the basic state against which small perturbations develop is a function of $\vec{v}$ and r only, and the equilibrium condition, namely $\partial\Phi_0/\partial r = r\Omega^2$, is satisfied.

In order to find the perturbed distribution function f_1, it is possible to integrate Eq. (1) over t, using the unperturbed orbits of stars. In the postepicyclic approximation, the orbits are given by Grivnev (1988), Griv and Peter (1996), and Griv *et al.* (1999a),

$$r = -\frac{v_\perp}{\kappa}\left[\sin\left(\phi_0 - \kappa t\right) - \sin\phi_0\right]; \quad v_r = v_\perp\cos\left(\phi_0 - \kappa t\right); \tag{2a}$$

$$\varphi = Dv_\perp^2 t + \frac{2\Omega}{\kappa}\frac{v_\perp}{\kappa r_0}\left[\cos\left(\phi_0 - \kappa t\right) - \cos\phi_0\right]; \tag{2b}$$

$$v_\varphi \approx r_0\frac{d\varphi}{dt} + r_0\frac{v_\perp}{\kappa}\frac{d\Omega}{dr}\sin\left(\phi_0 - \kappa t\right) \approx \frac{\kappa}{2\Omega}v_\perp\sin\left(\phi_0 - \kappa t\right), \tag{2c}$$

where $v_\perp$, ϕ_0 are constants of integration, $|v_\perp/\kappa r_0| \approx \rho/r_0 \ll 1$, $\kappa = 2\Omega[1 + (r/2\Omega)(d\Omega/dr)]^{1/2}$, and in galaxies $|(r/2\Omega)(d\Omega/dr)| \ll 1$ and $\kappa \sim \Omega$. Also, ρ is the mean epicyclic radius. The total azimuthal velocity of the stars was represented as a sum of the small random velocity v_φ and the circular velocity $r_0\Omega$; v_r is the random velocity in the radial direction. These expressions are written in the local rotating frame.

In Eqs. (2), the quantity involving D determines the systematic drift along the circular trajectory with radius r_0, and $|D|v_\perp^2 \ll \Omega$. The drift term $Dv_\perp^2 \approx (\rho^2/2r_0)(d\Omega/dr)$ leads to a new effect discussed in the paper; as a rule, in galaxies $(\rho/r_0)^2 \ll 1$. The minor drifting motion of a star is analogous to the gradient B drift of electrically charged particles of a plasma and is due to the nature of the nonuniform rotation of galactic disks: in differentially rotating stellar systems, as in the motion of a charged particle in a nonuniform magnetic field, the curvature of the epicycle circle is greater on one side than on the other. The latter should lead to a gradient Ω drift perpendicular to both $\vec{\Omega}$ and $\vec{\nabla}\Omega$. In galaxies, $D < 0$ (Griv and Peter, 1996; Griv *et al.*, 1999a). The gradient Ω drift represents a possible source of Landau instability of small-amplitude gravity perturbations, since this allows possible energy transfer between the system and the wave at a resonance in a hydrodynamically (Jeans-) stable particulate disk.

446 GRIV ET AL.

We choose the basic Schwarzschild (anisotropic Maxwellian) distribution $f_0 = f_0(r_0, v_\perp^2)$ satisfying the unperturbed part of the kinetic equation (1). Such a distribution function for the unperturbed system is particularly important because it provides a fit to observations (Shu, 1970; Morozov, 1980; Griv and Peter, 1996). The function f_0 is a function of the two constants of epicyclic motion $\mathcal{E} = v_\perp^2/2$ and $r_0^2 \Omega(r_0)$, where $v_\perp^2 = v_r^2 + (2\Omega/\kappa)^2 v_\varphi^2$ and $r_0 = r + (2\Omega/\kappa^2)v_\varphi$.

3. Resonant Excitation of Waves

Making use of unperturbed orbits (2), it is straightforward to show that the perturbed distribution function can be expressed as

$$f_1 = -\Phi_1(r_0)\left[\kappa\frac{\partial f_0}{\partial\mathcal{E}}\sum_{l=-\infty}^{\infty}\sum_{n=-\infty}^{\infty}l\frac{e^{i(n-l)(\phi_0-\zeta)}J_l(\chi)J_n(\chi)}{\omega_* - mDv_\perp^2 - l\kappa}\right.$$

$$\left.+\frac{2\Omega}{\kappa^2}\frac{m}{r}\frac{\partial f_0}{\partial r}\sum_{l=-\infty}^{\infty}\sum_{n=-\infty}^{\infty}\frac{e^{i(n-l)(\phi_0-\zeta)}J_l(\chi)J_n(\chi)}{\omega_* - mDv_\perp^2 - l\kappa}\right], \tag{3}$$

where $k_r = k\cos\psi$, $k_\varphi = m/r = k\sin\psi$, $\tan\zeta = (2\Omega/\kappa)\tan\psi$, $\tan\psi = m/rk_r$ is the pitch angle, and $f_1 = 0$ at $t \to -\infty$, so we neglected the effects of the initial conditions (Griv et al., 2000a,b). In Eq. (3), $J_l(\chi)$ is a Bessel function of the first kind with its argument $\chi = k_*v_\perp/\kappa$ and $k_* = [k_r^2 + (2\Omega k_\varphi/\kappa)^2]^{1/2} = k\{1 + [(2\Omega/\kappa)^2 - 1]\sin^2\psi\}^{1/2}$ is the effective wavenumber. In Eq. (3) the denominators vanish at resonances when $\omega_* - mDv_\perp^2 - l\kappa = 0$. The most important resonance is the corotation resonance, for which $l = 0$ and correspondingly $\omega_* - mDv_\perp^2 = 0$. Resonances of a higher order, in particular, the inner ($l = -1$) and outer ($l = 1$) Lindblad resonances, are dynamically less important.

The linearized Poisson equation in a suitable form is $(d^2/dz^2 - k^2)\Phi_1 = 4\pi G\sigma_1\delta(z)$, where $\delta(z)$ is the Dirac delta-function. In contrast to the original Lin–Shu expression, this equation includes effects of finite inclination of spiral arms ($k_\varphi \neq 0$). The linearized Poisson equation in such a form is readily solved to give the improved Lin–Shu type expression for the perturbed surface density of the two-dimensional disk (Lin and Lau, 1979; Bertin, 1980; Griv et al., 1999a, b).

We consider the galactic layer to be "tenuous." By "tenuous" we mean that the frequency of disk cooperative oscillations is smaller than the epicyclic frequency. In the opposite case of the high perturbation frequencies ($\omega_*^2 \gg \kappa^2$), the effect of the disk rotation is negligible and therefore not relevant to us.

Integrating Eq. (3) over velocity space $\int_0^\infty f_1 dv_\perp^2$ and equating the result to the perturbed density $\sigma_1 = -|k|\Phi_1/2\pi G$ given by the solution of the Poisson equation, the generalized dispersion relation in the plane $z = 0$ may be easily obtained (Griv et al., 2000a,b):

$$\frac{k^2 c_r^2}{2\pi G\sigma_0 |k|} = -\kappa \sum_{l=-\infty}^{\infty} l \frac{e^{-x} I_l(x)}{\omega_* - l\kappa} + 2\Omega \frac{m\rho^2}{rL} \sum_{l=-\infty}^{\infty} \frac{e^{-x} I_l(x)}{\omega_* - l\kappa} - i\gamma, \tag{4}$$

where $m|D|v_\perp^2 < |\omega_*| < \kappa$, $I_l(x)$ is a Bessel function of imaginary argument, $x = k_*^2 \rho^2$, $\rho = c_r/\kappa$, $|L| \approx \left|\partial \ln(2\Omega\sigma_0/\kappa c_r^2)/\partial r\right|^{-1}$ is the radial scale of spatial inhomogeneity, and $|\rho^2/rL| \ll 1$. The last term on the right-hand side in Eq. (4) is assumed to be small in comparison with other terms; this term yields a Landau damping/growth of oscillations of a disk that would have been stable in the hydrodynamical limit.

In the newly derived dispersion relation (4), we have

$$\gamma = \sum_{l=-\infty}^{\infty} \left(2\Omega \frac{m\rho^2}{rL} - l\kappa\right) \frac{\pi J_l^2(\chi')}{2m|D|c_r^2} \exp\left(-\frac{\Re\omega_* - l\kappa}{2m Dc_r^2}\right) H\left(\frac{\Re\omega_* - l\kappa}{mD}\right),$$

$$\tag{5}$$

where H is a Heveaside function and $\chi' = (k_*\rho)\sqrt{(\Re\omega_* - l\kappa)/m Dc_r^2}$.

In the derivation of Eq. (4), the integral $\int f_1 dv_\perp^2$, containing a resonant denominator, was estimated following Landau's prescription. That is, for perturbations with $(\Re\omega_* - l\kappa)/mD > 0$, the Cauchy principal value of the integral yields the non-resonant dispersion relation, and the term with γ determines the slow, $|\Im\omega_*/\Re\omega_*| \ll 1$, Landau growth/damping of the hydrodynamically stable oscillations. This means that the term with γ should be used only when the frequency, defined by the non-resonant dispersion relation, is real: $(\Re\omega_*)^2 > 0$. In other words, the disk is Jeans-stable.

The resonance in the velocity integral above $\int f_1 dv_\perp^2$ is in fact not with the epicyclic motion of the stars but with the drift motion of the epicyclic (guiding) center $\propto m Dv_\perp^2$. Indeed, the term $m Dv_\perp^2$ describes the additional Doppler shift in the frame of the drifting guiding center. Once there are stars, the guiding centers of which can be in resonance with the wave, the corresponding growth/damping rate is nonzero. Near the resonances, the sign of the drift of the guiding center is determined by the sign of mD, so that the resonance is impossible when $(\omega_* - l\kappa)/mD < 0$. Individual "spread across the resonance" stars (whose epicyclic orbits make them cross the resonance, while their guiding centers do not) take no part in the resonant Landau interaction.

The local dispersion relation (4) generalizes the standard Lin–Shu–Kalnajs one for *nonaxisymmetric* perturbations propagating in a spatially *inhomogeneous* disk including the *resonance* zones. The relation (4) is complicated: it is highly nonlinear in the wavefrequency ω_*. Therefore, in order to deal with the most interesting oscillation types, let us study only various limiting cases of perturbations described by some simplified variations of Eq. (4).

First, considering low-frequency perturbations, $|\Re\omega_*| < \kappa$, the terms in series (4) for which $|l| \geqslant 2$ may be neglected. In this event, consideration will be limited

448   GRIV ET AL.

to the transparency region between the inner and outer Lindblad resonances. Obviously, the $l = 0$ harmonic that defines the corotation resonance will dominate the series. Secondly, we consider the weakly inhomogeneous disk and the most important for the problem of spiral structure low-m perturbations: from now on in all equations $2\Omega(m\rho^2/rL) \ll 2\Omega \sim \kappa$ and $m \sim 1$. Therefore, in small terms proportional to L^{-1} and γ we include only $l = 0$ harmonics. Finally, analyzing the dispersion relation (4), it is useful to distinguish between two limiting cases: the cases of epicyclic radius that is small compared with wavelength ($x \lesssim 1$) and of epicyclic radius that is large compared with wavelength ($x \gg 1$).

As a result of these simplifications, the dispersion relation reads

$$\omega_*^3 - \omega_*\omega_J^2 + \omega_{\mathrm{grad}}\kappa^2 - i\omega_*\omega_{\mathrm{res}}\kappa = 0, \tag{6}$$

where $\omega_J^2 \approx \kappa^2 - 2\pi G\sigma_0|k|F(x)$ is the squared Jeans frequency and $F \approx 2\kappa^2 e^{-x} I_1(x)/k^2 c_r^2$ is the so-called reduction factor. In the long-wavelength limit, $F(x) \approx (k_*/k)^2[1 - x + (3/4)x^2]$ and $x \lesssim 1$, and in the short-wavelength limit, $F(x) \approx (1/k\rho)^2[1 - (1/2\pi x)^{1/2}]$ and $x \gg 1$. The aperiodic Jeans instability (gravitational collapse) occurs when $\omega_J^2 < 0$ (the real part of the frequency of this unstable motion vanishes in a rotating reference frame). Such instabilities, which imply the displacement of macroscopic portions of a gravitating system, may be also analyzed through the use of relatively simple hydrodynamic equations (e.g. Lin and Lau, 1979). Also in Eq. (6),

$$\omega_{\mathrm{grad}} = 2\Omega e^{-x} I_0(x)\frac{2\pi G\sigma_0|k|}{k^2 c_r^2}\frac{m\rho^2}{rL} \quad \text{and} \quad |\omega_{\mathrm{grad}}| \ll \kappa$$

is the frequency of the low-frequency "gradient oscillations" and

$$\omega_{\mathrm{res}} = \Omega\frac{2\pi^2 G\sigma_0|k|}{k^2 c_r^2}\frac{\kappa\rho^2}{rL}\frac{J_0^2(k_*\rho\sqrt{\Re\omega_*/mDc_r^2})}{|D|c_r^2}\exp\left(-\frac{\Re\omega_*}{2mDc_r^2}\right)H\left(\frac{\Re\omega_*}{mD}\right)$$

is the frequency of the "resonant oscillations." To repeat ourselves, in the simplified dispersion relation (6) we treat the term involving ω_{grad} as well as the term involving ω_{res} as small corrections.

Equation (6) has three roots which describe three branches of oscillations: two classical Jeans branches (long-wavelength, $x \lesssim 1$, and short-wavelength, $x \gg 1$, branches) modified by the spatial inhomogeneity, and an additional gradient one modified by the Jeans mode. If the condition $|\omega_*|^3 \sim |\omega_J|^3 \gg |\omega_{\mathrm{grad}}|\kappa^2$ (and $|\omega_J| \lesssim \kappa$), holds (we have made use of the fact that this is typically for flat galaxies), from Eq. (6) one defines the dispersion law of the most important for a gravitating medium low-m ($m < 10$) Jeans oscillations

$$\omega_{*1,2} = \pm p|\omega_J| - \omega_{\mathrm{grad}}\frac{\kappa^2}{2\omega_J^2} + is\omega_{\mathrm{res}}\frac{\kappa}{\omega_J} \tag{7}$$

or $\omega_{*1,2} = \pm p|\omega_J| - \omega_{\mathrm{grad}}(\kappa^2/2\omega_J^2)$, depending on whether ω_J/mD is positive or negative, respectively. Here $p = 1$, $s = 1$ for Jeans-stable ($\omega_J^2 > 0$)

perturbations and $p = i$, $s = 0$ for Jeans-unstable ($\omega_J^2 < 0$) perturbations. Accordingly, a spatial inhomogeneity will not influence the stability condition of Jeans modes. In the weakly inhomogeneous disk, from the equations above it follows that the correction to the Jeans frequency for the disk inhomogeneity is small, $\sim \kappa(m\rho^2/rL) \ll \kappa$, and hence $|\Re\omega_{*1,2}| \approx |\omega_J|$ for the low-m spiral patterns we are interested in. It follows from Eq. (7) that sufficient random motion can suppress the instability of arbitrary Jeans perturbations (see Griv and Peter (1996) and Griv et al. (1999b) for an explanation).[2] We are interested only in such hydrodynamically (Jeans-) stable rapidly rotating disks.

On the other hand, the result (7) means that small-amplitude spiral ($\psi \neq 0$) modes in a differentially rotating disk with allowance of the gradient Ω drift ($D \propto c_r^2(d\Omega/dr) \neq 0$) may be unstable oscillatory ($\Re\omega_{*1,2} \neq 0$, $\Im\omega_{*1,2} > 0$, and $|\Im\omega_{*1,2}/\Re\omega_{*1,2}| \ll 1$), i.e. wave excitation occurs. The existence of gradient Ω drift produces an oscillatory growing instability in Jeans-stable waves. The necessary conditions for this excitation of spiral density waves are $\Re\omega_{*1,2}/D \approx \omega_J/D > 0$, $\omega_J L > 0$, and $\omega_J^2 > 0$. Thus, Jeans-stable disks with radially dependent densities, angular velocities, and temperatures are nevertheless unstable to resonant Landau-drift instabilities. The larger the angular velocity gradient, the more unstable the system, because in Eq. (7) $D \propto d\Omega/dr$. This microinstability leads to oscillatory growing waves propagating through the fluidlike system with phase velocities ω_J/k_φ, which resonate with stars that drift at the same velocities. Because $|\Im\omega_{*1,2}/\Re\omega_{*1,2}| \ll 1$, contrary to the ordinary aperiodic Jeans instability, wave propagation can now occur. This is the Landau excitation, or collisionless resonant excitation of oscillations in a Jeans-stable system. To emphasize, $\Im\omega_{*1,2} \to 0$ as $D \propto d\Omega/dr \to 0$ and/or $m \to 0$, i.e. in the limit of a small gradient Ω and/or radial perturbations respectively, the growth rate of the waves approches zero. The latter means that the free kinetic energy associated with the differential rotation is one possible source for the growth of the spiral wave energy. On the other hand, spatial inhomogeneity ($L^{-1} \neq 0$) and the finite temperature ($c_r^2 \neq 0$) are also crucial factors in the effect.

Such a branch of oscillation in stellar disks has been expected (Griv and Peter, 1996; Griv, 1998). The effect was missed in all earlier theories either because (a) the second-order drift terms proportional to the temperature of a system as well as (b) the pitch-angle dependent effects or (c) the spatial inhomogeneity were ignored in a kinetic description of stellar disks (Lin and Shu, 1966; Julian and Toomre, 1966; Shu, 1970; Lynden-Bell and Kalnajs, 1972; Mark, 1977; Bertin, 1980; Morozov, 1980), or a gasdynamic (fluidlike) description was used (Goldreich and Lynden-Bell, 1965; Lovelace and Hohlfeld, 1978; Lin and Lau, 1979; Drury, 1980; Narayan et al., 1987; Papaloizou and Savonije, 1991) to study the prop-

---

[2] Jeans instabilities are thought to heat disks up to Toomre's Q-parameter values of $2 - 2.5$ (Lin and Lau, 1979; Bertin, 1980; Morozov, 1980; Griv and Peter, 1996; Polyachenko and Polyachenko, 1997; Griv, 1998; Griv et al., 1999b; Griv et al., 2000b,c). This is large enough to turn off the Jeans instability in a stellar disk.

450   GRIV ET AL.

erties of particulate systems. In the former case sufficiently far from resonances $|\omega_* - l\kappa| \gg m|D|v_\perp^2$ the minor drift term $\propto D$ may be omitted. But it is clear that in the resonant parameter regime $\omega_* - mDv_\perp^2 - l\kappa = 0$ the drift term becomes the principal one, and, therefore, should be retained along with other terms of order $\rho/r_0 \ll 1$ in the linearized equations of motion. In the latter case the equations of a theory lose microstructure, or particle effects.

Thus, it is shown that spiral gravity perturbations (e.g. those produced by a bar or oval structure in a galactic center, a spontaneous spiral perturbation and/or a companion galaxy) can be unstable even under conditions where the disk is Jeans stable; oscillatory unstable waves can be emitted at the corotation resonance. The instability has the kinetic origin due to the presence of poles in the dispersion relation, and can develop only if there is spatial inhomogeneity of the nonuniformly rotating dynamically hot disk. The source of free energy for the instability appears to be nonuniform rotation itself, and is presumably released when angular momentum is transfered outward. In the present theory, in essence, the oscillatory unstable waves with $(\Re\omega_*)^2 > 0$, $\Im\omega_* > 0$, and $|\Im\omega_*/\Re\omega_*| \ll 1$ are generated in the localized zone near the corotation resonance and propagate in the transparency region between the inner and outer Lindblad resonances in a disk. Gas dynamic, axisymmetric treatments, and kinetic treatments that ignore drifts and/or inhomogeneity overlook the effect. It seems likely that by these oscillatory unstable vibrations, one can explain a type of recurrent spiral modes discovered in collisionless N-body simulations by Sellwood and Lin (1989) and Donner and Thomasson (1994).

4. Astronomical Implications

As a demonstration of the astronomical implication of Eqs. (5) and (7), the development of dynamic Landau microinstabilities can result directly in the formation of different observable structural features in galaxies. We speculate that the Landau-type drift excitation of propagating nonaxisymmetric density waves may be related to the development and existence of the weak spiral structures in Jeans-stable subsystems of relatively old stars with ages $\gtrsim 10^9$ yr in flat galaxies.

Intrestingly, spiral arms are smoother in images of galaxies in the near IR (Schweizer, 1976; Rix and Zaritsky, 1995), indicating that the old disk stars participate in the pattern. Block and Puerari (1999) pointed out that to derive a coherent physical framework for the excitation of spiral structure in galaxies, one must consider the co-existence of two different dynamical components: a gas and young-stars dominated Population I disk and an evolved stellar Population II component. The classical Hubble classification scheme has as its focus, the morphology of the Population I component only. Our major conclusion is that the dynamical behaviour of gas and young OB stars is decoupled from that of Population II disk: different subsystems of a galaxy may prefer different types of instabilities (e.g. Jeans or Landau instabilities) for spiral generation. The latter may explain why spiral structures in Population I and Population II disks are often not coincide.

Apart from providing an account of weak spiral structures of galaxies, including the Milky Way, as such microinstabilities grow, it must lead to the slow increase of a dispersion of random velocities of stars ("heating") and the slow redistribution of mass by the collective collisionless mechanism, that is, to the slow dynamical relaxation of the system. Thus, disk galaxies can be viewed as time-dependent systems, where secular evolution, due to resonant effects, is transforming their morphologies and their kinematics over less a Hubble time. Martinet (1995) already reviewed observational facts as well as results of numerical simulations which suggest long-term evolution of galaxies.

References

Bertin, G.: 1980, *Phys. Rep.* **61**, 1.
Binney, J. and Tremaine, S.: 1987, *Galactic Dynamics*, Princeton Univ. Press, Princeton, NJ.
Block, D.L. and Puerari, I.: 1999, *Astron. Astrophys.* **342**, 627.
Chandrasekhar, S.: 1960, *Principles of Stellar Dynamics*, Dover, New York.
Donner, K.J. and Thomasson, M.: 1994, *Astron. Astrophys.* **290**, 785.
Drury, L.O'C.: 1980, *Mon. Not. R. Astron. Soc.* **193**, 337.
Goldreich, P. and Lynden-Bell, D.: 1965, *Mon. Not. R. Astron. Soc.* **130**, 125.
Griv, E.: 1998, *Astrophys. Lett. Commu.* **35**, 403.
Griv, E. and Peter, W.: 1996, *Astrophys. J.* **469**, 89.
Griv, E., Yuan, C. and Gedalin, M.: 1999a, *Mon. Not. R. Astron. Soc.* **307**, 1.
Griv, E., Rosenstein, B., Gedalin, M. and Eichler, D.: 1999b, *Astron. Astrophys.* **347**, 821.
Griv, E., Gedalin, M., Eichler, D. and Yuan, C.: 2000a, *Phys. Rev. Lett.* **84**, 4280.
Griv, E., Gedalin, M., Eichler, D. and Yuan, C.: 2000b, *Astrophys. Space Sci.* **271**, 21.
Griv, E., Gedalin, M., Eichler, D. and Yuan, C.: 2000c, *Planet. Space Sci.* **48**, 679.
Grivnev, E.M.: 1988, *Soviet Astron.* **32**, 139.
Julian, W.H. and Toomre, A.: 1966, *Astrophys. J.* **146**, 810.
Lin, C.C. and Shu, F.H.: 1966, *Proc. Natl. Acad. Sci.* **55**, 229.
Lin, C.C., Yuan, C. and Shu, F.H.: 1969, *Astrophys. J.* **155**, 721.
Lin, C.C. and Lau, Y.Y.: 1979, *Stud. Appl. Math.* **60**, 97.
Lovelace, R.V.E. and Hohlfeld, R.G.: 1978, *Astrophys. J.* **221**, 51.
Lynden-Bell, D. and Kalnajs, A.J.: 1972, *Mon. Not. R. Astron. Soc.* **157**, 1.
Mark, J.M.-K.: 1977, *Astrophys. J.* **212**, 645.
Martinet, L.: 1995, *Fund. Cosmic Phys.* **15**, 341.
Morozov, A.G.: 1980, *Soviet Astron.* **24**, 391.
Narayan, R., Goldreich, P. and Goodman, J.: 1987, *Mon. Not. R. Astron. Soc.* **228**, 1.
Papaloizou, J.C. and Savonije, G.J.: 1991, *Mon. Not. R. Astron. Soc.* **248**, 353.
Polyachenko, V.L. and Polyachenko, E.V.: 1997, *JETP* **85**, 417.
Rix, H.-W. and Zaritsky, D.: 1995, *Astrophys. J.* **447**, 82.
Schweizer, F.: 1976, *Astrophys. J. Suppl.* **31**, 313.
Sellwood, J.A. and Lin, D.N.C.: 1989, *Mon. Not. R. Astron. Soc.* **240**, 991.
Sellwood, J.A. and Kahn, F.D.: 1991, *Mon. Not. R. Astron. Soc.* **250**, 278.
Shu, F.H.: 1970, *Astrophys. J.* **160**, 99.
Toomre, A.: 1964, *Astrophys. J.* **139**, 1217.

Toomre, A.: 1990, in: *Dynamics and Interactions of Galaxies*, p. 292, ed. R. Wielen, Springer-Verlag, Heidelberg.

Toomre, A. and Kalnajs, A.J.: 1991, in: *Dynamics of Disc Galaxies*, p. 341, ed. B. Sandelius, Göteborg Univ. Press, Göteborg.

CONSTRUCTING OF THE STELLAR VELOCITY FIELD USING THE HIPPARCOS DATA

ALEXANDER TSVETKOV and MARIA BABADZHANYANTS
Astronomical Institute of St.- Petersburg University St. Petersburg, 198504, Russia

Abstract. To investigate the velocity field in the solar vicinity we have to know three components of stellar velocities. Since the Hipparcos catalogue does not contain the radial velocities we cannot find the spatial velocity of each star. The individual parallaxes allow us to arrange the selections of stars, which can form any configuration in space. The solution of the Airy-Kowalsky equations using such selections yields the components of the solar motion with regard to a group of stars. The differences between the various values are the main idea to construct the stellar velocity field.

1. The Method of Constructing Velocity Field

The influence of the Solar motion on a proper motion of star is described by the Airy-Kowalski equations (Kulikovsky, 1991):

$$\mu_l \cos b = \frac{V_x}{kr} \sin l - \frac{V_y}{kr} \cos l \tag{1}$$

$$\mu_b = \frac{V_x}{kr} \cos l \sin b + \frac{V_y}{kr} \sin l \sin b - \frac{V_z}{kr} \cos b \tag{2}$$

These equations were solved over the Hipparcos data using standard least square method.

2. Constructing Velocity Field - Selection of the Stars

We consider the galactic orthogonal coordinate system and chose a net 400pc size with the distances 50 pc from one knot to another. Every knot was a center of a sphere of radius 200 pc. The combined solution of equation (1) and (2) was derived for stars belonged to a sphere. We denote the solution as $V_i = (V_x, V_y, V_z)$ for i-sphere. The central solution is V_0. The vector $DV_i = V_0 - V_i$ can be interpreted as relative motion of the center of i-sphere with respect to the Sun. This algorithm was applied to three groups of stars: O-B, A-F and K-M giants. The accuracy of parallaxes better than 25% and proper motions less than 30 "/cy were used. Figures 1a and 2a, b show the results for O–B stars.

Pretka-Ziomek et al./Dynamics of Natural and Artificial Celestial Bodies, 453–454, 2001.
© 2001 *Kluwer Academic Publishers.*

3. Artificial Velocity Fields Based on the Oort-Lindblad Model

We compared the derived fields with the model ones based on the Oort-Lindblad model, which contains only the plane Galactic rotation. The artificial proper motions were calculated according to this model. The result is represented by Figure 1 b. The velocity field for XZ and YZ planes are equal to zero.

4. The Main Results

In the XY-plane the velocity field of O-B stars is in a good correspondence to the Oort-Linblad model. The XZ and YZ planes indicate significant rotation in non-galactic plane. It means that the rotation vector of the nearby O-B stars is non-orthogonal to the Galactic plane. The kinematics of the solar neighborhood is more complicated than one described of standard kinematical models.

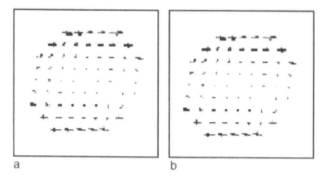

Figure 1. a. Original velocity vector field for O–B stars in XY plane and b. model velocity vector field for O–B stars in XY plane.

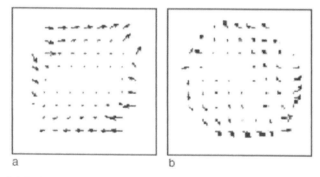

Figure 2. Original velocity vector field for O–B stars a. in XZ plane and b. in YZ plane.

References

Kulikovsky, P.G.: 1991, *Stellar Astronomy*, Nauka, Moskva (in Russian).

TIDAL AND ROTATIONAL EFFECTS IN THE EVOLUTION OF HIERARCHICAL TRIPLE STELLAR SYSTEMS

TAMÁS BORKOVITS (borko@electra.bajaobs.hu)
Baja Astronomical Observatory of Bács-Kiskun County, H-6500 Baja, Szegedi út,
P.O. Box. 766., Hungary

A new numerical integrator was developed by the author for studying together the orbital and spin evolution of hierarchical triple stellar systems. The stars are considered as liquid, viscous bodies. The initial orientation of the axis of rotation of each component can be arbitrary. The angular velocity of the rotation can differ from the Keplerian angular velocity. The equations of the orbital motion and the (Eulerian) equations of the star rotations are integrated simultaneously. The dynamical tides are treated as a force term calculated as the divergence of the viscous stress tensor (see eg. Kopal, 1978). Only the first order terms are integrated. For the more distant third companion only a mass–point model is applied. This means the effect of the third body is calculated only as a point mass–point mass interaction. The numeric integrator is a seventh order Runge-Kutte-Nyström one (Fehlberg, 1974). Regularization is not applied.

In the following, I present the first results applying the method for an Algol-like system. I integrated this system on several ways with the same parameters which Harrington (1984) used for the Algol. Here I illustrate two of them. First a mass–point approximation was used (as a reference). The evolution of the orbital elements is shown in Fig. 1a. The presence of the Kozai resonance (Kozai, 1962) is evident. This makes the system unstable.

In the other presented run (Figure 1b) an $n = 3$ politropic model was used for 'Algol A', and an $n = 1.5$ one for 'Algol B'. The primary's initial angular velocity was two times larger than the Keplerian one, while the secondary's equator was inclined with $10°$ to the orbital plane. Dissipation was also considered. In that case the picture changed dramatically. The system became clearly stable (cf. Söderhjelm, 1984, Kiseleva et al., 1998) in the sense of Kozai resonance. The eccentricity of the close orbit shows only lower amplitude fluctuations on at least two different time scales (from decades to centuries). Nevertheless, the visible inclination (and the longitude of the node) of the binary vary on two different ways. The large amplitude variation due to the rotation of the orbital plane of the binary caused by the distant third body which would yields easily observable effects on a time scale of a century (eg. the change of the eclipse depth in eclipsing binaries, and a virtual period change). There is also a few tenths degree of fluctuation (which cannot be seen in the figure) produced by the precession of the aligned secondary star, which illustrates the interaction between the spin and the orbital angular momentum.

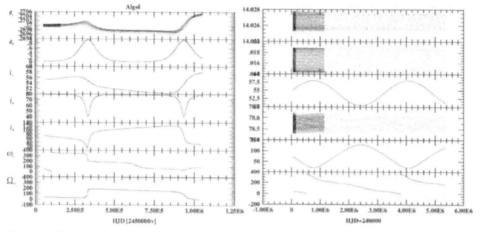

Figure 1. The evolution of some orbital elements of the 'Algol' system according to two different physical models (see text). (Left panel (from top): e_1, e_0, i_1, i_{mut}, i_0, ω_0, Ω_0; right: a_0, e_0, i_1, i_{mut}, i_0, and Ω_0. Subscript 0 refers to the close orbit, while 1 refers to the wide one.)

This work is in its first stage yet. In the following I plan several runs with different initial conditions in the geometrical and physical parameters. The studies will concentrate on the following three main topics: (*a*) Short term observable effects produced by the interaction between rotation and tides, as well as the presence of the third body, (*b*) the effectiveness of the dissipation in the synchronization of such systems, (*c*) finally, the formation, and stability of triple stellar systems.

Acknowledgements

This work was supported by the National Grant OTKA F030147.

References

Fehlberg: 1974, *NASA Technical Report R-432*.
Harrington, R. S.: 1984, *ApJ* **277**, L69.
Kiseleva, L. G., Eggleton, P. P. and Mikkola, S.: 1998, *MNRAS* **300**, 292.
Kopal, Z.: 1978, *Dynamics of Close Binary Systems*, D. Reidel, Dordrecht.
Kozai, Y.: 1962, *AJ* **67**, 591.
Söderhjelm, S.: 1984, *A&A* **141**, 232.

ANALYSIS OF CATALOG CORRECTIONS WITH RESPECT TO THE HIPPARCOS REFERENCE FRAME

F.J. MARCO and J.A. LÓPEZ
Departamento de Matemáticas Univ Jaume I de Castellón, Spain
E-mail:lopez@mat.uji.es, marco@mat.uji.es

M.J. MARTÍNEZ
Departamento de Matemática Aplicada Univ Politecnica de Valencia, Spain
E-mail:mjmartin@mat.upv.es

Abstract. An study of different analytical procedures to compare catalogues in spherical domains is presented to obtain analytical expressions for the necessary reductions. We consider the case of catalogues with bounded declination limits, near of which the errors must be carefully modeled.

Key words: reference frames, catalogues.

1. Introduction

This paper deals with two complementary but equally necessary aspects to determine the analytical signals that reflect the discrepancies between different catalogues. The first of them refers to the establishment of a theoretical model and the second one deals with the practical realization of the theoretical model.

2. Continuous Functional Model and Discretization of the Problem

Let us consider $D = [0, 2\pi] \times [-\gamma, \gamma]$ with boundary $\Gamma = \Gamma^+ \cup \Gamma^-$ with $\Gamma^\pm = [0, 2\pi] \times \{\pm\gamma\}$. We search an $u \in L^2(D)$ with $u|_\Gamma \equiv g^\pm \neq 0$ for a given g. We put $u = u_0 + u_g$ where $u_0|_\Gamma = 0$ (which may be developed in an adequate series) and $u_g|_\Gamma = g$ arbitrary. To calculate u_g we suppose $g^\pm(\alpha) = \frac{a_0^\pm}{2} + \sum_{k \geqslant 1}[a_k^\pm \cos k\alpha + b_k^\pm \sin k\alpha]$. As we stated in [3] it suffices to take $u_g(\alpha, \delta) = \sum_{k \geqslant 0}\{y_k^+(\delta)g_k^+(\alpha) + y_k^-(\delta)g_k^-(\alpha)\}$ where $y_k^\pm$ verify the Legendre equation with boundary conditions $y_k^+(\gamma) = 1, y_k^+(-\gamma) = 0$ and $y_k^-(-\gamma) = 1, y_k^-(\gamma) = 0$ (Courant and Hilbert, 1989). We have a discrete set of points, so we must approximate g and u_0. We suppose $\{P_k(\sigma)\}$ an orthonormal basis to develop u_0 as $u_0 = \sum_k c_k P_k(\sigma)$ and we put $v = u - u_g$. We search $c_k, 0 \leqslant k \leqslant R$ so that $\int_D[v - u_0]^2 d\sigma = Min$, given the equations $\langle v - \sum_k c_k P_k, P_l \rangle = 0 \ 0 \leqslant l \leqslant R$ where $\langle, \rangle$ is the inner product in $L^2(D)$. If $\langle, \rangle_h$ denotes the product in the discretized space, the equations result $\langle v, P_l \rangle_h = \sum_k c_k \langle P_k, P_l \rangle_h$ If the points that determine the discretization h are equally spaced

Pretka-Ziomek et al./Dynamics of Natural and Artificial Celestial Bodies, 457–458, 2001.
© 2001 *Kluwer Academic Publishers.*

458   F.J. MARCO ET AL.

and the discrete product $\langle v, w \rangle_h = h \sum_i v_i w_i$ is chosen, the previous equations coincide with the ones obtained from the minimum squares method (statistic). If we have that the vectors $\{P_k(\sigma_i)\}$ are algebraically orthonormal, the values for c_k are numerically stable. In other case, the validity of the method is more uncertain.

3. Approximations in the Boundary

Given the lack of data in the fictitious boundary chosen, we purpose to consider a little band $B^\pm$ around each $\Gamma^\pm$ where $u_g \approx u$ ($u_0 \approx 0$). We take a Fourier development of $\widetilde{g}$ around ξ and we define $g^\pm(\alpha) = \widetilde{g}^\pm(\alpha, 0)$ through the minimization of $\int_D [u - \widetilde{g}(\alpha, \xi)]^2 d\alpha d\xi$. We obtain condition equations with coefficients in the (α, ξ)-space. Let $\xi = \eta(\alpha)$ a regression curve for the data. We approximate $u(\alpha, \xi) \approx u(\alpha, \eta(\alpha)) + \frac{\partial u}{\partial \xi}(\alpha, \eta(\alpha))(\xi - \eta(\alpha))$ and analogously for $\widetilde{g}$. We denote $u_\alpha = u(\alpha, \eta(\alpha))$ and $\widetilde{g}_\alpha = \widetilde{g}(\alpha, \eta(\alpha))$ and the condition equations are the same if and only if $\frac{\partial u}{\partial \xi} \eta \equiv 0$ (In a real case, the conditions are reduced to conditions in mean). So, if the discrete data (α_i, ξ_i, u_i) accomplish that the equation $\xi = \eta(a)$ is $\xi = 0$ (quadratic mean), we can consider the data as if they were $(\alpha_i, 0, u_i)$. It is not possible to assure a stable computation of the coefficients of g and u_0, due to the particular distribution of the data, unless we assure the algebraic orthogonality of the basis on the discrete points. The regression non-parametric methods may be useful to be used directly to obtain estimations of a function. Afterwards, other approximation to this function is obtained, using the infinite norm.

References

Courant, R. and Hilbert, D.: 1989, *Methods of Mathematical Physics*, Wiley Classic Library.
Martinez, M.J., Marco, F.J. and Lopez, J.A.: 1997, 'Hogenization of Astrometrical Catalogs in a band', *Journees Systemes de reference spatio-temporels*, Prague.
Simonoff, J.S.: 1998, *Smoothing Methods in Statistics*, Springer.
Wang, M.P. and Jones, M.C.: 1995, *Kernel Smoothing*, Chapman & Hall.

Printed by Publishers' Graphics LLC